Early Modern Zoology

Intersections

Yearbook for Early Modern Studies

VOLUME 7 – 2007

Early Modern Zoology

The Construction of Animals in Science, Literature and the Visual Arts

Edited by

Karl A.E. Enenkel and Paul J. Smith

BRILL

LEIDEN • BOSTON
2007

Illustration on the cover: Ulisse Aldrovandi, *De quadrupedibus digitatis viviparis libri tres* [...], Bologna: 1637.

Brill has made all reasonable efforts to trace all rights holders to any copyrighted material used in this work. In cases where these efforts have not been successful the publisher welcomes communications from copyrights holders, so that the appropriate acknowledgements can be made in future edition, and to settle other permission matters.

This book is printed on acid-free paper.

A C.I.P. record for this book is available from the Library of Congress.

ISSN 1568-1181
ISBN 978 90 04 13188 0

PRINTED IN THE NETHERLANDS

CONTENTS

VOLUME ONE

THE CONSTRUCTION OF THE ANIMAL IN ZOOLOGICAL ILLUSTRATIONS

VOLUME TWO

THE CONSTRUCTION OF THE ANIMAL IN ITS EARLY MODERN USE AND ABUSE

LIST OF ILLUSTRATIONS

Colour plates I–XXXI can be found following page 12.

Figures 1–6 (belonging to the article by Karl A.E. Enenkel):

Figures 1–9 (belonging to the article by Paul J. Smith):

Figures 1–6 (belonging to the article by Vibeke Roggen):

Figures 1–6 (belonging to the article by Eric Jorink):

Figures 1–16 (belonging to the article by Karin Leonhard):

Figures 1–13 (belonging to the article by Thea Vignau-Wilberg):

Figures 1–7 (belonging to the article by Florike Egmond):

Figures 1–14 (belonging to the article by Rebecca Brienen):

Figures 1–4 (belonging to the article by Suzanne J. Walker):

Figures 1–10 (belonging to the article by Almudena Pérez de Tudela and Annemarie Jordan Gschwend):

Figures 1–16 (belonging to the article by Johan Koppenol):

Figures 1–7 (belonging to the article by Vincent Buyens):

Figures 1–17 (belonging to the article by Paulette Choné):

Figures 1–6 (belonging to the article by Sarah Cohen):

NOTES ON THE EDITORS

Karl A.E. Enenkel is Professor of Neo-Latin Literature at Leiden University and teaches classical Latin and Neo-Latin in the Department of Classics. His research focuses on international Humanism, the reception of Classical Antiquity, the history of ideas, the description of literary genres and emblem studies. He is the author of *Francesco Petrarca: De vita solitaria, Buch 1. Kritische Textausgabe und ideengeschichtlicher Kommentar* (1991) and of *Kulturoptimismus und Kulturpessimismus in der Renaissance* (1995). He has co-edited and co-authored *Lipsius in Leiden* (1997), *Modelling the Individual. Biography and Portrait in the Renaissance* (1998), *Recreating Ancient History* (2001), *Mundus Emblematicus. Studies in Neo-Latin Emblem Books* (2003), *Cognition and the Book. Typologies of Formal Organisation of Knowledge in the Printed Book of the Early Modern Period* (2004), *The Manipulative Mode. Political Propaganda in Antiquity* (2005), *Petrarch and his Readers in the Renaissance* (2006) and several collections of studies on the reception of Classical Literature. Recently, he wrote a book on Neo-Latin autobiography (*Die Erfindung des Menschen. Die Autobiographik des frühneuzeitlichen Humanismus von Petrarca bis Lipsius*; forthcoming 2007). He is member of the editorial board of *Imago figurata* and *Proteus. Studies in Early Modern Identity Formation* and the general editor of *Intersections*.

Paul J. Smith is Professor of French Literature at Leiden University. His research focuses on 16th-, 17th- and 20th-century French literature, its reception in the Netherlands, French and Dutch fable and emblem books and literary rhetoric. He is the author of *Voyage et écriture. Etude sur le* Quart Livre *de Rabelais* (1987), *Het schouwtoneel der dieren. Embleemfabels in de Nederlanden (1567–ca. 1670)* (2006) and co-author of *Francis Ponge. Lectures et méthodes* (2004). He has edited *Editer et traduire Rabelais à travers les âges* (1997) and co-edited *Lectures de René Char* (1990), *Le paradoxe en linguistique et en littérature* (1996), *Fabuleux La Fontaine* (1997), and *Montaigne and the Low Countries (1580–1700)* (2007). He is member of the editorial board of *Intersections*, *Montaigne Studies* and *Neophilologus*.

LIST OF CONTRIBUTORS

Rebecca P. Brienen is Assistant Professor of Art History at the University of Miami. Recently, she published *Visions of Savage Paradise: Albert Eckhout, Court Painter in Colonial Dutch Brazil* (2006).

Vincent Buyens is preparing his Ph.D. thesis at the Department of Literary Studies at the Catholic University of Leuven. His research focuses on the seventeenth-century Brussels writer Willem van der Borcht.

Paulette Choné is Professor of Modern Art and Research Director at the University of Bourgogne (Dijon). She specializes in the symbolic meaning of art in the 16th and 17th centuries, including emblem studies, and in art theory. She recently published *Flore au paradis. Emblématique et vie religieuse aux XVIe et XVIIe siècles* (2004).

Sarah R. Cohen is Associate Professor of Art History at the University at Albany, State University of New York. She has published *Art, Dance, and the Body in French Culture of the Ancien Régime* (2000) and numerous articles on representations of the human body in early modern European culture. She is currently completing a book on portrayals of animals in European art and natural philosophy in the early modern era.

Pia Cuneo is Professor of Art History at the University of Arizona. She has published on the art and city chronicle of the Augsburg painter and printmaker Jörg Breu the Elder, as well as on various facets of hippological culture in early modern Germany.

Louise Hill Curth is Senior Lecturer in Health Studies at Bath Spa University, England. Her major research interests are in human and veterinary medical beliefs and practices in early modern England. She is the author of *English Almanacs, Astrology & Popular Medicine, 1550–1700* (2006) and the editor of *From Physick to Pharmacology: Five hundred years of British Drug Retailing*.

Florike Egmond has taught history at various Dutch universities and currently works as a researcher for the Clusius Project, based at Leiden University. Her main research topic is Renaissance natural history. Recently she co-edited *The Whale Book. Whales and other marine animals as described by Adriaen Coenen in 1585* (2003) and *Correspondence and Cultural Exchange in Early Modern Europe* (2007).

Susanne Hehenberger is lecturer at the Department of History at the University of Vienna. Her research interests include history of criminal law, criminality, sexuality and religion in early modern period, and gender studies. She is currently working on the provenance of reliquaries collected in the Kunsthistorisches Museum of Vienna and on the prosecution of blasphemy in early modern Austria. She is co-editor of the magazine *Frühneuzeit-Info*.

Annemarie Jordan-Gschwendt is working as an independent scholar (Ph.D. 1994, Brown University). Her areas of specialization include patronage, collecting and curiosity cabinets at the Renaissance courts in Austria, the Netherlands, Spain and Portugal. She is the author of *Retrato de Corte em Portugal. O legado de António Moro (1552–1572)* (1994) and of numerous articles and exhibition catalogues.

Eric Jorink is Researcher at the Huygens Institute (Royal Dutch Academy of Arts and Sciences), The Hague. He is the author of *Het Boeck der Natuere. Nederlandse geleerden en de wonderen van Gods Schepping, 1575–1715* (2006). Currently, he is working on a biography of the Dutch microscopist and anatomist Jan Swammerdam (1637–1680).

Johan Koppenol is professor of Dutch literature at the Vrije Universiteit Amsterdam. He is a specialist in Dutch Renaissance literature and has published on Jan van Hout, Jacob Cats and Joost van den Vondel, and has edited works by P.C. Hooft and Jacob Westerbaen.

Karin Leonhard is Research Assistant at the Institute of Art History at the University of Eichstätt/Ingolstadt. She specializes in seventeenth-century Dutch painting and the scientific revolution of the seventeenth century.

Almudena Pérez de Tudela is an art historian in the Department of Art History at the Universidad Complutense de Madrid and curator of the Real Monasterio de San Lorenzo of the Escorial, Madrid.

Vibeke Roggen is Associate Professor of Classics at Oslo University. Her scholarly interests are Neo-Latin literature and textual criticism. She has published *Intellectual play – word and picture. A study of Nils Thomassøn's Latin rebus book* Cestus sapphicus (2002).

Franziska Schnoor has worked at the Department of Medieval Studies at Göttingen University, and is currently completing her Ph.D. on Latin poetry about animals written in the medieval and early modern eras.

Thea Vignau-Wilberg is Curator emeritus of Dutch and Flemish Prints and Drawings at the Staatliche Graphische Sammlung in Munich. She specializes in sixteenth-century drawing and printing. She has published monographies on Joris Hoefnagel, Christoph Murer, Rembrandt (*Rembrandt auf Papier – Werk und Wirkung*, 2001) and Netherlandish artists in the Munich collections (2005).

Suzanne J. Walker is Assistant Professor for Art History at Tulane University, New Orleans. She is currently working on a book on Rubens and violence.

INTRODUCTION

Karl A.E. Enenkel – Paul J. Smith

Early modern zoology

'The cervix of the lion does not consist of cervical vertebras, but of a single bone', states Conrad Gesner in his *Historia animalium*, volume I, of 1551, when investigating the intriguing anatomy of the mammals. This short remark contains in a nutshell the dazzling complexity of early modern zoology. It obviously reflects a vivid interest in anatomy, one of the new scientific methods of research which are generally associated with the scientific progress ascribed to the early modern period. This progress is thought to be connected with the replacement of 'bookish' knowledge with empirical study. But what was the basis of Gesner's anatomy of the lion? The lion was present in quite a number of 16th-century menageries – did Gesner, who was a medical doctor and a zoologist at the same time, dissect a lion? Or, did he eventually use the dissection reports of others?

The invention of zoology is only one of the fascinating features that the re-definition of the animal and animal species in the early modern period brought about. The discovery of the new world confronted intellectuals with surprising and hitherto unknown species, which challenged the existing systems of animal classification. By the expeditions into the New World, new information was gathered about a strange and exotic fauna. Exotic and European wild animals, as well, were brought together in new places of knowledge – e.g. courtly menageries, curiosity cabinets or academic collections. Artistic progress in the visual arts offered new options for animal description, which were reinforced by the printing press and the reproduction of identical graphic illustrations and their dissemination among larger audiences.

To us, these features may seem very familiar. They suggest close connections between the early modern and modern periods. Nowadays, zoology, as a main branche of biology, is totally established in the system of sciences. Modern zoology is based on empirical methods and furnished with highly developed technical means. By these means, zoology participates in the spectacular progress modern science achieves almost on a daily basis. Furthermore, a large popular interest in zoology exists. Books on animals and animal encyclopedias, both enriched with attractive illustrations, belong to the stock of general book stores. Zoos are spread all over the world and form an indispensable part of the modern entertainment sector. Animal films are popular to a degree that television stations came into being which are exclusively dedicated to nature programmes.

Nevertheless, these striking familiarities may be highly deceptive. It is not a given fact that modern zoology did indeed develop from its early modern counterpart. Its general aims, interests, research methods, organisation of knowledge etc. may differ fundamentally. It seems questionable, for instance, whether its scope was identical. Modern zoology, for example, studies animals for their own sake, and it pays much attention to the endangered status of many species. It has become clear that the animals' most dangerous enemy is man, who rules all over the world. Early modern zoology shows a completely different approach: it departs from the dogma that man has the full right to exploit animals – that their only *raison d'être* is that they serve man, providing food, clothes, means of transport, medicines, entertainment and so on. Furthermore, in early modern zoology one may detect a vivid interest in classification which at first sight seems to be similar to the modern science's approach – but are the principles of classification compatible? As a result of Darwin's evolutionary theory, the understanding of classification has changed profoundly. In addition, it is not clear whether early modern anatomy and physiology, especially their methods and organisation of knowledge, really correspond to their modern counterparts.

In the last decades, an increasing interest in the history of science has come into being, which also affects biology. The history of biology has been the scope of a number of larger projects, e.g. Ilse Jahn's *Geschichte der Biologie*[1] (first edition 1982) or Änne Bäumers[2] (1991) and Ch. Hünemörder's[3] works with the same title. Erik Jorink has researched the approach to nature by Dutch scholars and scientists 1575–1715

[1] Hamburg: 2004 (following the second, corrected special issuing Heidelberg – Berlin: 2002 of the third edition Jena: 1998).

[2] Frankfurt a.M.: 1991.

[3] Stuttgart: 1985.

(2006).[4] Paula Findlen has analysed the representation of 'nature' in early modern collections (1994).[5]

In the studies on the history of biology, botany has received considerably more attention than zoology. In his recent monograph, *The Science of Describing. Natural History in Renaissance Europe* (2006), which is almost entirely dedicated to botany, Brian Ogilvy ascribes this emphasis to the 'preeminence of botany' over early modern zoology: 'My periodization has focussed on botany, the chief focus of natural history in the Renaissance. [...] In fact, zoology generally lagged behind botany in the Renaissance; developments that occurred in botany often took a decade or two to appear in zoology'.[6]

Nevertheless, it does not seem adequate to consider early modern zoology only as a kind of appendix to early modern botany. Zoology deals with different topics and the early modern zoologist struggled with problems, difficulties and challenges that differ considerably from the botanist's. Animals are more complex organisms than plants and they are much more difficult to be analysed and observed. Their mobility and their tendency to hide make observation complicated, to say the least. Their much more complex interior organisation requires advanced research methods, e.g. on the fields of anatomy and physiology. It is also very tricky to captivate and keep animals, especially in the case of exotic and higher organised species. In captivity, they often show a different and atypical behaviour. Empirical anatomy and physiology were comparatively new research interests that required special technical understanding and skills. Just how difficult the understanding of the processes of life was is illustrated by the fact that e.g. the circulation of blood was discovered not earlier than in 1628, by William Harvey.[7]

Already from this small number of arguments it seems rewarding to pay separate attention to zoology. The second volume of Änne Bäumer's *Geschichte der Biologie*, which offers a first description of the

---

[4] Jorink E., *Het Boek der Natuere. Nederlandse geleerden en de wonderen van Gods schepping 1575–1715* (Leiden: 2006).

[5] Findlen P., *Possessing Nature. Museums, collecting, and scientific culture in early modern Italy* (Berkely: 1994).

[6] Ogilvie B., *The Science of Describing. Natural History in Renaissance Europe* (Chicago – London: 2006) 49.

[7] *Exercitatio anatomica de motu cordis et sanguinis in animalibus* (Frankfurt, G. Fitzer: 1628).

zoological treatises of the 'Renaissance', is a pioneering publication in the field.[8]

The editors would like to make clear that the present publication does not have the slightest ambition to compete with this compendium, or the related parts in other compendia, let alone to replace them. It does, however, have the intention to highlight other aspects, to focus attention on different tendencies, to shed light on certain intriguing details and to investigate early modern zoology in connection with the use of animals in various fields, theory and practice alike, such as hunting, horsemanship, veterinary medicine, courtly life, jurisdiction, literature and the visual arts. These aspects are studied by Suzanne Walker, Pia Cuneo, Louise Hill Curth, Annemarie Jordan-Gschwend and Almudena Pérez de Tudela, Susanne Hehenberger, Johan Koppenol, Franziska Schnoor, Vincent Buyens, Paulette Choné and Sarah Cohen. Various aspects of the science of zoology are investigated by Karl Enenkel, Paul Smith, Vibeke Roggen, Erik Jorink, Karin Leonhard, Thea Vignau-Wilberg, Florike Egmond and Rebecca Parker Brienen. The last three articles focus on a specific part of the science, the zoological illustration.

If one looks at general tendencies and methods of approach, the studies in this volume differ in important points from previous works. Ever since Boas's *The Scientific Renaissance 1450–1630*[9] there has been a general tendency to interpret the achievements of early modern science in a teleological way, viz. as forerunners of modern science. Many a time, the attention was focused in a one-sided way on aspects that could be used to demonstrate the scientific progress or, vice versa, the failure to effectuate scientific progress.

With this respect, the history of zoology is no exception. The teleological approach is prominently present in Änne Bäumer's *Geschichte der Biologie*. In general, she interprets the early modern zoologists as forerunners of their modern counterparts. She states that 'die Entwicklung der modernen zoologischen Disziplinen' 'nimmt in der humanistischen Zoologie ihren Anfang'.[10] The 'bewusste Rückbesinnung auf antike methodische Ansätze' directly leads to scientific progress, among others to the differentiation of the 'zoologischen Spezialgebiete

[8] *Zoologie der Renaissance – Renaissance der Zoologie*, in eadem, *Geschichte der Biologie*, vol. 2 (Frankfurt a.M. – Bern – New York – Paris: 1991).

[9] (New York: 1962).

[10] Cf. her own summary of her monograph in "Zoologie der Renaissance – Renaissance der Zoologie", in Döring K. – Wöhrle G. (eds.), *Antike Naturwissenschaft und ihre Rezeption*, Bd. I und II (Bamberg: 1992) (275–295) 279.

Embryologie, Physiologie, Vergleichende Anatomie'.[11] In her opinion, the early modern zoological compendia (Pierre Gilles, Michael Herr, Adam Lonitzer, Edward Wotton, Conrad Gesner etc.) 'wiesen in ihren methodischen Ansätzen der zukünftigen Forschung den Weg'.[12] Their attention for anatomy provided the 'entscheidende Kriterien zur Einteilung (classification) der Tiere [...]. Damit waren alle Ansätze für eine adäquate Behandlung des Tierreichs gegeben, die neuzeitliche Zoologie war begründet'.[13] Bäumer identifies Renaissance zoology with the early modern scientific revolution: 'Die Renaissance der zoologischen Forschung ist hier gleichzusetzen mit dem, was man gemeinhin als wissenschaftliche Revolution bezeichnet'.[14]

Progress, however, although one of the best accepted dogmas of modern science, is not an adequate tool for the understanding of the functioning and organisation of *early modern* science. The authors of this volume, therefore, do not depart from a teleological conception of the history of science. They make an effort to investigate the various discourses in which animals are discussed in zoological treatises and other genres of literature and art. Their approach is characterised by awareness that these discourses may differ fundamentally from modern ones, and that they are connected to specific historical contexts, interests, needs, and literary, theological, philosophical and artistic traditions. If one carefully studies these parameters, a rich and complex picture of early modern zoology appears, with a broad spectrum of varieties and differentiations. It is a sparkling and intriguing picture, which is characterized above all by *its striking alterity and discontinuity from modern science.*

The authors of this volume do not think that in early modern zoology there is a clear chronological development from bookish to empirical knowledge, from "traditional" to "objective" description, from symbolical to "realistic" illustration etc. Various methods of animal description may occur at the same time or in "reverse" order. Most important are the specific historical contexts, interests, needs and the literary, theological, philosophical and artistic discourses. By investigating within those discourses and contexts, it appears that early modern men did not so

[11] Ibidem.
[12] Ibidem 280.
[13] Ibidem.
[14] Ibidem 285.

much objectively describe or depict animals; rather, they *construct* animals according to the above mentioned parameters.

Karl Enenkel analyses the discourses in which the mammals were constructed in two most important zoological compendia, Edward Wotton's *De differentiis animalium* and Gesners *Historia animalium*, volume I, with special reference to the species *Panthera leo*; Paul Smith investigates the remarkable descriptions of the exotic toucans and hornbills, which appear to be based to a much higher degree on intertextuality than one may suppose at first sight. Erik Jorink demonstrates that "advanced" miscroscopical research and "traditional" emblematic understanding of insects existed side by side, at the same time and sometimes by the same person. Suzanne Walker investigates the puzzling construction viz. deconstruction of the stag in early modern hunting treatises such as Jacques du Fouilloux's *La Vénerie* or George Gascoigne's *The Noble Art of Venerie or Hunting*. Pia Cuneo shows how a "traditional" animal like the horse was fundamentally re-constructed in early modern German theory and practice. Karin Leonhard demonstrates in which way the knowledge of shells was organized in early modern curiosity cabinets and choncological treatises. Almudena Pérez de Tudela and Annemarie Jordan Gschwend shed light on the role exotic animals played at the Habsburg courts of Iberia and central Europe.

In general, the editors have made an effort to have represented in this volume the most important categories of animals. *Mammals* are treated in the studies of Karl Enenkel, Rebecca Brienen, Suzanne Walker, Pia Cuneo, Louise Hill Curth, Johan Koppenol, Franziska Schnoor, Sarah Cohen and Vincent Buyens; *birds* by Paul Smith, Vibeke Roggen, Rebecca Brienen, Johan Koppenol and Vincent Buyens; *fishes* by Florike Egmond; *insects* by Erik Jorink and Thea Vignau-Wilberg; lower organised animals by Karin Leonhard.

Because of the discovery of the New World it is of course tempting to understand early modern zoology by the descriptions of exotic species. Indeed, in recent publications exotic animals have received much more attention than species from the "Old World". In this collection of essays, however, we tried to find a kind of equilibrium between exotic and traditionally known species. Indigenous animals like the horse (Cuneo, Hill Curth, Koppenol), the cow (Hill Curth, Hehenberger), the crane, swan and beaver (Buyens), the dog, cat, sheep (Koppenol), the deer (Walker, Cohen, Koppenol) get as much attention as exotic species from the New World (Pérez de Tudela – Jordan Gschwend, Smith, Brienen, Koppenol) or the Old World (Enenkel, Koppenol, Leonhard, Smith). It appears that the species of the New World do not offer us a key to understand the systems of animal classification used in the early modern period. Much more so, the zoological discourses of the early modern period were dominated by the literary tradition of classical antiquity. The authorisation of zoological knowledge by Aristotle, Pliny, Athenaeus and others even in cases in which empirical knowledge *was* available, forms in fact one of the fascinating features of early modern zoology.

Zoology and the arts

Most studies of animals in early modern literature and the visual arts tend to foreground the symbolic significance of the animals represented, neglecting somehow their zoological reality or their natural-historical conceptualisation. An example of this 'symbolist' approach is the recently published *Lexikon der Tiersymbole* (2004) by Sigrid and Lothar Dittrich.[15] Although this impressive work does not totally neglect the animals' reality and early modern reflection on them, its main accent – as indicated by its title – is on their symbolism. Studies that follow a more 'naturalistic' perspective, such as Arianne Faber Kolb's recent monograph on Jan Brueghel's *Entry of the Animals into Noach's Ark*,[16] which looks at the painting as 'a visual catalogue of animals and birds function[ing] as a type of microencyclopedia',[17] are indeed rare.

The art historical and literary contributions of the present volume of *Intersections* aim to continue along this naturalistic line of approach: how are the zoological reality and early modern reflection on natural history thematised in the literary and visual arts? How are animals de- and re-constructed, as it were, in the arts? This emphasis on the zoological reality and early modern reflection on the natural world comes to the fore in the article by Johan Koppenol on animals in Dutch poetry of the sixteenth and seventeenth centuries. Although Koppenol also deals with the more traditional symbolic and moralising aspects of animals, his main focus is on the seventeenth-century poets' vision of 'wild' and exotic animals and domestic animals in everyday life.

The contributions to the present volume show that the subject of the de- and re-construction of the animal from natural history to the field

[15] Dittrich S. – Dittrich L., *Lexikon der Tiersymbole. Tiere als Sinnbilder in der Malerei des 14.–17. Jahrhunderts* (Petersberg: 2004).
[16] Kolb A.F., *Jan Brueghel the Elder. The Entry of the Animals into Noach's Ark* (Los Angeles: 2005).
[17] Kolb, *Jan Brueghel* 27.

of the arts is highly problematic. Indeed, literature and painting are so much closely connected with zoology that it is often impossible to arrive at a clear-cut distinction between the fields. This is, for instance, the case – studied by Paul Smith – with the French eighteenth-century zoologist Buffon, *for whom the ideal zoological description is in fact a literary one*, based upon the predominance of style and stylistic variation. The same confusion can be observed with respect to the borderline between zoological illustrations (i.e. illustrations made for a zoological publication) and more 'artistic' paintings. In her contribution, Rebecca Brienen shows how Albert Eckhout and Georg Marcgraf, both appointed to document the newly found *exotica* during Johan Maurits's expedition to Brazil, come up with very different results: Eckhout with highly artistic oil and chalk sketches, Marcgraf with rather flat but precise watercolours that were intended for publication, but did not end up in print as it should be, because of the inadequacy of the anonymous engraver.

These examples show us how important it is to study in detail the interconnections between watercolours, engravings and zoological texts. The general connections between illustrations and texts are more or less known, since for instance the well documented catalogues by Jan Balis, *Van diverse pluimage. Tien eeuwen vogelboeken* (1968) and Laurent Pinon, *Livres de zoologie de la Renaissance* (1995).[18] These interconnections changed with the growing technical perfection of printing and illustrating – from the rather crude and clumsy woodcuts of the beginnings to the refined ones of the second half of the sixteenth century, and their gradual replacement by other techniques: copper engravings (seventeenth century) and lithography (eighteenth century). In this development the zoological discourse leans more and more on the illustration. This growing dependency can even imply a reversal: the illustration becomes more important than the text – a development also visible in other scientific fields, such as medicine and botany.

From a teleological perspective, this development has sometimes had a negative influence on the history of zoology. There are indeed some well-known examples of inaccurate or fictitious illustrations, such as Dürer's rhinoceros, which is uncritically reproduced in zoological works until the eighteenth century, or the illustration of the sloth, represented

[18] Balis J., *Van diverse pluimage. Tien eeuwen vogelboeken* (Antwerp – The Hague – Brussels: 1968); Pinon L., *Livres de zoologie de la Renaissance. Une anthologie (1450–1700)* (Paris: 1995).

since Thevet's *Singularitez de la France antarctique* (1557), not as climbing or hanging upside-down in a tree, but as standing on its four legs. Other examples, studied in the present volume, are the imaginary toucan that Gesner constructed out of a bill and some incomplete descriptions of the bird he received from his informants, and the strange 'tattooed tuna', a fish with painted ships on its body, studied in detail by Florike Egmond.

All these instances point to the dangers of quick generalisation and the aforementioned 'teleological' perspective. Indeed, from a *non*-teleological perspective, the case of a zoologist like Pierre Belon shows us in his *Histoire de la nature des oyseaux* (1555) the complexity of the problem of intermediation in zoological works. On the one hand, Belon confesses that he is totally dependent on the work of the painters of his watercolours: 'il n'y a description ni portrait d'oyseau en tout cest oeuvre, qui n'est esté devant les yeux des peintres'.[19] On the other hand, the engravings, based on the watercolours, do nothing but confirm the text. According to Philippe Glardon, in his edition of the *Histoire de la nature des oyseaux* 'tout se passe comme si Belon, et le ou les illustrateurs [...], conscients de l'incomplétude foncière de leur mode de représentation, éprouvaient le besoin de mettre l'accent sur ce qui est déjà perceptible au niveau du texte, les marques, pour éviter que l'oiseau ne s'échappe du cadre qui lui est assigné'. This explains the conclusion of Belon, which will make a curious and paradoxical impression on the (modern) reader: 'admonestons le Lecteur, qu'il ait plus desgard à la description que luy en baillerons, qu'aux couleurs du peinctre, et traicts du tailleur'. In the last instance, for Belon the printed word seems to be superior to the painted or printed image.

The case of Belon argues in favour of more detailed attention to the role of sketches and watercolours in the process of writing and printing. In her contribution, Thea Vignau-Wilberg studies the illustration of insects in the different media of watercolour, painting and print from Dürer to the generation of Joris Hoefnagel, Ulisse Aldrovandi, Thomas Muffet (or Moufet) and their successors, while Karin Leonard studies with the same precision the interconnections between conchological sketches, printed illustrations and painted still-lives.

This attention is of recent date: one thinks for instance of the beautiful watercolours on which Aldrovandi based his illustrations, preserved

[19] All quotations are from Pierre Belon du Mans, *L'Histoire de la nature des oyseaux. Facsimilé de l'édition de 1555*, ed. P. Glardon (Geneva: 1997), "Introduction" LXV–LXVI.

in the form of albums in the Aldrovandi Museum in Bologna and only recently made available in print and on internet.[20] Aldrovandi's albums, however, have another function too: they complete his own cabinet of curiosities, because a collection of *naturalia* alone can never do justice to the richness of Nature. Moreover, these albums are intended to document the collection, which is often composed of very transitory zoological material, due to the imperfections of contemporary taxidermy. The album of watercolours can function as a catalogue, as in the case of the watercolours of the Museum of Rudolph II[21] and the paintings made by Anselmus De Boodt.[22] In some instances these catalogues were put into print, as in the cases of Albert Seba (who had all the printed copies coloured by hand) and Ole Worm. In the present volume the contributions of Rebecca Brienen and Florike Egmond deal with the multifunctionality of this kind of album. Florike Egmond studies in detail the *libri picturati* painted in the Southern Netherlands during the 1560s under the patronage of the aristocratic collector Charles de Saint Omer and the 'Fish Books' made by Adriaen Coenen,[23] whereas Rebecca Brienen focuses on the *libri picturati* of Marcgraf and Eckhout.

Another important lacuna which the present volume tries to fill concerns the 'naturalistic' approach to the arts – an approach that is clearly different from more iconographical approaches. This volume does not pretend to cover the subject in its whole breadth, but it does present some interesting case studies, which will certainly stimulate other research in the field. Moreover, our point of departure differs from most other approaches, because we do not focus on the presence and symbolism of one particular animal species in one or more genres (like H.W. Janson's study of apes and ape lore in the arts, or Michael Bath's monograph on the iconography of the stag),[24] nor do we provide an

---

[20] Biancastella A. (ed.), *Les animaux et les créatures monstrueuses d'Ulisse Aldrovandi* (Arles: 2005). Also available on Internet: http://www.filosofia.unibo.it/aldrovandi/default.htm.

[21] See Haupt H. – Vignau-Wilberg T. – Irblich E. – Standinger M., *Le Bestiaire de Rodolphe II. Cod. min. 129 et 130 de la Bibliothèque nationale d'Autriche* (Paris: 1990).

[22] Maselis M.-C. – Balis A. – Marijnissen R.H., *De albums van Anselmus De Boodt (1550–1632). Geschilderde natuurobservatie aan het Hof van Rudolf II te Praag* (Tielt: 1989).

[23] See also Egmond F. – Mason P. (eds.), *The Whale Book. Whales and other marine animals as described by Adriaen Coenen in 1585* (London: 2003) and Egmond F., *Het Visboek. De wereld volgens Adriaen Coenen 1514–1587* (Zutphen – The Hague: 2005).

[24] Janson H.W., *Apes and Ape Lore in the Middle Ages and the Renaissance* (London: 1976); Bath M., *The Image of the Stag. Iconographic Themes in Western Art* (Baden-Baden: 1992).

exhaustive survey of the presence of all of the animals in the work of one particular artist (like Colin Eisler's study of Dürer's animals)[25] or author (like Lazare Sainéan's book on the natural history in Rabelais' novels).[26] Our starting point lies more in the work or works themselves and in the way they reflect upon animals.

In her contribution Paulette Choné studies the role of 'natural antipathy' in an illustrated architectural treatise by the French architect Joseph Boillot. This treatise deals with the ways in which the descriptive parameter of antipathy (the relationship of enmity between different zoological species, as explained by Gesner in section D of most of his zoological descriptions) can be used in the decoration of architectural orders.[27] Vincent Buyens deals with the perception of natural history in Willem van der Borcht's *Sedighe Sinne-beelden* (1642), a Dutch collection of animal emblems that goes back not to the medieval animal allegories, but to a Latin medical encyclopaedia: the *Historia medica* (1639) by the Brussels doctor Willem van den Bossche. Franziska Schnoor studies how knowledge about, and more generally human contact with animals are thematised in collections of proverbs of the fifteenth and sixteenth centuries, and her conclusion is that human relationships confer on the genre of the proverb its *raison d'être*.

One of the surprising new themes presented in two of the contributions to this volume concerns the much discussed hierarchy between man and animal. In his provocative *Apologie de Raymond Sebond*, the French sceptical philosopher Michel de Montaigne problematises the supposed superiority of humans to animals. Humans and animals are much alike: 'il se trouve plus de difference de tel homme à tel homme que de tel animal à tel homme'.[28] In this context Montaigne proclaims a certain empathy with animals: 'Quand je me jouë à ma chatte, qui sçait si elle passe son temps de moy plus que je ne fay d'elle?'[29] These provocative statements aroused reactions from, among others, Descartes, who considered the animal as a 'beast-machine'. In her contribution

---

[25] Eisler C., *Dürer's Animals* (Washington – London: 1991).

[26] Sainéan L., *L'histoire naturelle et les branches connexes dans l'oeuvre de Rabelais* (Paris: 1921).

[27] For the use of the same parameter in the work of Jan Brueghel the Elder, see Smith P.J., "Sympathy in Eden. On Rubens's and Brueghel's *Paradise with the Fall of Man*", forthcoming in *Intersections* 9.

[28] Michel de Montaigne, *Oeuvres complètes*, eds. Thibaudet A. – Rat M. (Paris: 1962) 444 (*Essais*, Book II, chap. 12).

[29] Montaigne, *Oeuvres* 430.

Paulette Choné demonstrates how, in presenting his animals as powerful, upstanding creatures, but all at man's service, Boillot enters into discussion with Montaigne's ideas on the resemblance between men and beasts. Sarah Cohen demonstrates that the Descartes-Montaigne controversy on animals is echoed in the Netherlandish game piece as practised by painters like Frans Snyders, Jan Fyt and Jan-Baptist Weenix.

The editors hope that the present volume offers a glimpse of the intriguing variety of discourses on animals among early modern scientists, writers and artists and that it will stimulate further research in this fascinating field.

COLOUR PLATES
I – XXXI

[PLATE I. JORINK – Fig. 4, p. 171]

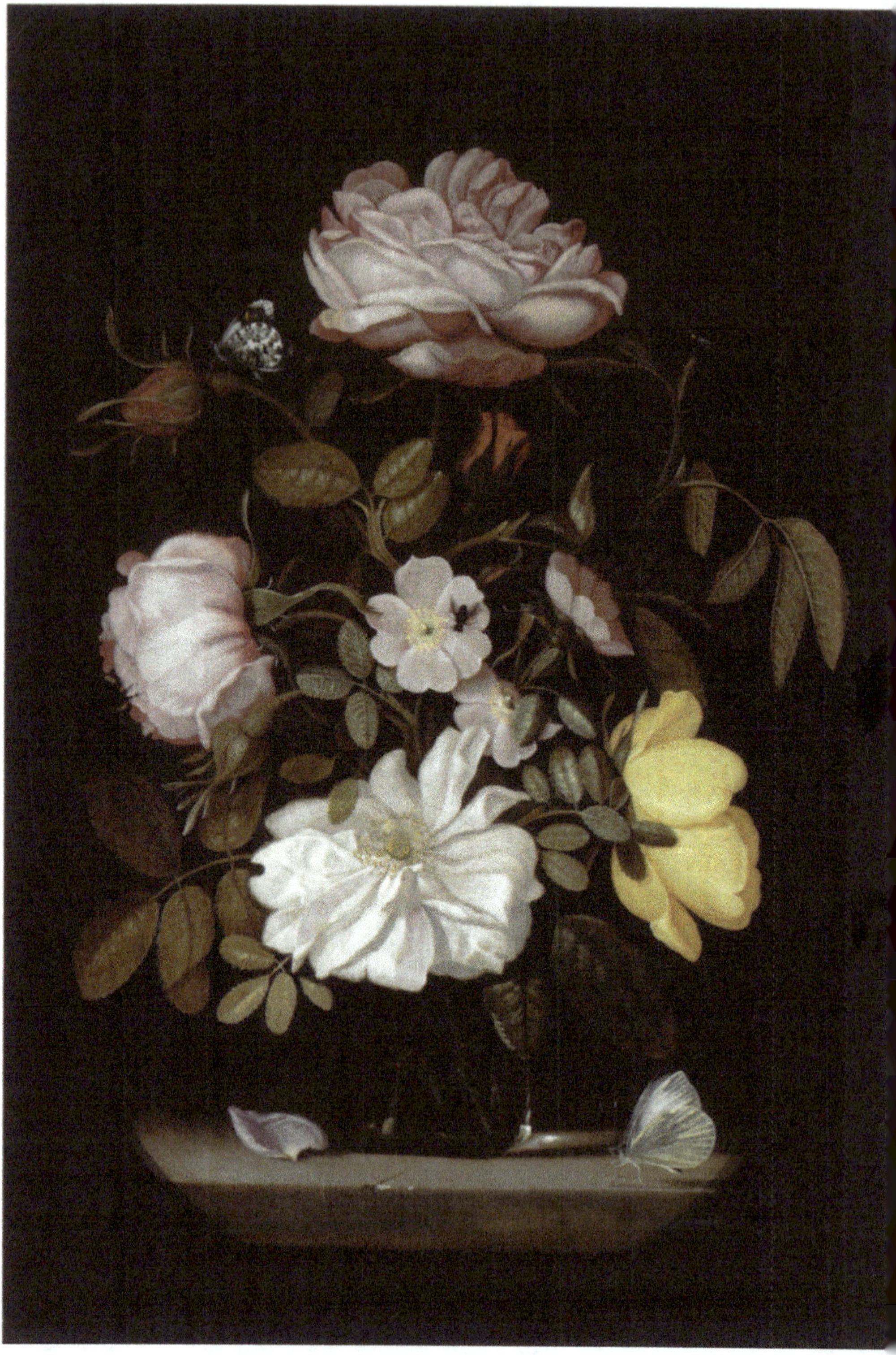

[Plate II. Leonhard – Fig. 1, p. 178]

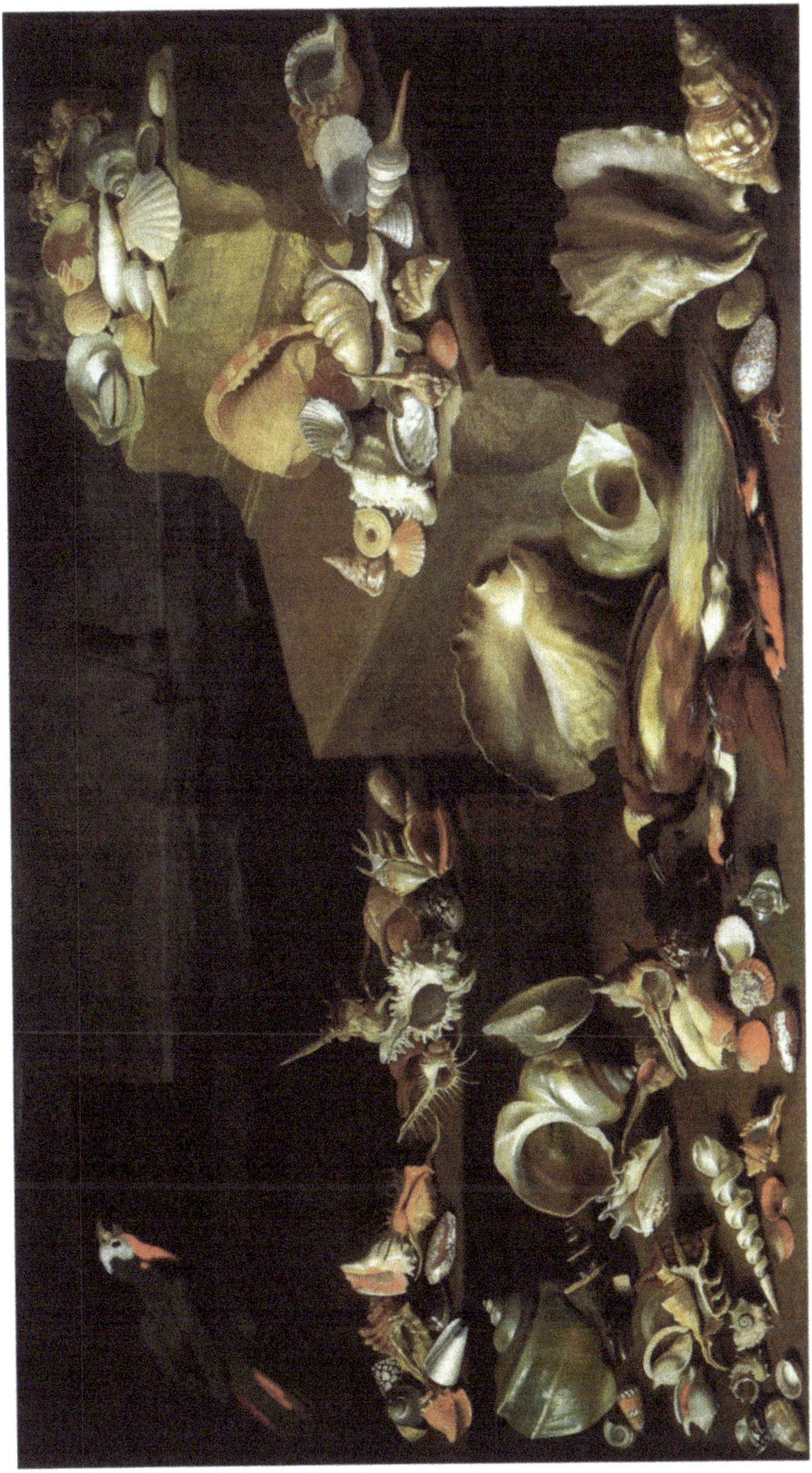

[PLATE III. LEONHARD – Fig. 3, p. 184]

[Plate IV. Leonhard – Fig. 5, p. 187]

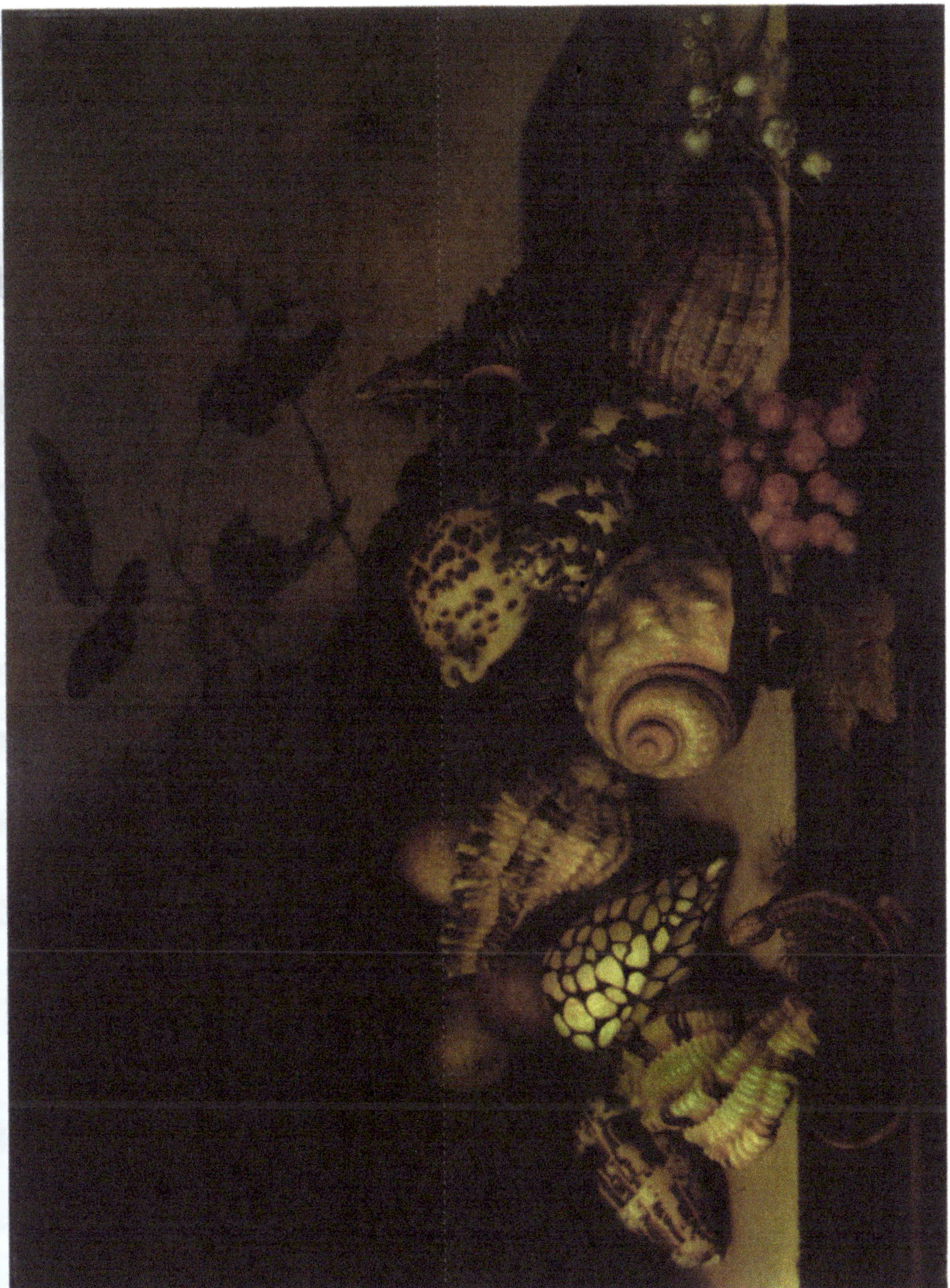

[PLATE V. LEONHARD – Fig. 6, p. 189]

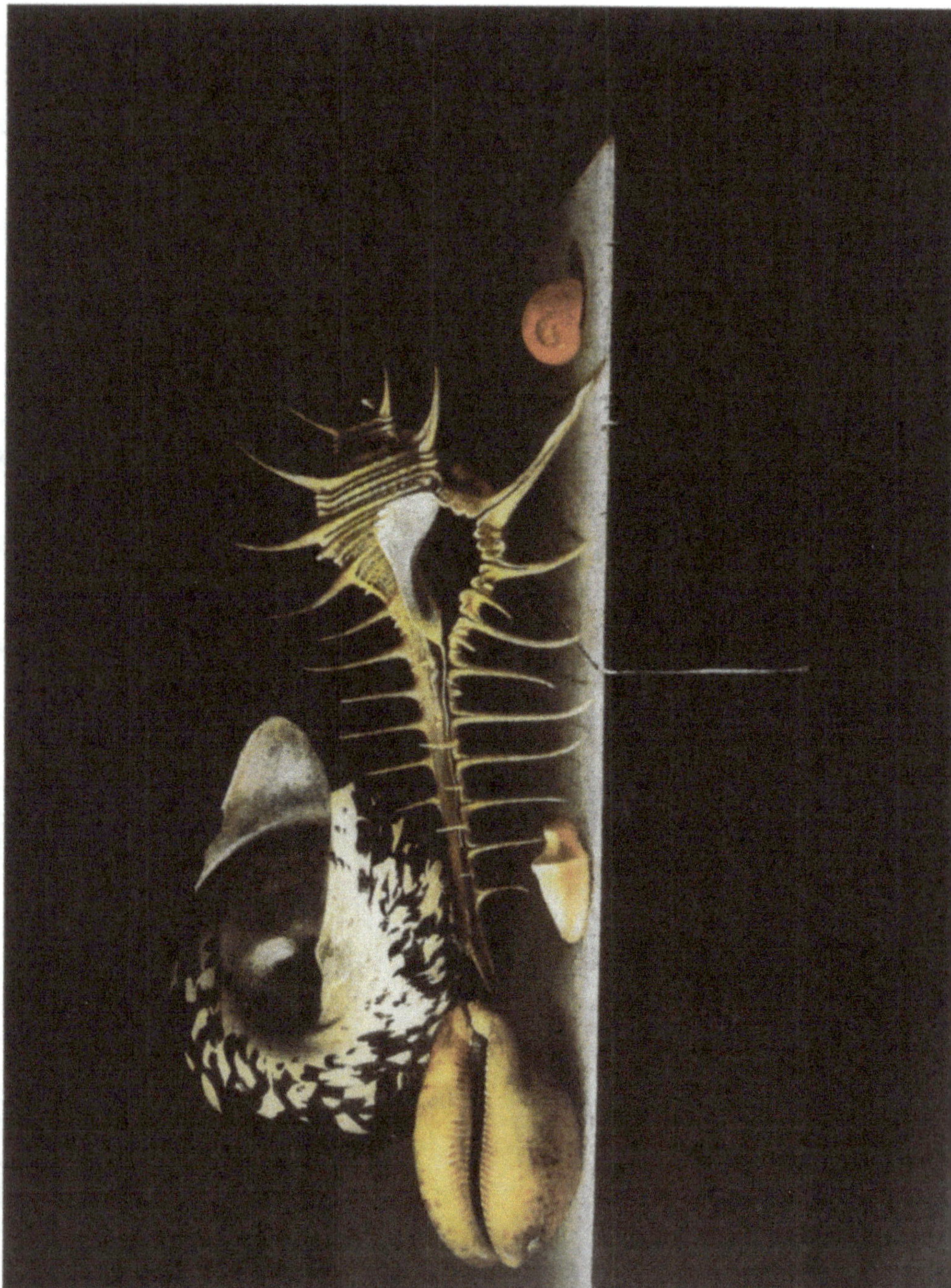

[PLATE VI. LEONHARD – Fig. 13, p. 206]

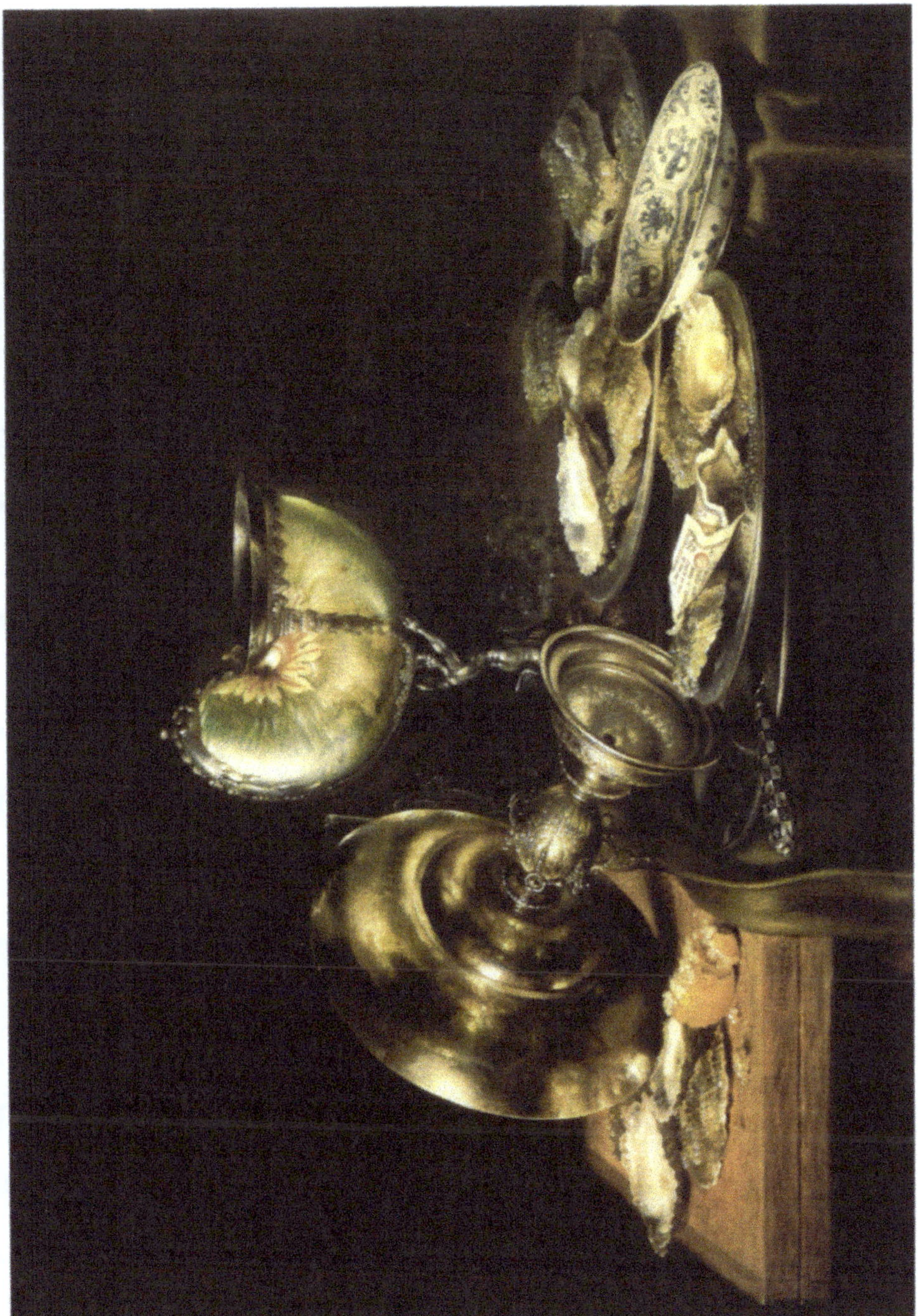

[Plate VII. Vignau-Wilberg – Fig. 3, p. 224]

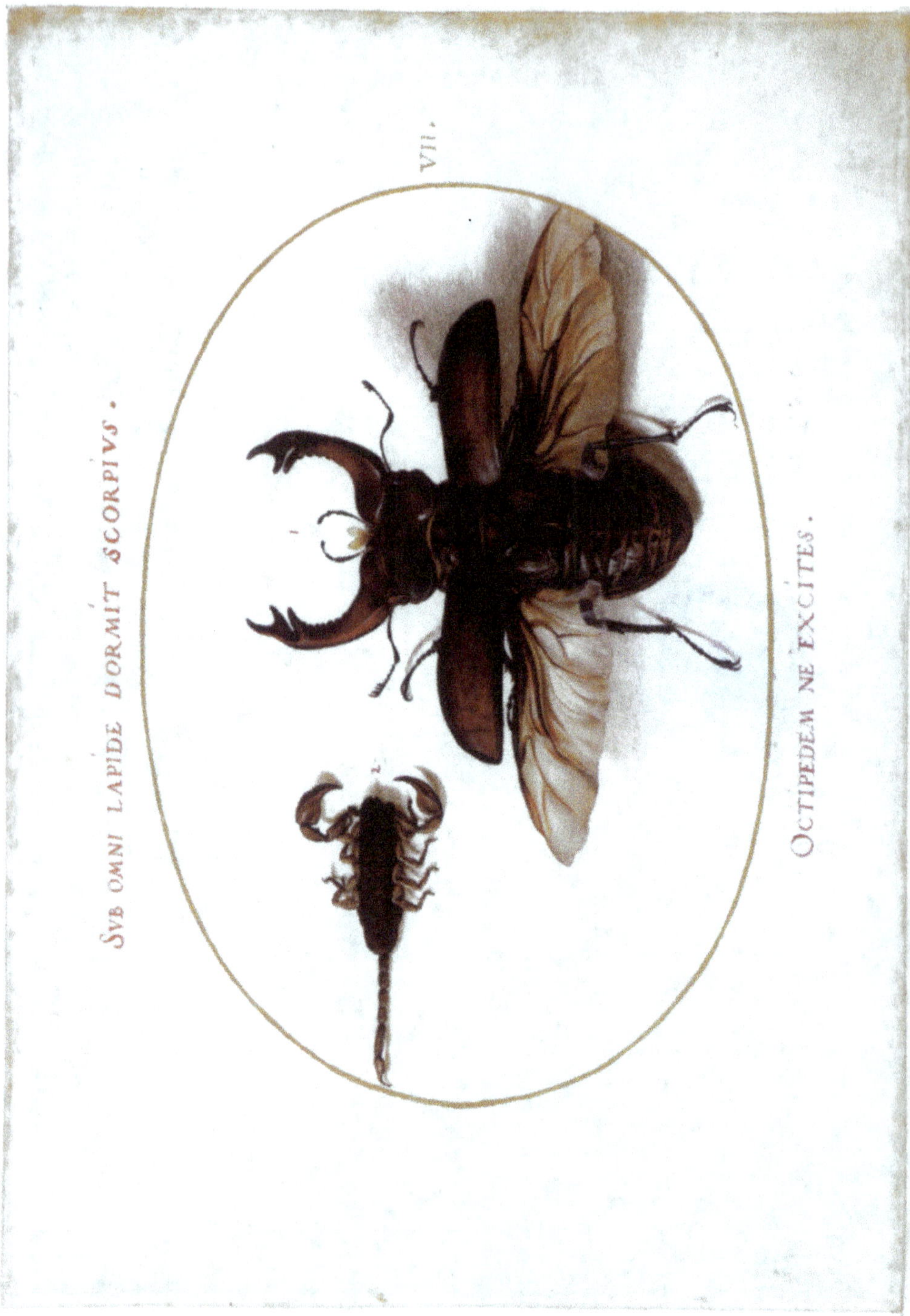

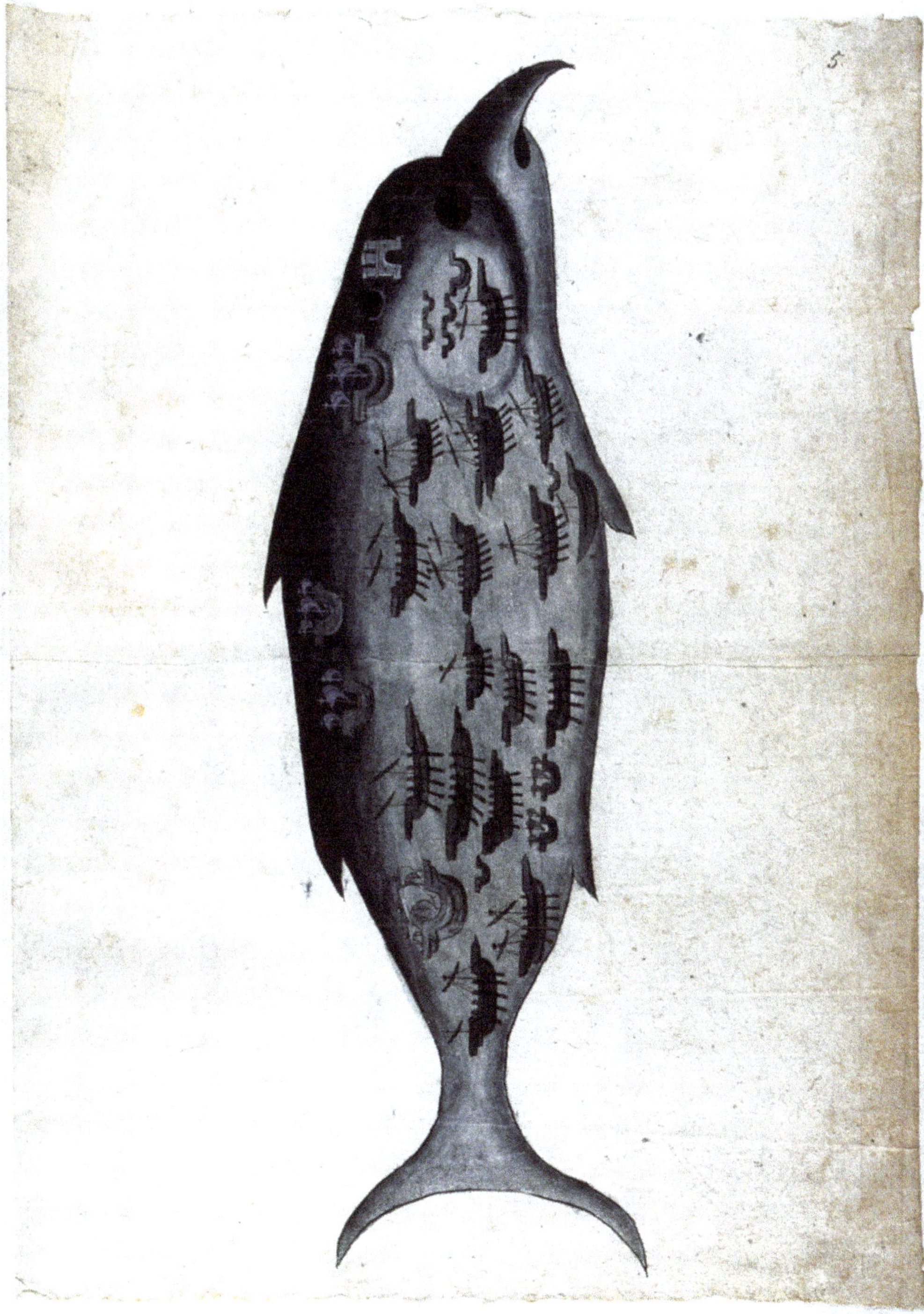

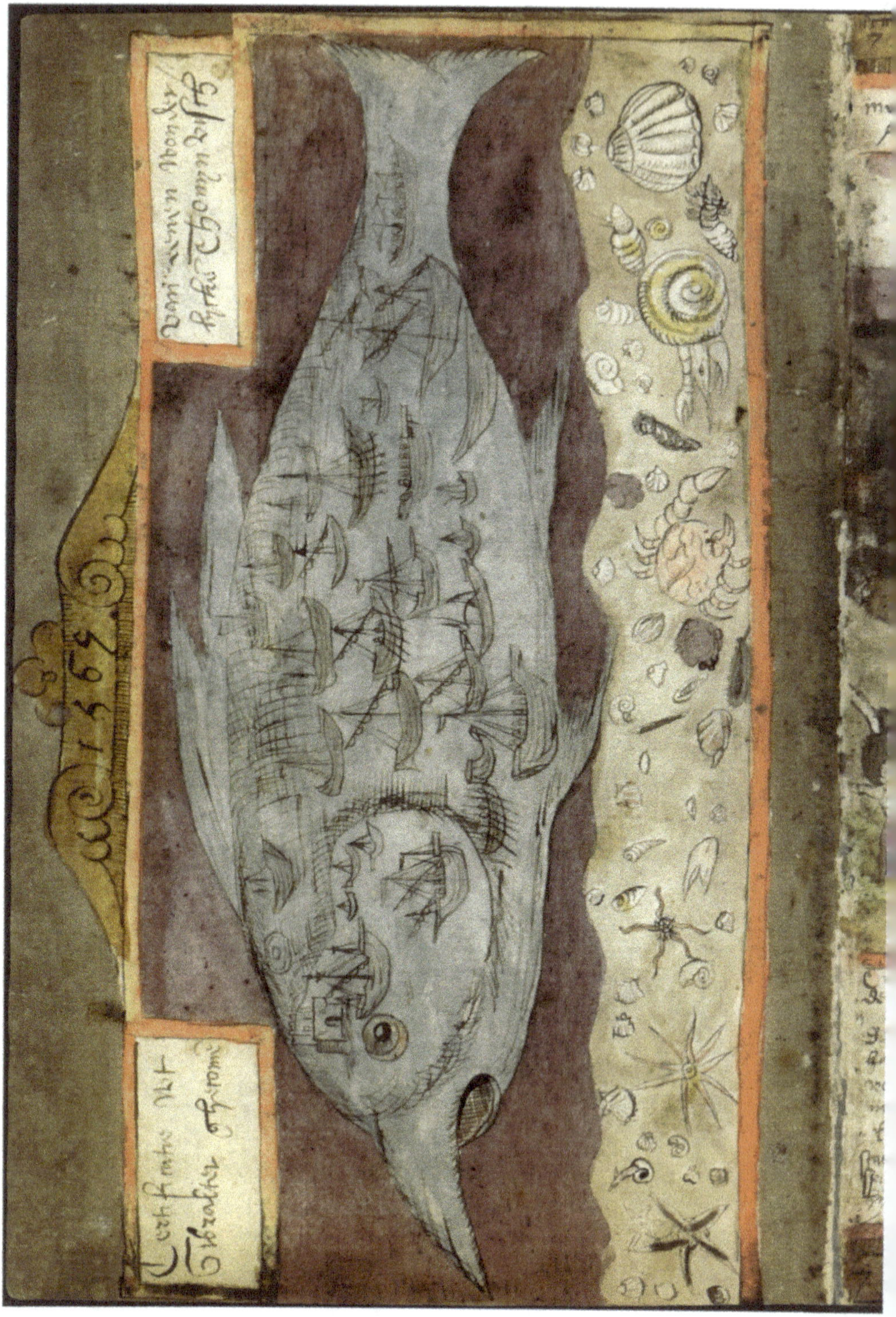

[Plate X. Egmond – Fig. 3, p. 251]

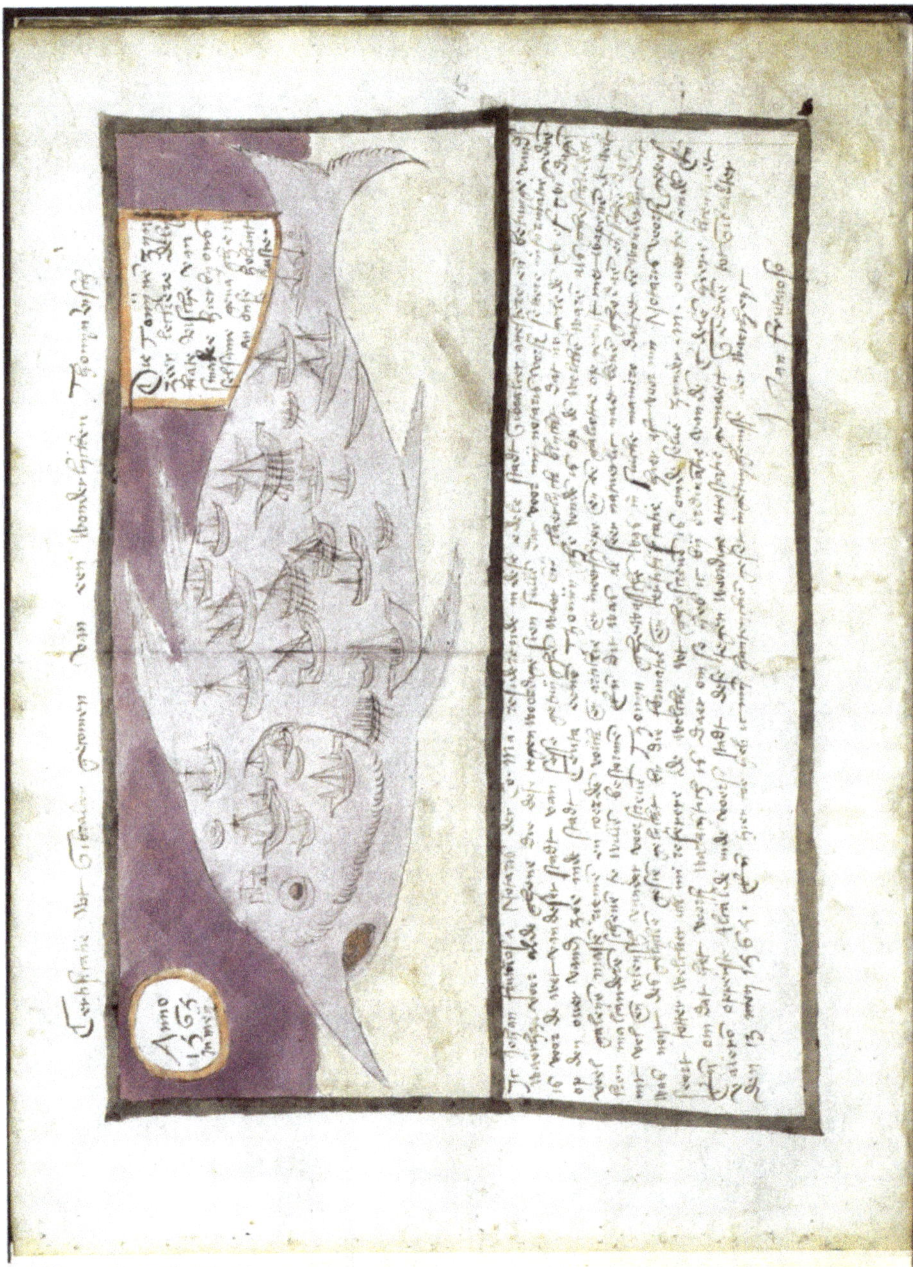

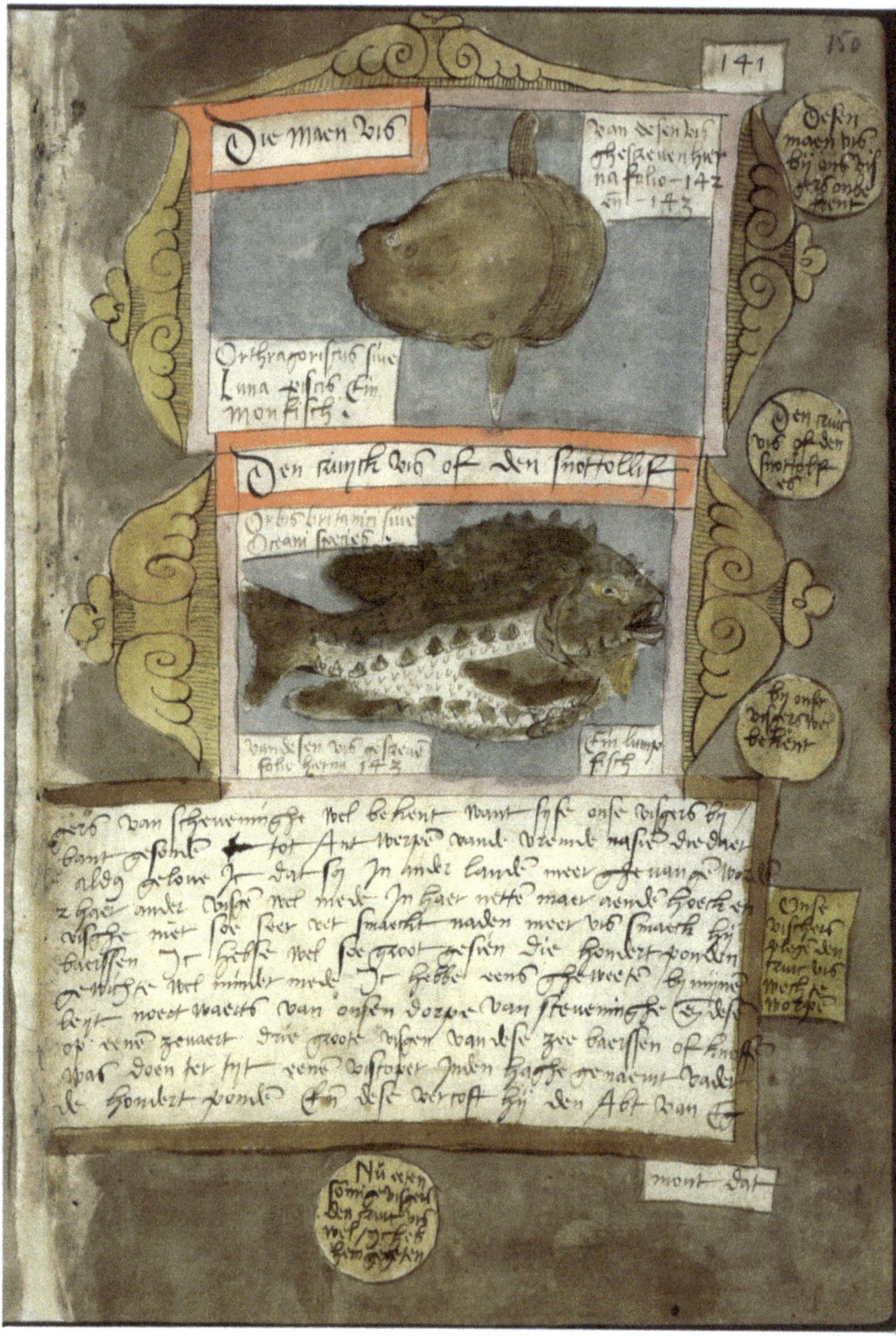

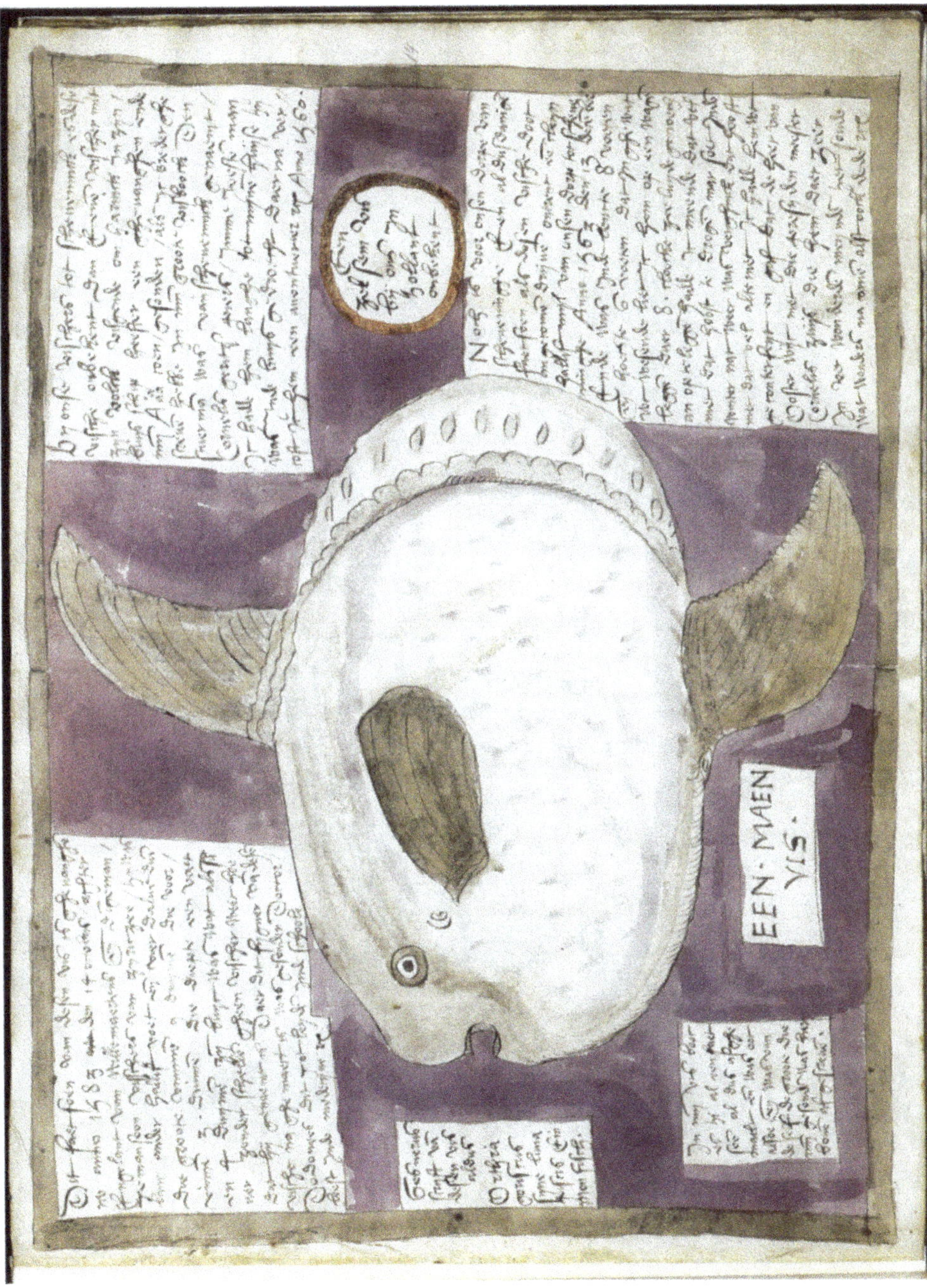
EEN· MAEN VIS·

[Plate XIII. Brienen – Fig. 1, p. 283]

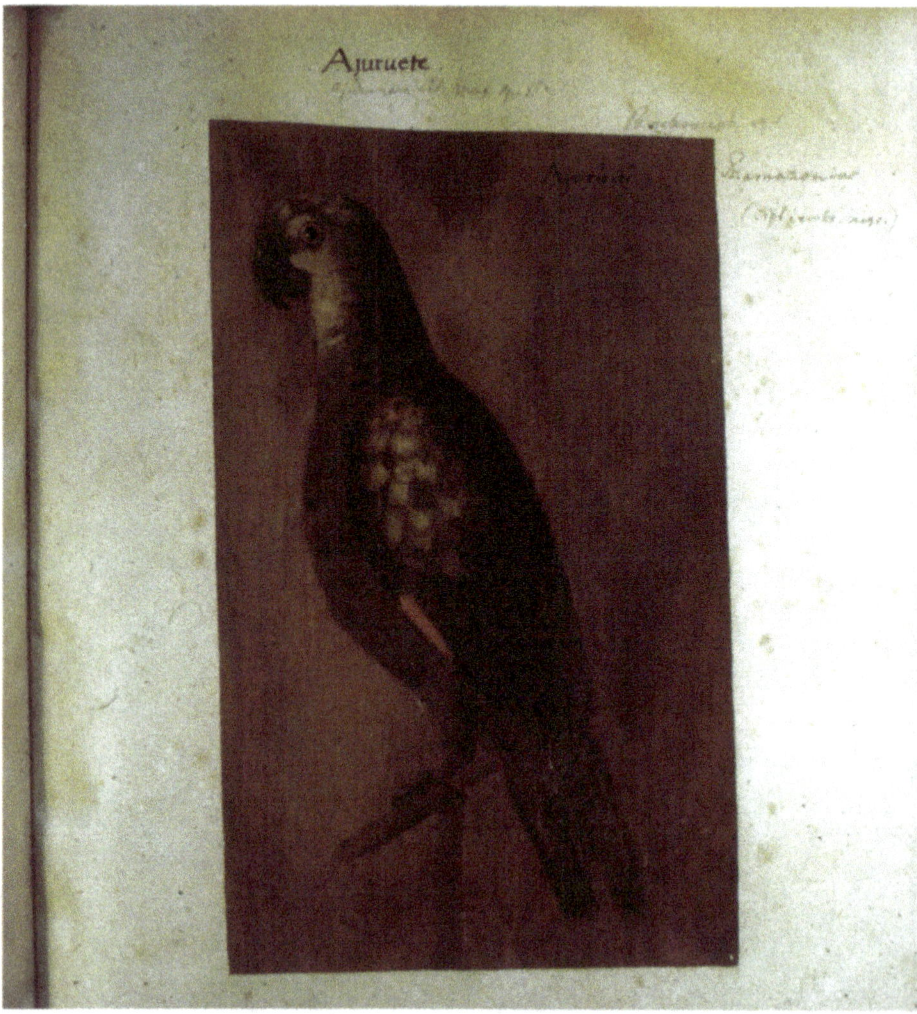

[PLATE XIV. BRIENEN – Fig. 2 (detail), p. 284; Fig. 12 (detail), p. 304]

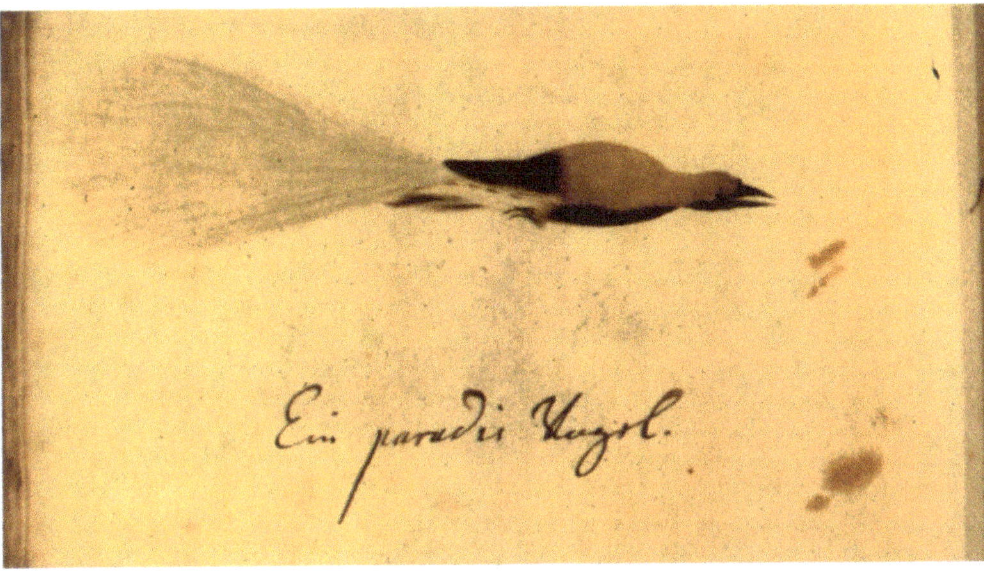

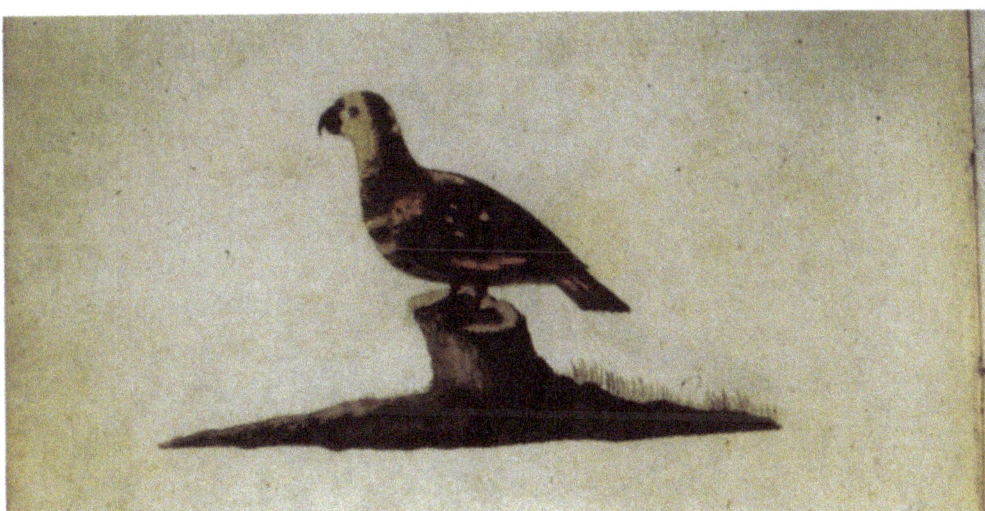

[PLATE XV. BRIENEN – Fig. 4 (detail), p. 287; Fig. 5 (detail), p. 289]

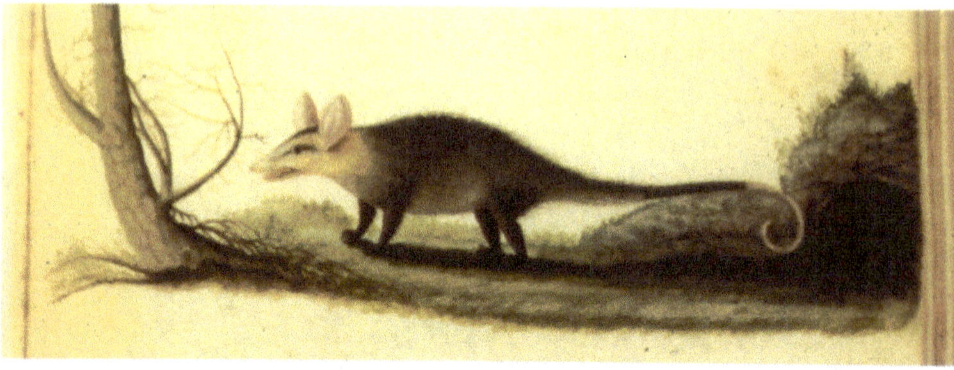

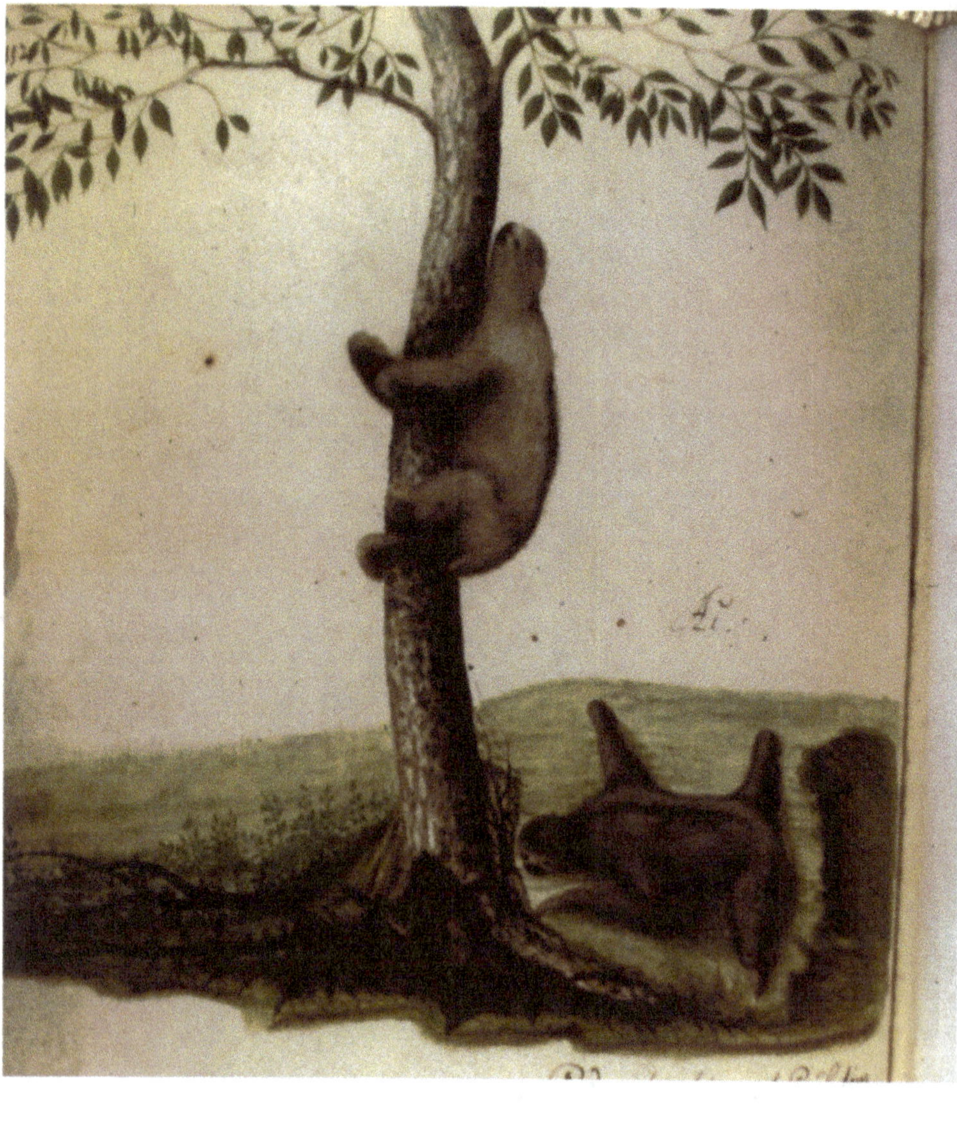

[PLATE XVI. BRIENEN – Fig. 6 (detail), p. 290]

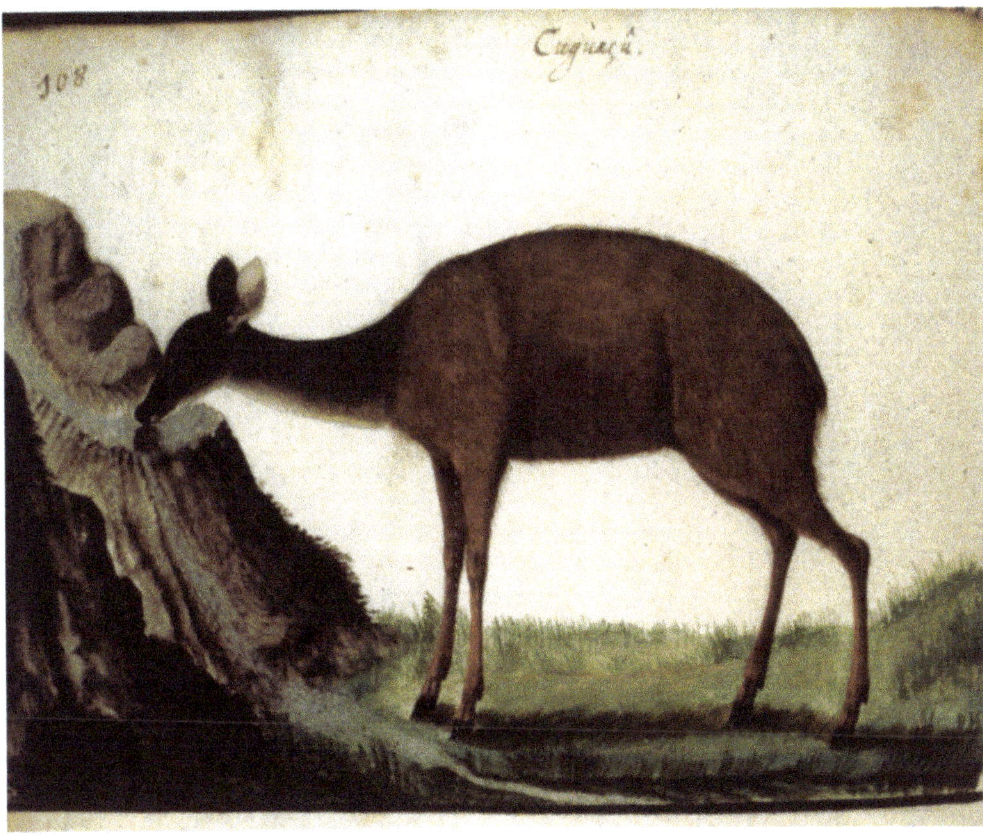

[PLATE XVII. BRIENEN – Fig. 7, p. 293]

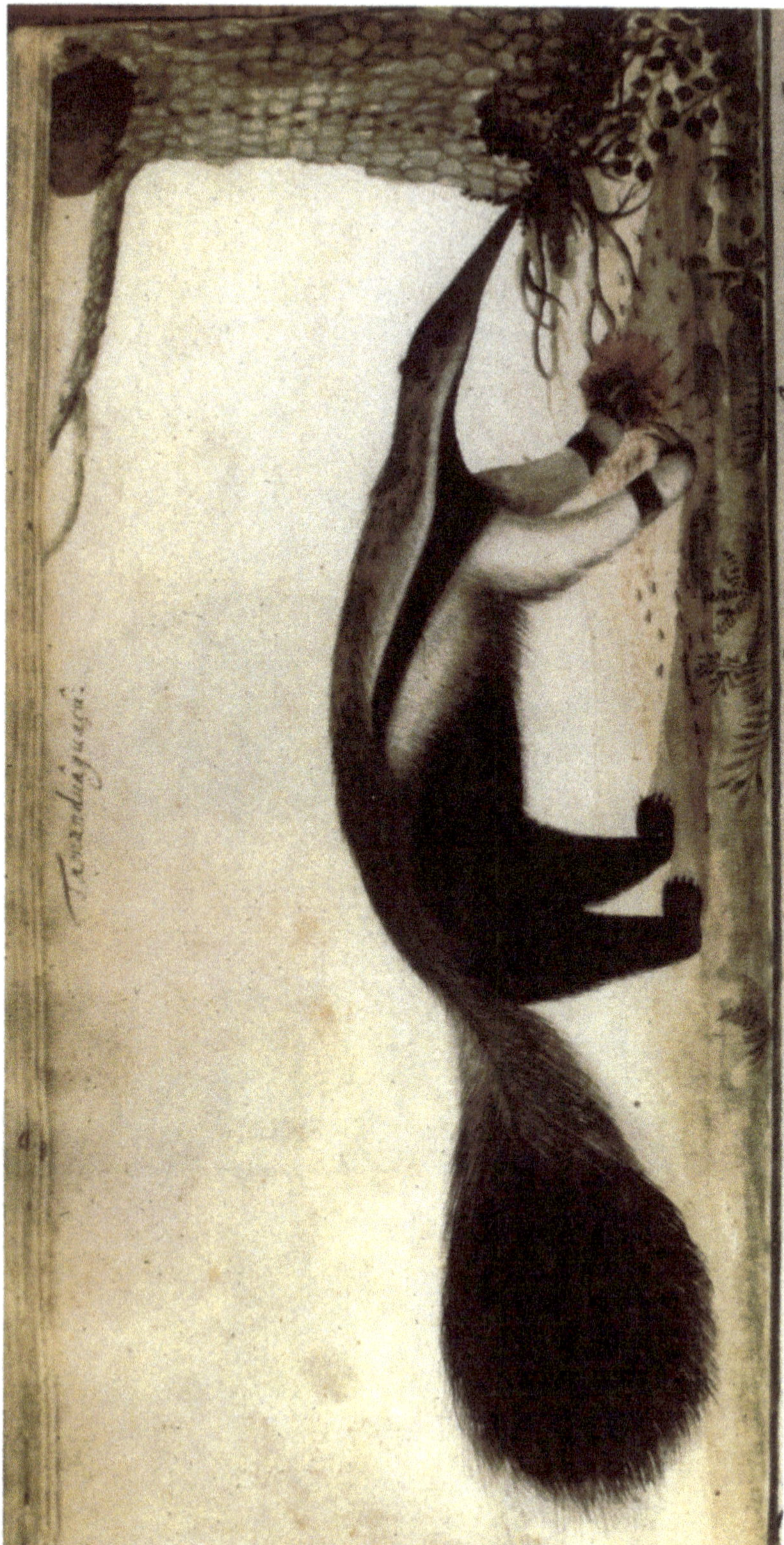

[PLATE XVIII. BRIENEN – Fig. 9, p. 300]

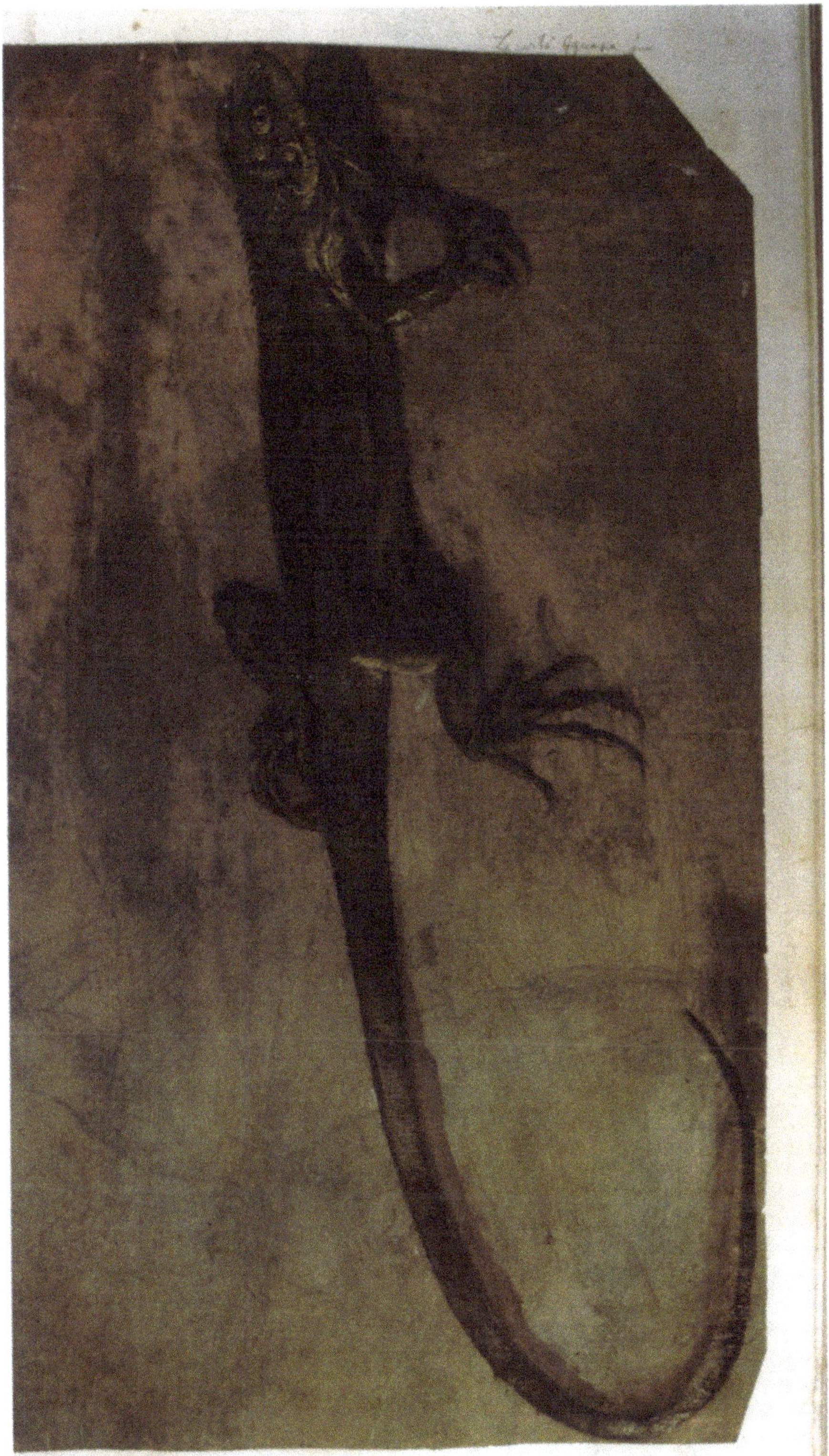

[Plate XIX. Brienen – Fig. 10, p. 301]

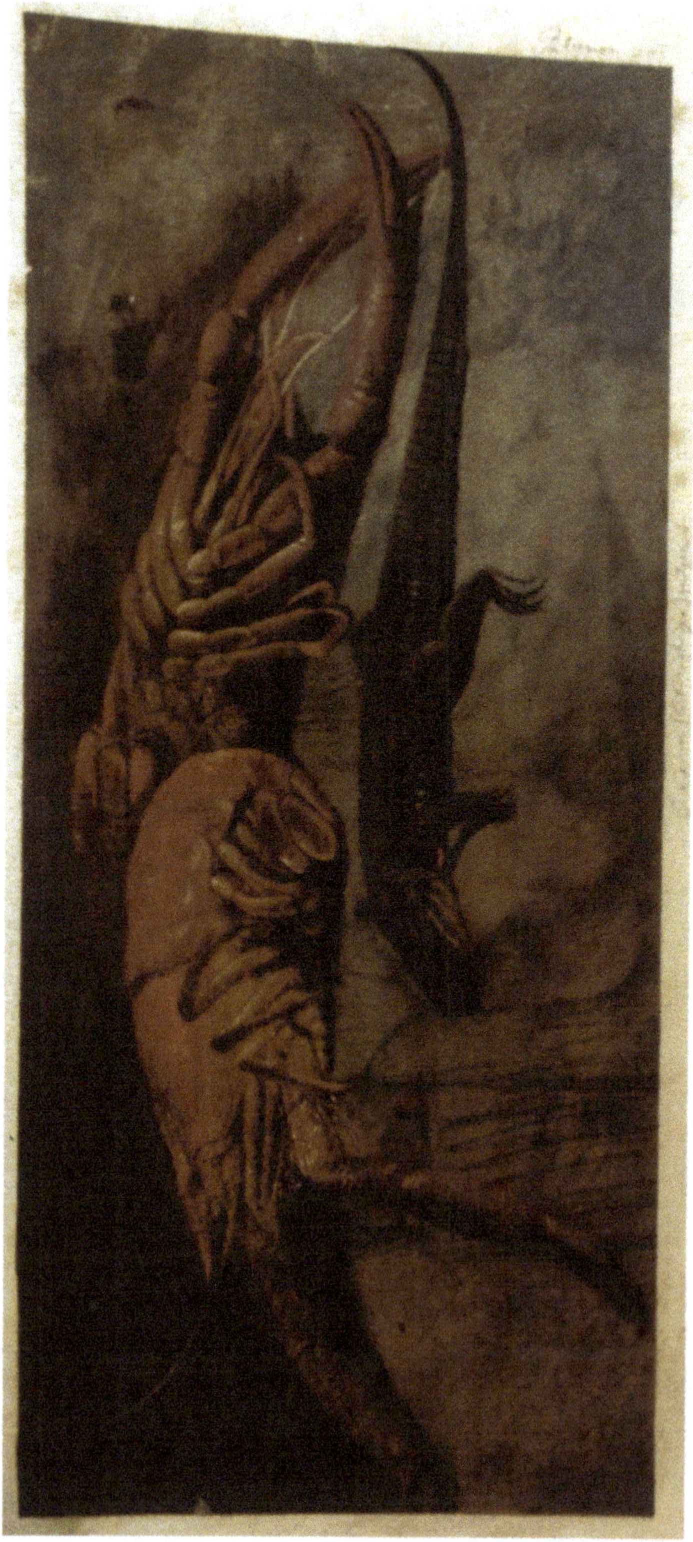

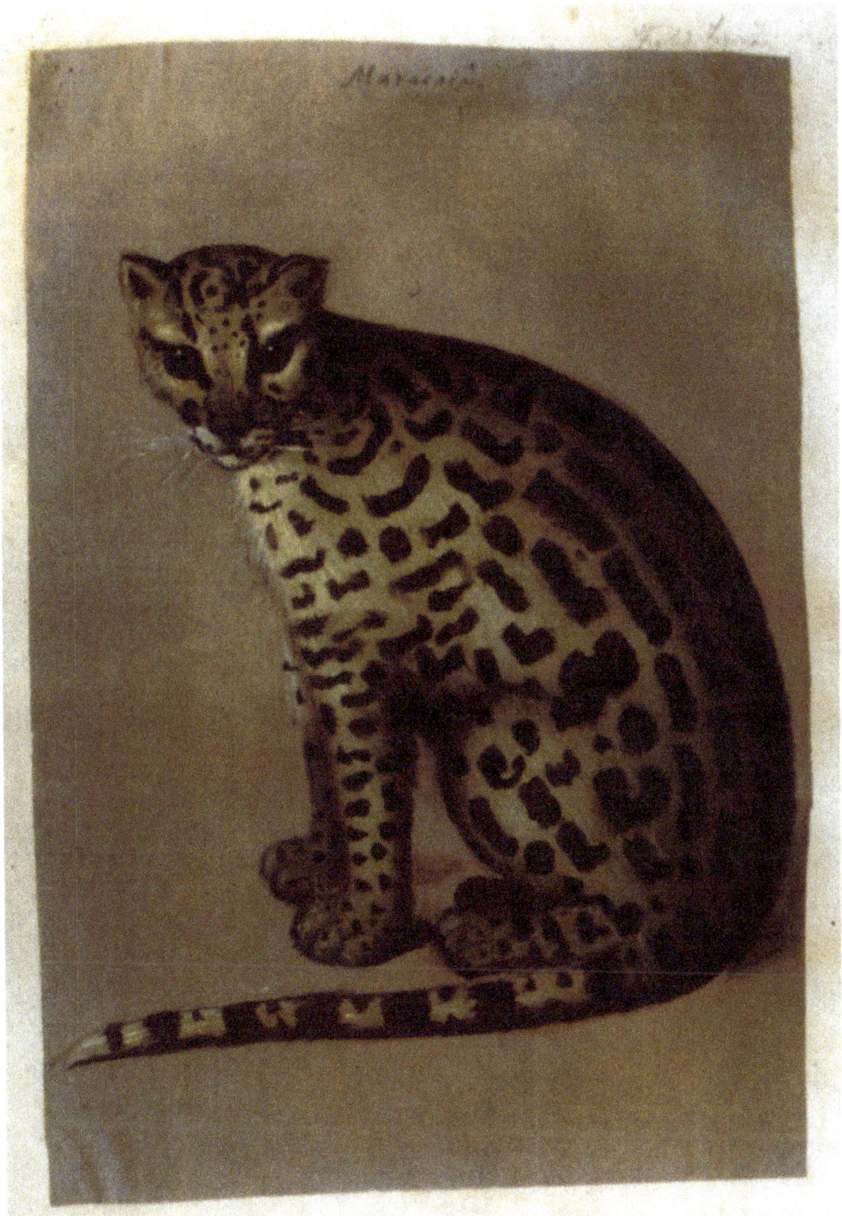

[PLATE XXI. PÉREZ DE TUDELA & JORDAN GSCHWEND – Fig. 1, p. 427]

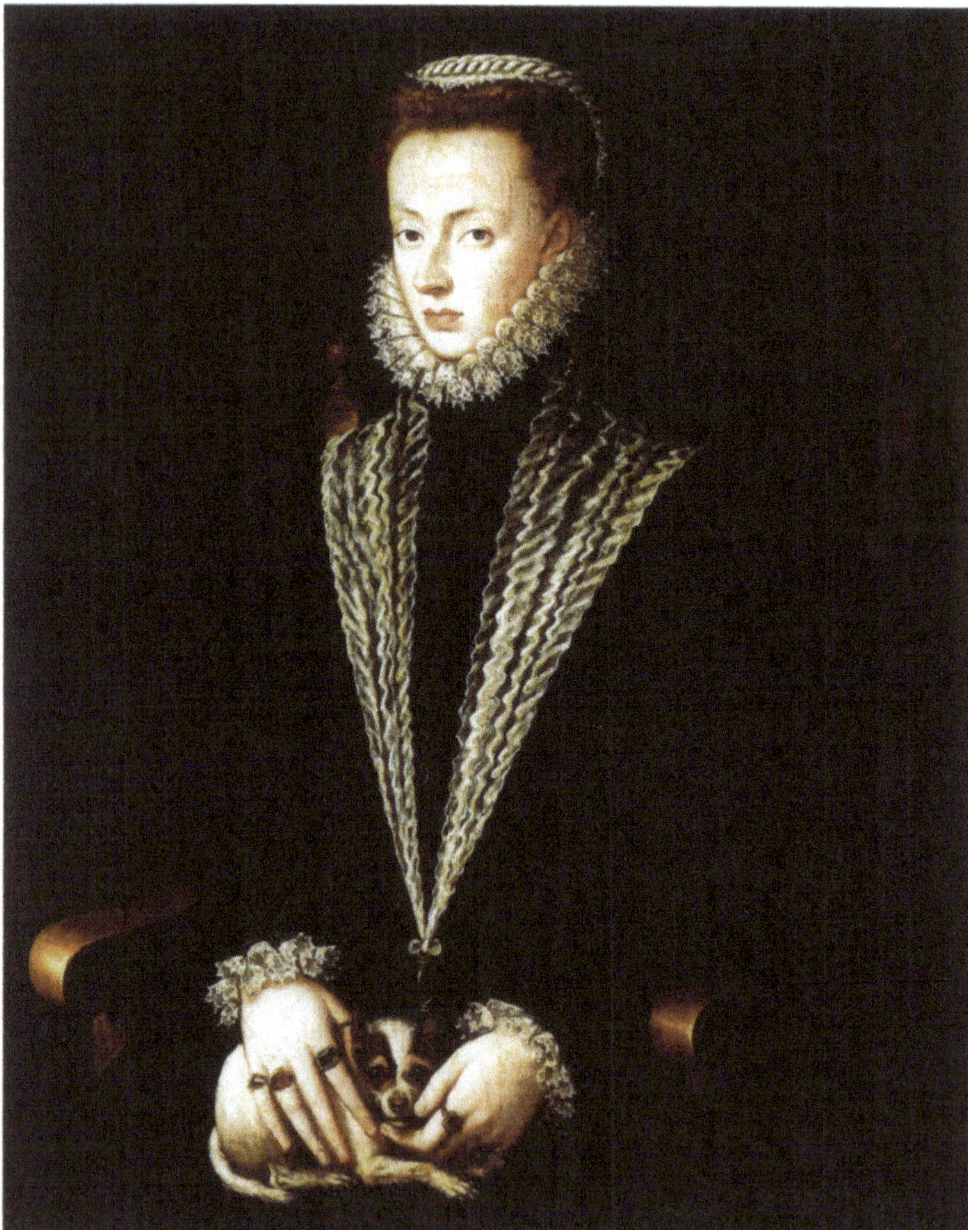

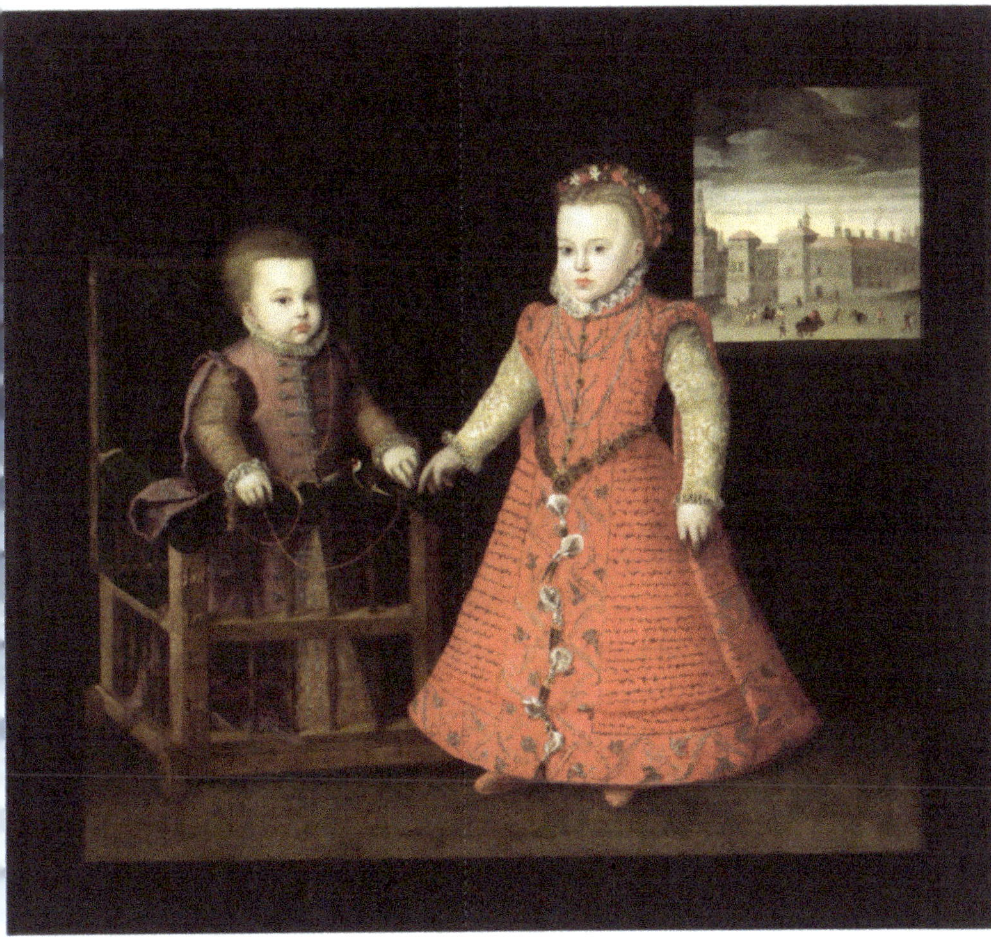

[PLATE XXIII. PÉREZ DE TUDELA & JORDAN GSCHWEND – Fig. 3, p. 430]

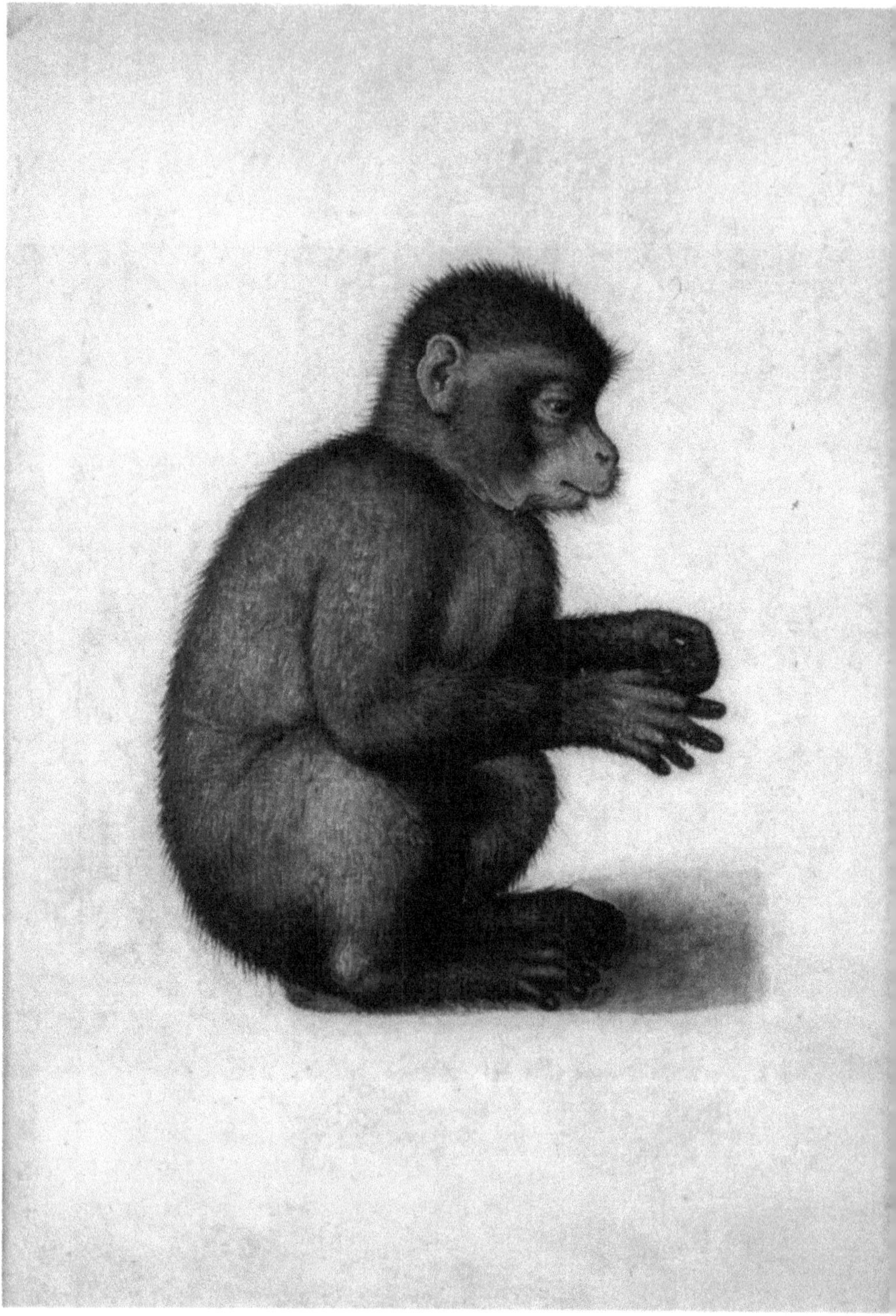

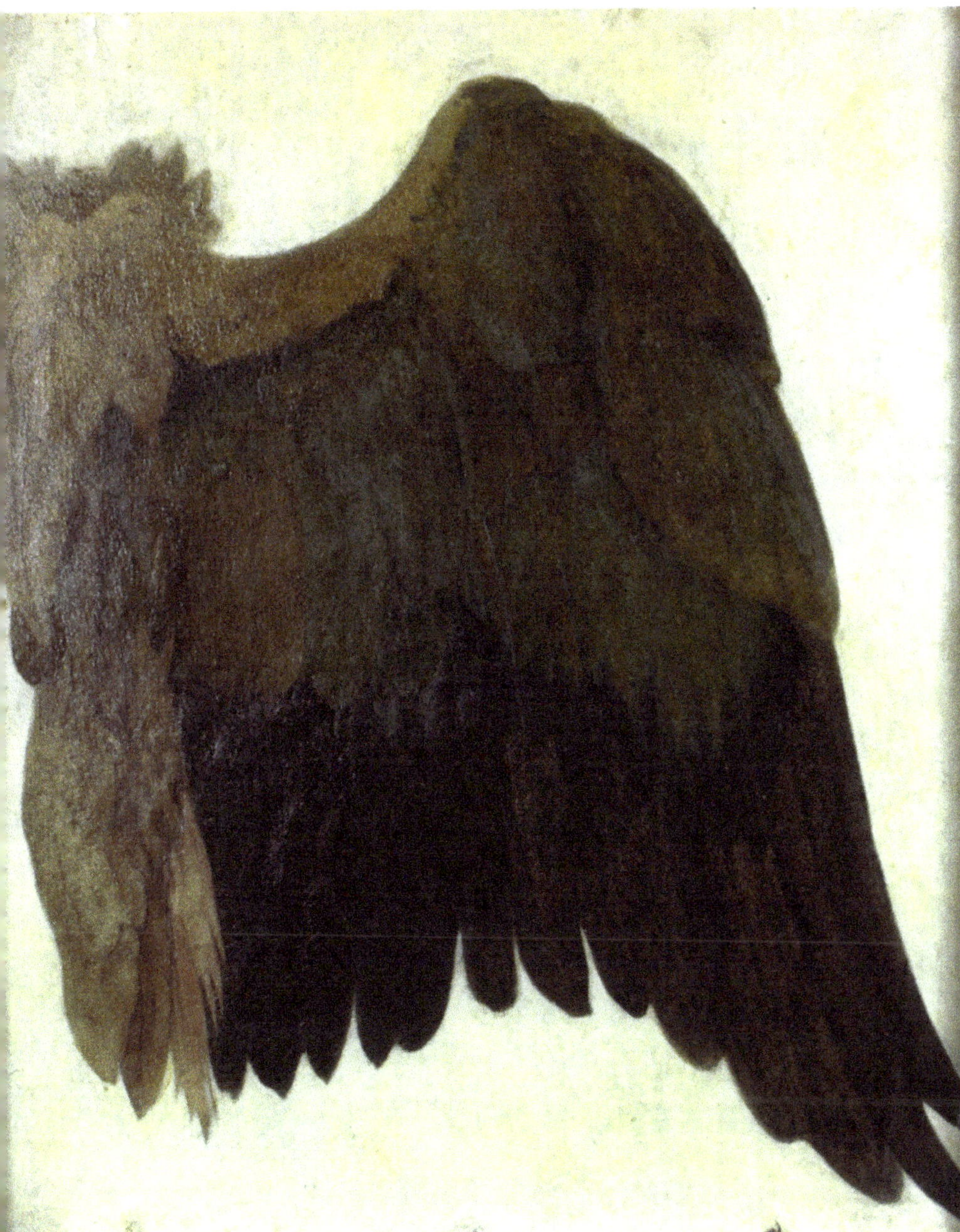

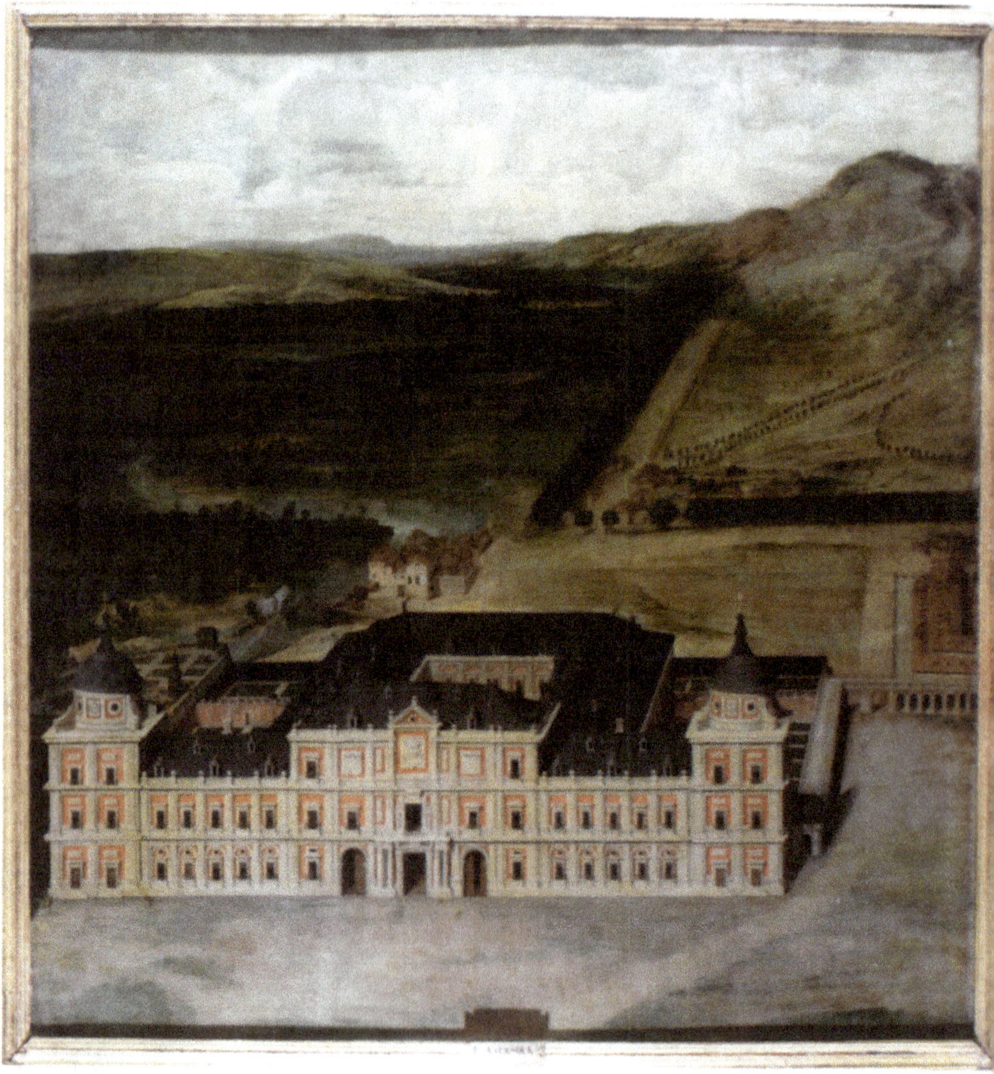

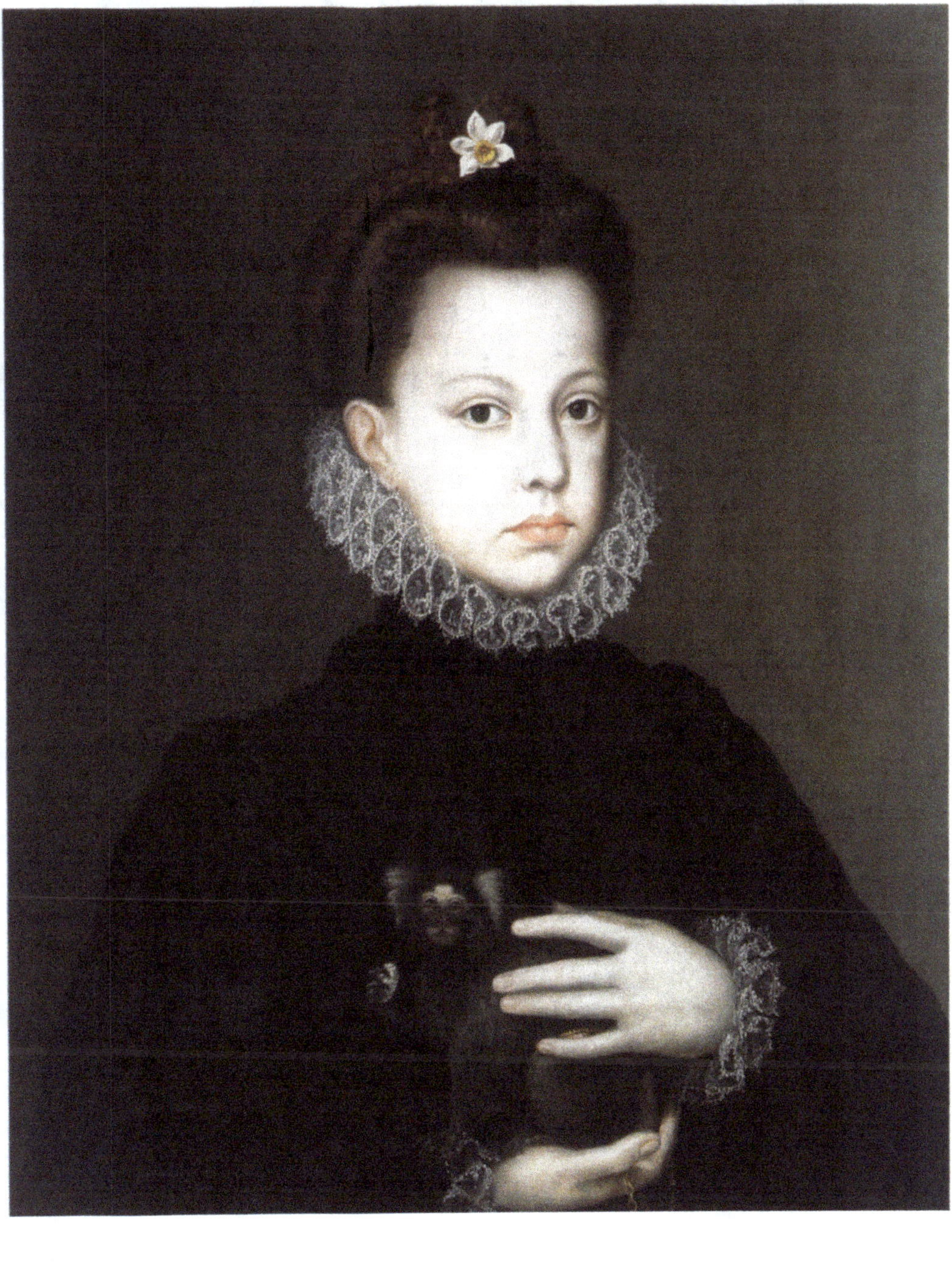

[PLATE XXVII. PÉREZ DE TUDELA & JORDAN GSCHWEND – Fig. 9, p. 442]

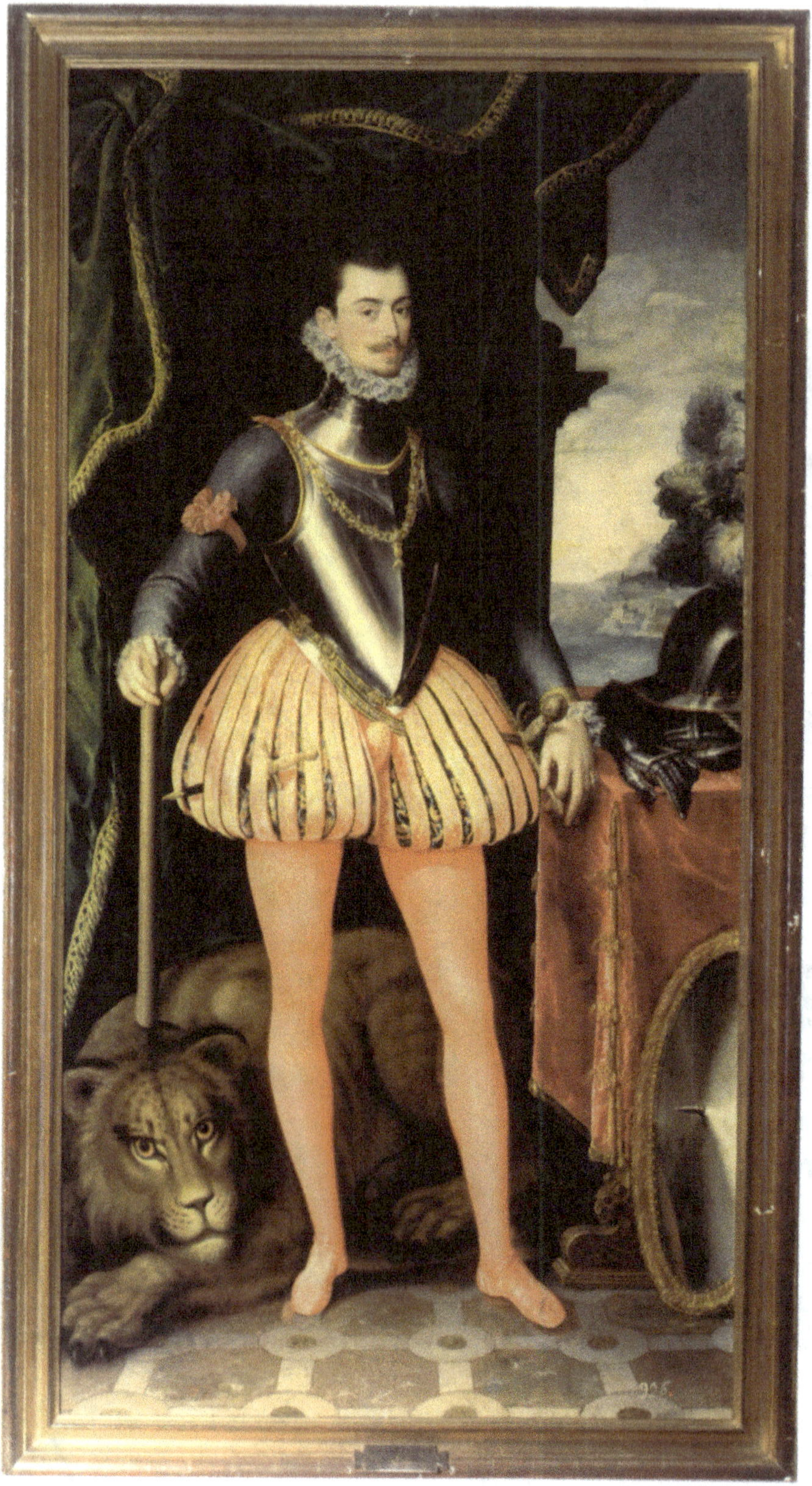

[Plate XXVIII. Pérez de Tudela & Jordan Gschwend – Fig. 10, p. 444]

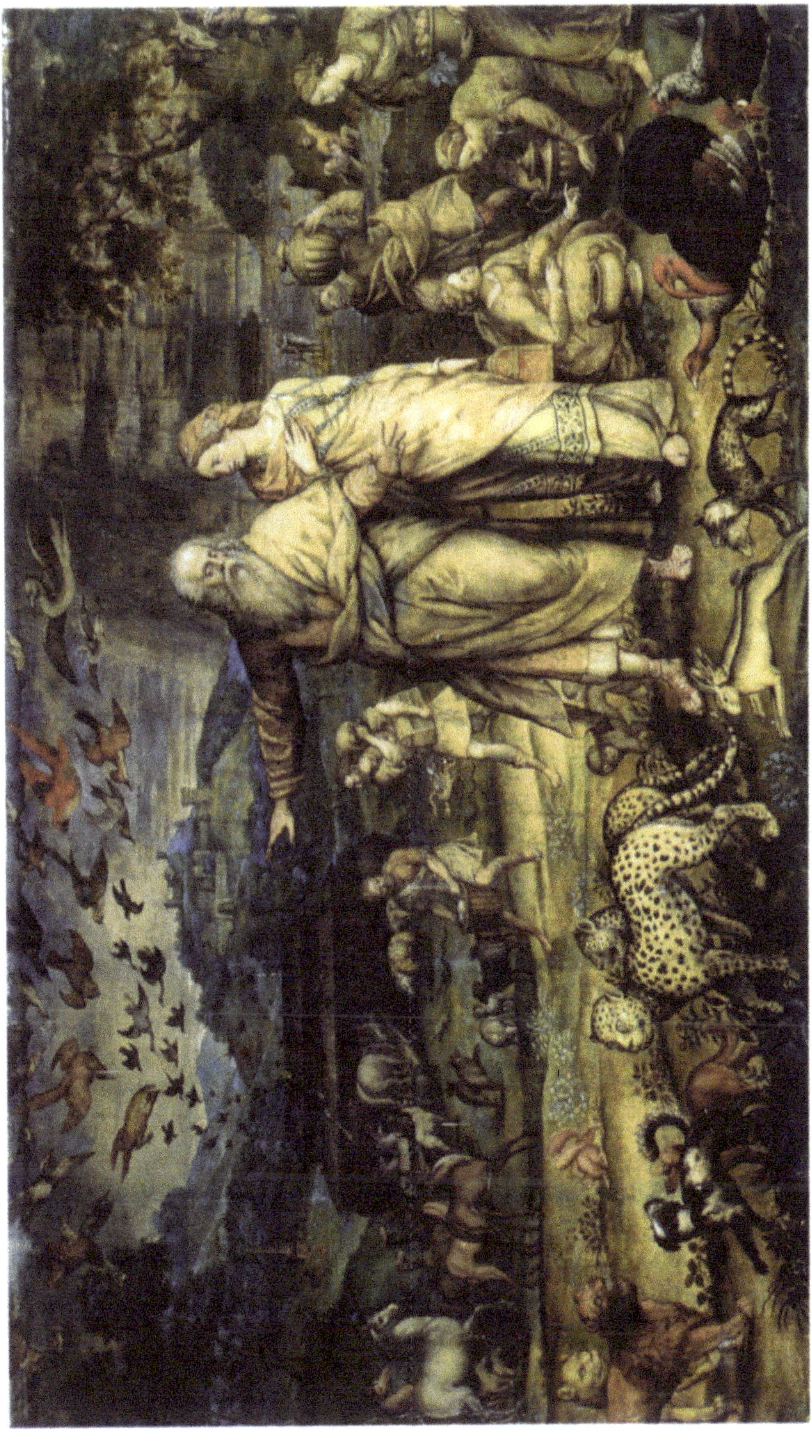

[PLATE XXIX. KOPPENOL – Fig. 5a, p. 477]

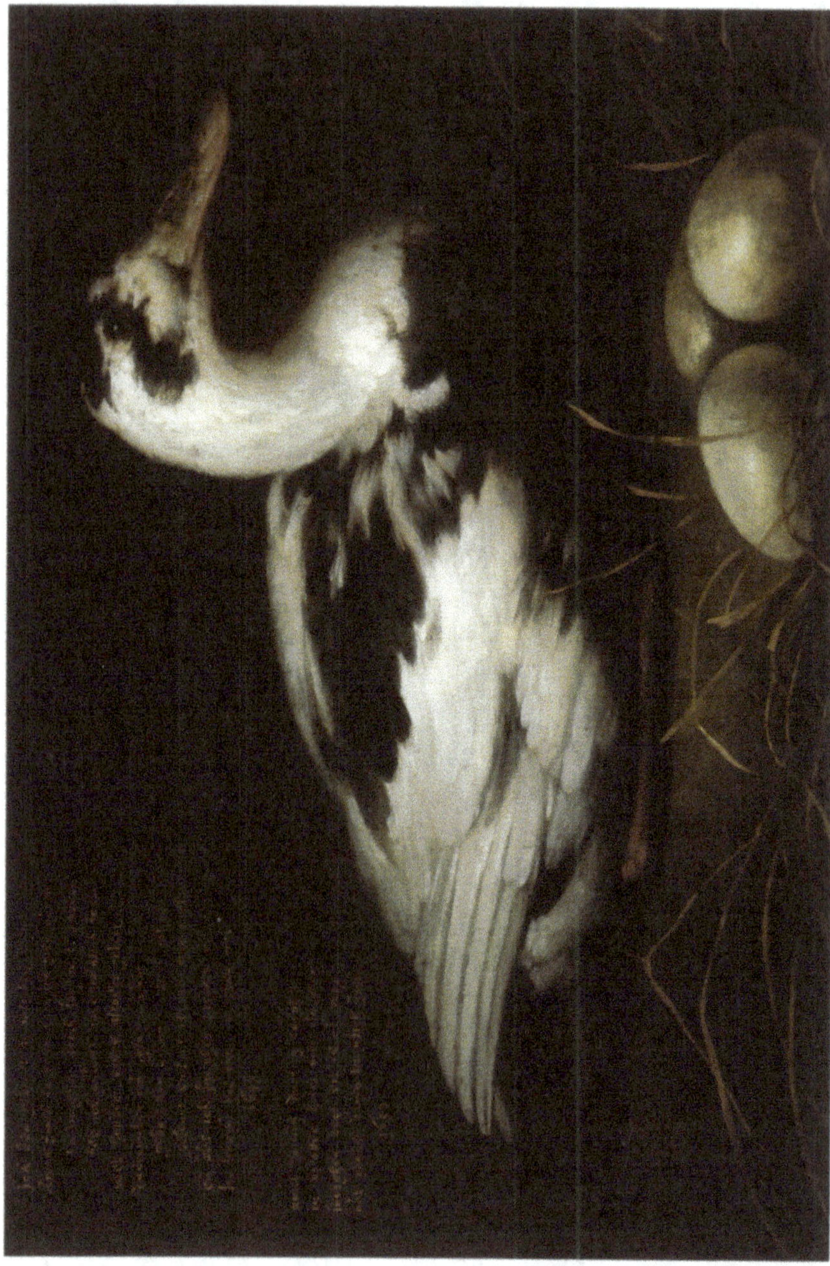

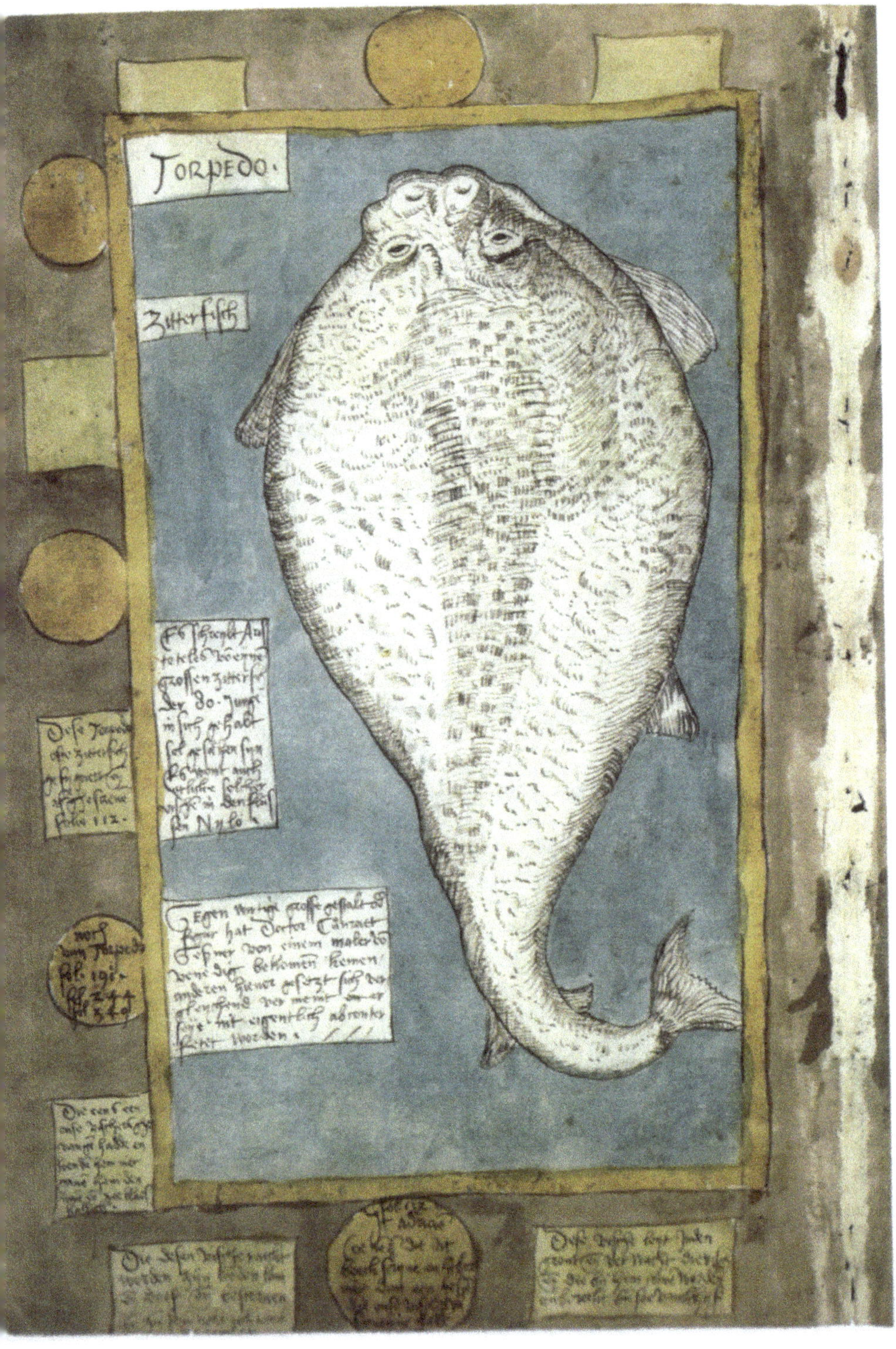
Torpedo.
Zitterfisch

[Plate XXXI. Cohen – Fig. 4, p. 627]

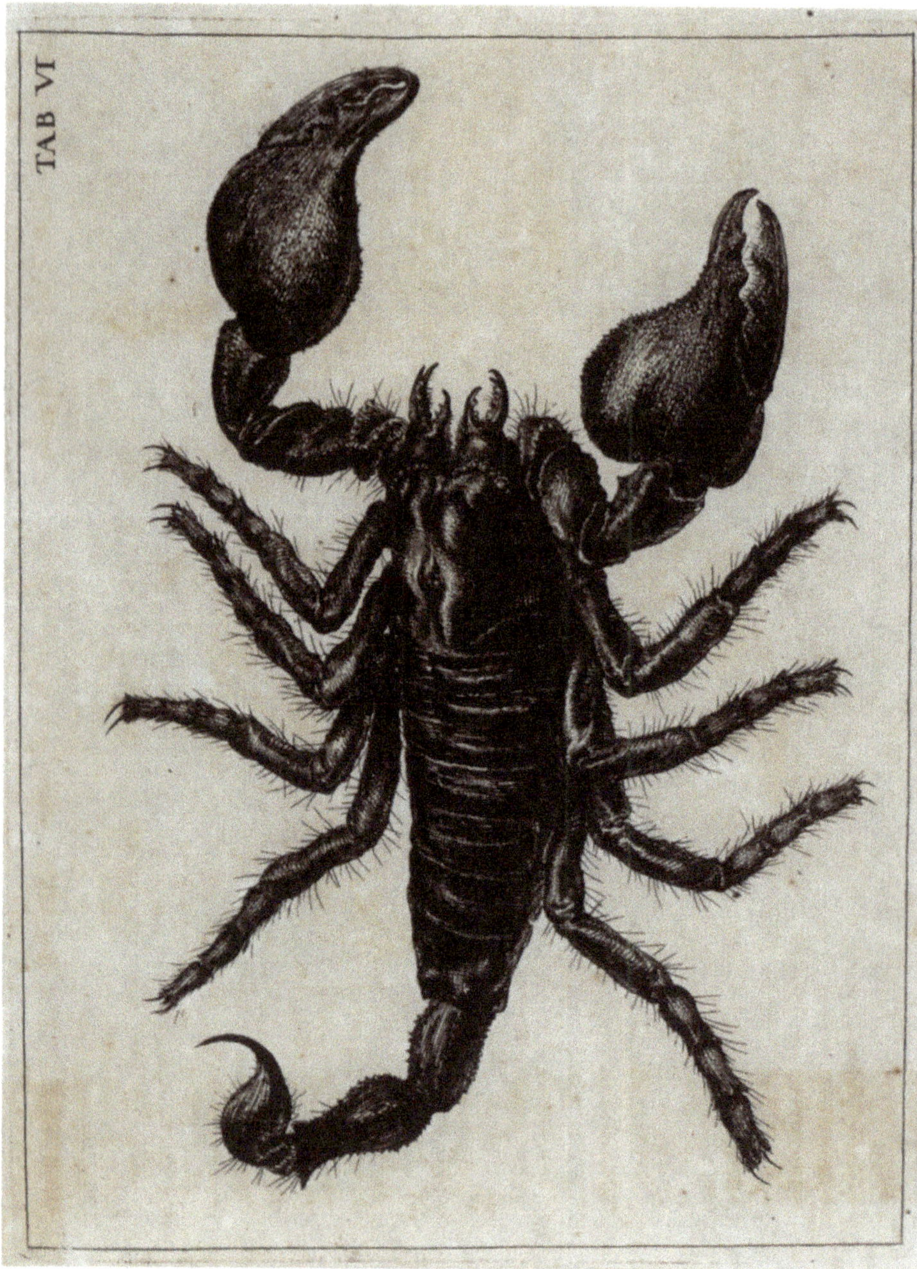

ZOOLOGY: THE CONSTRUCTION OF AN EARLY MODERN SCIENCE

ZUR KONSTITUIERUNG DER ZOOLOGIE ALS WISSENSCHAFT IN DER FRÜHEN NEUZEIT: DISKURSANALYSE ZWEIER GROSSPROJEKTE (WOTTON, GESNER)

Karl A.E. Enenkel

Einleitung

Conrad Gesner, der Verfasser einer der ersten frühneuzeitlichen Zoologien, leitet seine *Historia animalium* (1551) mit einer umfassenden Verteidigungsrede ein,[1] in der er sich für den entlegen und merkwürdigen Gegenstand im Zusammenhang mit dem Umfang des Werkes entschuldigt, jedoch dieses mit dem Hinweis auf den vielfältigen Nutzen, den es dem Leser bringe, rechtfertigt. Ins selbe Horn bläst die Titelseite der *Historia animalium*, auf der hervorgehoben wird, dass die Zoologie Intellektuellen unterschiedlicher Bereiche nütze: Philosophen, Ärzten, Grammatiklehrern, Philologen sowie Dichtern [Abb. 1].

Kein moderner Zoologe würde wohl auf den Gedanken kommen, sein Werk auf eine derartige Weise einzuleiten. Die Zoologie ist als Hauptzweig der Biologie im Wissenschaftssystem der Moderne seit dem 19. Jahrhundert etabliert.[2] Ob oder inwiefern das Tier einen sinnvollen und epistemologisch gerechtfertigten Forschungsgegenstand abgebe, ist keine Frage, die im Wissensdiskurs der Moderne eine Rolle spielt. Es liegt kein Grund vor, die Berechtigung der Wissenschaft mit dem Hinweis auf ihre Nutzanwendung nachzuweisen, etwa in dem Sinn, dass Schriftsteller daraus ihren Stoff beziehen oder Grundschullehrer daraus ihren Unterrichtsstoff zusammenstellen könnten. Nicht einmal der Popularisierungsfall würde eine solche Rechtfertigung erfordern. Bereits im 19. Jahrhundert, als sich die Zoologie als Wissenschaft zuerst voll entfaltete, wurde ihr ein so reges Interesse zuteil, dass sie gleich

---

[1] Conrad Gesner, *Historiae animalium I de quadrupedibus vivparis* [...] (Zürich, Christoffel Froschauer: 1551) fol. ß ff. (Vorwort an den Leser: „Conradus Gesnerus candidis lectoribus").

[2] Vgl. Jahn I., *Grundzüge der Biologiegeschichte* (Jena: 1990).

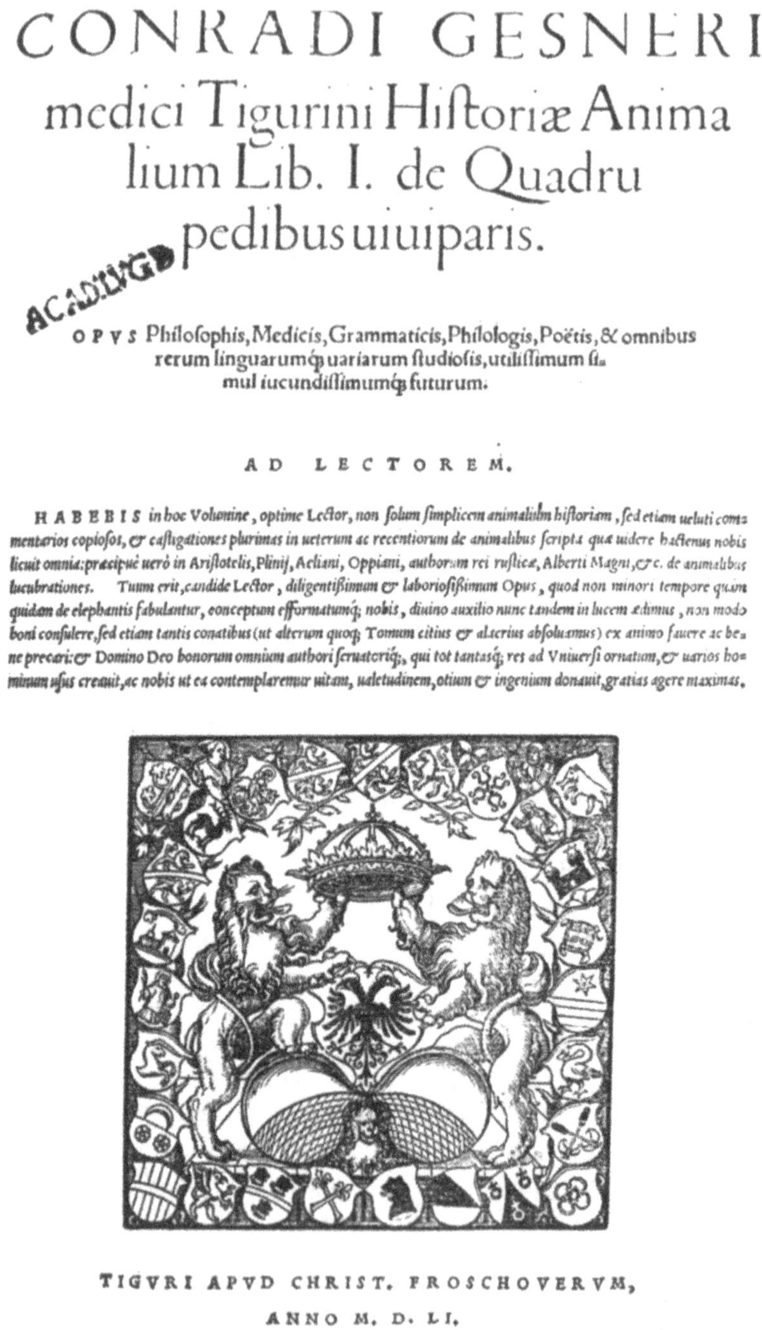

CONRADI GESNERI

medici Tigurini Hiſtoriæ Anima lium Lib. I. de Quadru pedibus uiuiparis.

OPVS Philoſophis, Medicis, Grammaticis, Philologis, Poëtis, & omnibus rerum linguarumq; uariarum ſtudioſis, utiliſſimum ſimul iucundiſſimumq; futurum.

AD LECTOREM.

HABEBIS in hoc Volumine, optime Lector, non ſolum ſimplicem animalium hiſtoriam, ſed etiam ueluti commentarios copioſos, & caſtigationes plurimas in ueterum ac recentiorum de animalibus ſcripta quæ uidere haſtenus nobis licuit omnia: præcipuè uerò in Ariſtotelis, Plinij, Aeliani, Oppiani, authorum rei ruſticæ, Alberti Magni, &c. de animalibus lucubrationes. Tuum erit, candide Lector, diligentiſſimum & laborioſiſſimum Opus, quod non minori tempore quàm quidam de elephantis fabulantur, conceptum efformatumq; nobis, diuino auxilio nunc tandem in lucem ædimus, non modo boni conſulere, ſed etiam tantis conatibus (ut alterum quoq; Tomum citius & alacrius abſoluamus) ex animo fauere ac bene precari: & Domino Deo bonorum omnium authori ſeruatoriq;, qui tot tantasq; res ad Vniuerſi ornatum, & uarios hominum uſus creauit, ac nobis ut ea contemplaremur uitam, ualetudinem, otium & ingenium donauit, gratias agere maximas.

TIGVRI APVD CHRIST. FROSCHOVERVM,

ANNO M. D. LI.

Fig. 1. Conrad Gesner, *Historiae animalium I de quadrupedis viviparis* (Zürich, Chr. Froschauer: 1551), Titelseite. Universiteitsbibliotheek Leiden

auf mehreren Gebieten erfolgreich popularisiert werden konnte. Im deutschsprachigen Bereich ist Alfred Brehms *Illustriertes Thierleben* in sechs Bänden (zuerst Leipzig: 1864–1869) nur ein Beispiel einer erfolgreichen popularisierenden zoologischen Buchproduktion.[3] Nebenher kam es im 19. Jahrhundert zum Aufstieg des Zoos als öffentliche Einrichtung und Zeitvertreib der Massen, während naturkundliche Museen dem Publikum die Tore öffneten.[4] Das Popularisierungspotential der Zoologie ist in der Gegenwart noch angewachsen: Wissenschaftlich fundierte Tierfilme sind ein fester Bestandteil der Unterhaltungsbranche geworden. Es vergeht kein Tag, an dem in der international vernetzten Televisionswelt nicht gleich mehrere prächtige, technisch hochwertige und informationsreiche Tierfilme zur Auswahl stehen, ja es gibt sogar Fernsehsender, die sich ausschließlich auf Tier- und Naturfilme spezialisiert haben. Der Aufstieg der Tierfotographie mit dem damit verbundenen Anwachsen illustrierter popularisierender Buchproduktion sowie die Vernetzung aller Medien mit dem Internet haben ein Übriges dazu beigetragen, dass die Zoologie aus dem Geistesleben der Moderne kaum mehr wegzudenken ist.

Die Selbstverständlichkeit der Zoologie als Wissenschaft hängt mit dem spektakulären Aufstieg der Naturwissenschaften im 19. und 20. Jahrhundert zusammen. Die Forschungsdiskursivität, die damit einhergeht, ist fortschrittsorientiert: Eine Bezugnahme neuer Forschungsresultate auf zoologische Diskurse *vor* dem 19. Jahrhundert ist kein fachimmanentes Desideratum, was angesichts des rapiden empirischen Erkenntniszuwachses, der gewissermaßen täglich verbucht wird, der stets weiter fortschreitenden Spezialisierung und der Technikbezogenheit der modernen Forschungsmethoden verständlich ist.

Aus dieser Sachlage lässt sich ableiten, dass die Beschäftigung mit der Geschichte des Fachs am besten entweder als Spezialdisziplin oder als interdisziplinäre Forschungsrichtung organisiert werden kann. Gerade in den letzen Dezennien lässt sich ein stets wachsendes Interesse an der Geschichte der Naturwissenschaften wahrnehmen:[5] Lehr- und Forschungsaufträge, Planstellen, Professuren und Institute für die

---

[3] 2. Auflage (Leipzig: 1876–1879, 10 Bde.). Das Werk erlebte – vollständig oder gekürzt – eine Vielzahl von Auflagen, die, obwohl sein Inhalt völlig überholt ist, sich sogar noch im 21. Jahrhundert fortsetzen.

[4] Jahn I., „Zoologische Gärten – zoologische Museen. Parallelen ihrer Entstehung", *Bongo* 24 (1994) 7–30.

[5] Z.B. Brocke B. vom, *Wissenschaftsgeschichte als historische Disziplin* (Berlin: 1994).

Geschichte der Naturwissenschaften wurden ins Leben gerufen. Dieses verstärkte Interesse hat unter anderem dazu geführt, dass die Geschichte der Biologie in einigen groß angelegten Projekten in den Blick genommen wurde, z.B. in Ilse Jahns *Geschichte der Biologie* (Hrsg.; zuerst Jena: 1982),[6] die mittlerweile mehrfach aufgelegt worden ist, oder Änne Bäumers *Geschichte der Biologie* (Frankfurt a.M.: 1991).[7] Der große Fortschritt, den die Zoologie seit dem Anfang des 19. Jahrhunderts verbucht hat – man denke nur an die epochemachenden Leistungen Jean-Baptiste de la Lamarcks (*Recherches sur l'organisation des corps vivantes* [Paris: 1802]; *Histoire naturelle des animaux sans vertèbres* [Paris: 1815–1822]) oder Charles Darwins (*The Origin of Species* [London: 1859]) – hat dazu geführt, dass sich in der Biologiegeschichte eine schwerpunktmäßige Fokussierung auf das letzte Viertel des 18. Jahrhunderts, auf das 19. und die erste Hälfte des 20. Jahrhunderts erkennen lässt. Die Biologiegeschichte der frühen Neuzeit von ca. 1450–1650 ist zwar von diesem Elan ebenfalls erfasst worden, konzentrierte sich jedoch auf die Geschichte der Botanik. Änne Bäumers *Zoologie der Renaissance – Renaissance der Zoologie*, der wir eine erste Beschreibung der zoologischen Publikationen der frühen Neuzeit verdanken, bildet auf diesem Gebiet eine Pionierleistung.[8] Brian Ogilvies rezente Monographie *The Science of Describing. Natural History in Renaissance Europe* zeigt, dass die einseitige Gewichtung zur Botanik hin noch keineswegs überwunden ist: Die Arbeit ist fast zur Gänze der Botanik gewidmet. Ogilvie rechtfertigt dies mit der Vorrangstellung, welche die Botanik gegenüber der Zoologie gehabt habe („The Preeminence of Botany"): „My periodization has focussed on botany, the chief focus of natural history in the Renaissance. [...] In fact, zoology generally

---

[6] Hier zitiert nach der Auflage Hamburg: 2004 (nach der zweiten, korrigierten Sonderausgabe Heidelberg-Berlin: 2002 nach der dritten Auflage Jena: 1998).

[7] Vgl. weiter Findlen P., *Possessing Nature. Museums, collecting, and scientific culture in early modern Italy* (Berkely: 1994); Hünemörder Ch., *Geschichte der Biologie. Wesen und Aufgaben* (Stuttgart: 1985); Jorink E., *Het Boek der Natuere. Nederlandse geleerden en de wonderen van Gods schepping 1575–1715* (Leiden: 2006).

[8] In Dies., *Geschichte der Biologie*, Bd. 2. Weitere rühmliche Ausnahmen bilden Caroline Gmelig-Nijboers Monographie zu Gesner und einige Einzelstudien Änne Bäumers/ Bäumer-Schleinkofers (siehe Auswahlbibliographie). Laurent Pinon und A.J.J. van der Velde haben über die zoologische Buchproduktion der frühen Neuzeit buchgeschichtliche (Pinon) und bibliographische (van de Velde) Übersichten angelegt: Pinon L., *Livres de zoologie de la Renaissance. une anthologie (1450–1700)* (Klincksieck: 1995) und Velde A.J.J. van de, „Rond Gesner's Historiae animalium liber I van 1551", *Mededelingen van de Koninklijke Vlaamse Academie voor wetenschappen, letteren en schone kunsten van België. Klasse der Wetenschappen* XIII,17 (Brüssel: 1951).

lagged behind botany in the Renaissance; developments that occurred in botany often took a decade or two to appear in zoology".[9]

Während in keiner Weise an der Bedeutung der Botanik als frühneuzeitliche Wissenschaft gezweifelt werden soll, erscheint es verfehlt, die frühneuzeitliche Zoologie als schwachen Abklatsch der Botanik zu betrachten. Sie behandelt einen andersartigen Gegenstand und muss sich mit anderen Problemen als die Botanik auseinandersetzen. Tiere stellen weitaus komplexere, ungleich schwieriger erfassbare Organismen als Pflanzen dar, und ihre Beobachtung stellt durchaus andere, zum Teil viel höhere Ansprüche an den Forscher. Während sich Pflanzen ruhig dem Auge des Beobachters darbieten, entziehen sich ihm Tiere vielfach mit Hilfe des leistungsfähigen Bewegungsapparats, der evolutionsgeschichtlich ihre Überlebensgrundlage bildet, während sie in Gefangenschaft selbst mit hohem Aufwand entweder nur schwer gehalten oder nachgezüchtet werden können und selbst, wenn dies gelingt, andere Verhaltensmuster zeigen als in freier Wildbahn. Diese Probleme treten natürlich in verstärktem Maß bei exotischen Species auf. Desgleichen konfrontiert die analytische Untersuchung des tierischen Organismus den Zoologen sowohl mit anderen Problemen als auch anderen Möglichkeiten als den Botaniker die Analyse der Pflanzen.

Ein Beispiel bildet die Neukonstituierung der medizinischen Anatomie als empirische Wissenschaft um das Jahr 1520. Sie ist für den Botaniker unbedeutend, während sie dem Zoologen bisher ungekannte Möglichkeiten bietet, in das Innere des komplexen tierischen Organismus vorzudringen und die Lebensvorgänge (Physiologie) zu erforschen. Der Einstieg ins Innere des tierischen Organismus lässt sich mit der Entdeckung eines neuen Kontinents vergleichen. Die frühneuzeitliche Zoologie legte auf diesem Gebiet eine außerordentlich weite Strecke zurück, die unter anderem zur Entdeckung des tierischen Blutkreislaufs führte (William Harvey, *Exercitatio anatomica de motu cordis et sanguinis in animalibus*, 1628). Diese und andere Forschungen fanden jenseits der Botanik statt. Schon aufgrund dieser einfachen Tatsachen verdient die Zoologie der frühen Neuzeit eine Aufmerksamkeit, die nicht vom Interesse an der Botanik gesteuert wird.

Was die Herangehensweise betrifft, ist die Erforschung der Zoologie einem bestimmten Druck ausgesetzt, welcher den Gang der Forschung

---

[9] Ogilvie B., *The Science of Describing. Natural History in Renaissance Europe* (Chicago – London: 2006) 49.

steuert. Dieser wird von dem modernen Fortschrittsbewusstsein hervorgebracht, welches dazu führt, dass man explizit oder implizit die Leistungen der Vergangenheit als Entwicklungsschritte zur modernen Wissenschaft hin teleologisch einordnet. Diese Verstehensfigur trifft man auch in Bezug auf die Geschichte der Zoologie der Renaissance (1300–1600) an. Änne Bäumer, die in ihrer sonst äußerst wertvollen *Renaissance der Zoologie* schwerpunktmäßig *die Entwicklung zur modernen Wissenschaft* in den Blick nimmt und dabei ein Dreierschema ihres Lehrmeisters Fritz Kraft anwendet,[10] definiert die *Verdienste* der frühneuzeitlichen Zoologie vielfach im Hinblick auf die moderne Zoologie.

Ihrer Ansicht nach „nimmt die Entwicklung der modernen zoologischen Disziplinen in der humanistischen Zoologie ihren Anfang".[11] Die „bewusste Rückbesinnung auf antike methodische Ansätze" führt zum wissenschaftlichen Fortschritt, unter anderem zur Herausdifferenzierung der „zoologischen Spezialgebiete Embryologie, Physiologie, Vergleichende Anatomie" usw.[12] Die „zoologischen Kompendien" (Pierre Gilles,[13] Michael Herr,[14] Adam Lonitzer,[15] Edward Wotton, Conrad Gesner usw.) „wiesen in ihren methodischen Ansätzen der zukünftigen Forschung den Weg".[16] „Auch die Anatomie musste Bestandteil dieser Darstellung sein, da sie entscheidende Kriterien zur Einteilung der Tiere lieferte. Damit waren alle Ansätze für eine adäquate Behandlung des Tierreichs gegeben, die neuzeitliche Zoologie war begründet".[17] „Die

---

[10] Vgl. *Zoologie der Renaissance*, bsd. 405–422 „Übergang von der ersten zur zweiten Phase der humanistischen Zoologie" 405–407; „Die zweite Phase der humanistischen Zoologie" 407–411; „Übergang von der zweiten zur dritten Phase der humanistischen Zoologie" 411–414; „Die dritte Phase der humanistischen Zoologie" 414–417.

[11] Siehe die Zusammenfassung ihrer Monographie „Zoologie der Renaissance – Renaissance der Zoologie", in Döring K. – Wöhrle G. (Hrsg.), *Antike Naturwissenschaft und ihre Rezeption*, Bd. I und II (Bamberg: 1992) (275–295) 279.

[12] Ebd.

[13] Pierre Gilles, *Ex Aeliani historia per Petrum Gyllium latini facti itemque ex Porphyrio, Heliodoro, Oppiano, tum eodem Gyllio luculentis accessionibus aucti libri XVI de vi et natura animalium. Eiusdem Gyllii liber unus de Gallicis et Latinis nominibus piscium* (Lyon, Sebastian Gryphius: 1535).

[14] Michael Herr, *Gründtlicher Underricht, wahrhaffte und eygentliche beschreibung wunderbarlicher seltsamer art, natur, krafft und eygenschafft aller vierfüssigen thier, wild und zahm, so auff und in der erden oder wassern wonen* [...] (Straßburg, Balthasar Beck: 1546).

[15] Adam Lonitzer, *Naturalis historiae opus novum, in quo tractatur de natura et viribus arborum, fructicum, herbarum animantiumque terrestrium, volatilium et aquatilium, item gemmarum, metallorum succorumque concretorum* (Frankfurt a.M., Christoph Egenolph: 1551–1555).

[16] „Zoologie der Renaissance – Renaissance der Zoologie" 280.

[17] Ebd.

Renaissance der zoologischen Forschung ist hier gleichzusetzen mit dem, was man gemeinhin ‚als wissenschaftliche Revolution' bezeichnet".[18]

Ich kann mich dieser Herangehensweise nicht anschließen. Sie birgt die Gefahr in sich, reduktionistisch zu wirken, unsere Aufmerksamkeit einzuengen und uns den Blick auf die Eigenheiten der frühneuzeitlichen Wissenschaftsausübung zu verstellen. Gerade diese sollte m.E. den Mittelpunkt des Interesses bilden. Das Konzept der in groben Zügen linearen, progressiven Wissenschaftsentwicklung scheint mir gerade für die frühe Neuzeit wenig brauchbar, da es sich um eine Periode handelt, die von gesteigerter Diversifizierung und Pluralität des intellektuellen Lebens gekennzeichnet ist. Es gibt keine einheitliche Entwicklung, geschweige denn eine zukunftsorientierte homogene Entwicklung. Es existieren verschiedene Diskurse nebeneinander, die sich keineswegs in einem progressiven Entwicklungsverhältnis zur Moderne hin befinden.

Zum Beispiel scheint es mir nicht richtig, die „Entwicklung" der Zoologie in der frühen Neuzeit als kontinuierliche Hinwendung zur Empirie zu fassen. Wolfgang Franzius' *Historia animalium sacra* aus dem Jahre 1612[19] bietet keineswegs mehr Empirie als Edward Wottons *De differentiis animalium* aus dem Jahre 1552 oder Gesners *Historia animalium* aus dem Jahre 1551. Die letzten beiden Werke müssten, da sie ungefähr gleichzeitig entstanden sind, dieselbe wissenschaftliche ‚Entwicklungsstufe' repräsentieren. Sie unterscheiden sich jedoch, wie in diesem Aufsatz gezeigt werden soll, wesentlich von einander. In der vorliegenden Studie möchte ich daher von teleologischen und linear-fortschrittsbezogenen Herangehensweisen absehen. Es geht mir schlicht darum, die Diskurse nachzuzeichnen, die die Konstituierung der Zoologie als Wissenschaft in zwei Großprojekten, die um 1550 publiziert wurden, bestimmten.

Dabei werde ich übersichtsmäßige Erfassungen moderner zoologischer Forschungsresultate heranziehen,[20] die jedoch lediglich einen heuristischen Zweck haben. Der Vergleich soll helfen, *die diskursiven Eigenheiten*

---

[18] Ebd. 285.

[19] *Historia animalium sacra, in qua plerorumque animalium praecipuae proprietates in gratiam studiosorum Theologiae et ministrorum verbi ad usum eikonologikon breviter accommodantur, in Academia Witenbergensi* [...] *dictata* (Wittenberg, Zarcharias Schurer und Johannes Gormann: 1612). Zu diesem Werk vgl. in vorliegendem Band den Aufsatz von Vibeke Roggen „Biology and Theology in Franzius' *Historia animalium sacra* (1612)".

[20] Z.B. aus Macdonald D. (Hrsg.), *Enzyklopädie der Säugetiere* (Königswinter: 2003; urspr. englisch, Oxford: 2001). Gould E. – McKay G., *Enzyklopädie der Tierwelt. Säugetiere* (München: 2002; urspr. englisch, Sidney: 1990).

der frühmodernen Zoologie zu orten. Dabei schien es am zweckmässigsten, diese anhand einer bestimmten Tierklasse und des konkreten Beispiels einer Species herauszuarbeiten. Dafür wurde die Klasse der *Mammalia* und die Species *Panthera leo* (Löwe) ausgewählt. Der Vorteil dieser Species ist, dass sie die Hauptprobleme zu Tage fördert, welchen der frühmoderne Zoologe begegnente: Es handelt sich um ein hochentwickeltes exotisches Tier, das komplexe Verhaltensmuster aufweist, bei lebendigem Leib nicht leicht untersucht werden, zudem in freier Wildbahn nicht leicht beobachtet werden konnte, jedoch in verschiedenen Menagerien Europas gehalten wurde, wodurch es prinzipiell möglich war, neue empirische Daten zu erheben und eventuell auf anatomischem Weg sogar in das Innere des Tierkörpers vorzudringen.

Zur Heuristik der Säugetiere (Mammalia): einige Diskursmerkmale der modernen Zoologie

Die moderne Zoologie versteht sich als eine systematische Wissenschaft, für die die Taxonomie ein wichtiger Ausgangspunkt ist. Die Taxonomie ordnet die Species (Tierarten) innerhalb eines feinmaschigen Klassifizierungssystems ein, in welchem sie die Verwandtschaftsverhältnisse der unterschiedlichen Arten festlegt.[21] Die moderne Taxonomie ist kein Wissenschaftszweig, der sich als rein formale Festlegung der heute lebenden Arten versteht. Ihr Diskurs wird von der Evolutionstheorie bestimmt und ist mit einer Methodenpluralität verknüpft. Die Klassifizierung geschieht unter anderem mit Hilfe der Paläontologie, welche die Tierwelt der weit zurückliegenden Vergangenheit untersucht und die die in der Gegenwart lebenden Arten auf diese zurückführt. Z.B. werden die heute lebenden Mammalia (insgesamt ca. 4250–4700 Arten) in einen Stammbaum eingegliedert, der aus der Abspaltung der Theria von den eierlegenden Protheria vor ca. 200 Millionen Jahren emporwächst. Dasselbe von der Evolutionsbiologie und der Paläontologie gespeicherte Prinzip liegt der Einteilung der Mammalia in Unterklassen (2), Ordnungen (28), Familien (132–139) und Gattungen (ca. 1050–1100) zugrunde. Die Species Löwe/*Panthera leo* etwa wird der Unterklasse der *Eutheria*, der Ordnung

[21] Sowohl Macdonalds *Enzyklopädie der Säugetiere* (Vorwort, XII; XVI–XVIII) als auch Goulds und McKays *Säugetiere* (19–21) fangen bezeichnenderweise mit Ausführungen zur Taxonomie an.

Carnivora (Fleischfresser: 11 Familien;[22] ca. 100 Gattungen; ca. 240–270 Arten), der Familie der *Felidae* (Katzenartige: insgesamt 4 Gattungen; 37 Arten) und der Gattung Großkatzen-*Panthera* (5 Arten) zugeordnet. Die Katzenartigen leiten sich von den *Viveravines* her und spalteten sich im Oligozän, vor ca. 30 Millionen Jahren, von den Hyänen (*Hyaenidae*) ab. Die zoologische Taxonomie gilt keineswegs als abgeschlossen, sondern ist Gegenstand einer fortlaufenden Forschungsdiskussion. Da die Interpretation der gattungs- und artenspezifischen Merkmale subjektiv ist, variieren die taxonomischen Angaben. Diese Diskursorientierung spiegelt sich selbst in den überblicksmäßigen Erfassungen wider: Die *Carnivora* beispielsweise gliedern sich nach Goulds und McKays Enzyklopädie (1990) in 106 Gattungen und 270 Arten, nach Macdonalds *Enzyklopädie der Säugetiere* (2001) in 94 Gattungen und 239 Arten.

Die evolutionsbiologische Herangehensweise, die der modernen Zoologie zugrunde liegt, wird mit diversen Forschungsmethoden und -zweigen verknüpft, u.a. der *Vergleichenden Anatomie*, der *vergleichenden und experimentellen Physiologie*, der *Neuroendokrinologie, Mikrobiologie, Biochemie, Genetik, Gentechnologie und Verhaltensbiologie.*

Die von Methodenpluralität bestimmte Forschungsdiskursivität spiegelt sich z.B. in der modernen Definition der Säugetiere wider: Es handelt sich um warmblütige (Physiologie; Thermoregulation: Biochemie) Wirbeltiere (Vergleichende Anatomie), die ihre Jungen mit Milch säugen (Physiologie; Verhaltensbiologie; Biochemie), einen behaarten Körper (Thermoregulation: Biochemie, Zellbiologie/Mikrobiologie) und ein einteiliges Kiefergelenk besitzen, wodurch der Unterkieferknochen direkt mit dem Schädel gelenkt wird (Vergleichende Anatomie).

Der zoologische Deskriptionsdiskurs wird von einer wesentlich *funktionsanalytischen Methodik* gekennzeichnet. Das Körpermerkmal der Behaarung etwa wird nicht als rein morphologisches Fakt, sondern in Bezug auf seine physiologische Funktionalität – die Thermoregulation – beschrieben: Eine wesentliche Eigenschaft der Säuger ist, dass sie ihre Körpertemperatur von innen heraus (Endothermie) durch einen *biochemischen* Vorgang – durch Oxidation von Nahrung im Körper – regeln. Dies hat den Vorteil der Unabhängigkeit und der biochemischen Stabilität des Organismus, jedoch den Nachteil, dass dieses Energiesystem relativ kostspielig ist. Die Maßnahmen, die

[22] Nach Macdonald, *Enzyklopädie der Säugetiere* 3, werden die *Carnivora* in 9 Familien unterteilt.

erforderlich sind, um das System rentabel zu gestalten, werden evolutionsbiologisch als ‚*Anpassungen*' gedeutet: Solche ‚Anpassungen' der Säugetiere stellen Behaarung, Talgdrüsen (die beide der Wärmeerhaltung durch Isolation dienen) und Schweißdrüsen (die der Wärmeabfuhr durch Wasserausscheidung/Verdampfung/Kühlung dienen) dar. Die Anpassung der Behaarung beispielsweise kann durch die Forschungsmethoden der Mikrobiologie weiter erklärt werden. Eine Analyse der Zellstruktur des Haares zeigt, dass die innerste Schicht des Haares aus toten Zelltrümmern besteht, welche Luft enthalten. Die Zellstruktur des Haares ist also so angelegt, dass dieses eine starke isolierende und daher thermoregulierende Wirkung hat.

Ebenso fasst man die Reduktion auf einen Unterkieferknochen nicht einfach als morphologisches Fakt auf, sondern deutet sie evolutionsbiologisch als Anpassungsschritt zur Entwicklung höherer Lebensformen: Die Reduktion auf *einen* Unterkieferknochen ermöglicht, dass sich die Knochen, die das Gehirn umgeben, verstärken konnten – dies macht eine Vergrößerung des Gehirns und eine Ausdehnung des Nervensystems möglich, und bildet somit die Voraussetzung zur Entwicklung höherer Intelligenzen.

Ähnlich wird die Fortpflanzungsart sowohl evolutionsbiologisch als auch funktionsanalytisch beschrieben: Die meisten Säugetierarten sind Plazentatiere (*Eutheria*), die sich vor etwa 90 Millionen Jahren von den *Metatheria* (Beuteltieren) abgespaltet haben. Ein Vorteil der „Anpassung" der Plazentatiere ist, dass die Plazenta durch den intensiven Austausch von Nähr- und Abfallstoffen sowie Atemgasen zwischen dem Kreislaufsystem von Mutter und Kind längere Tragzeiten möglich macht. Diese ermöglichen ihrerseits eine längere und damit höhere Entwicklung des Kindes. Das Säugen mit Milch bewirkt sowohl eine größere Unabhängigkeit von der Umwelt (die Mutter produziert auch bei Nahrungsmangel weiter Milch) als auch eine Differenzierung in der Investition bei der Fortpflanzung. Z.B. können Männchen länger und besser gesäugt werden, wodurch sie sich rascher entwickeln: Dadurch wird die Voraussetzung zu einer Rangdifferenzierung der Männchen in Bezug auf ihre evolutionsbiologische Fitness geschaffen. Dies besitzt z.B. für die Species *Panthera leo* große Bedeutung, die sich durch einen ausgeprägten Geschlechtsdimorphismus auszeichnet: Die Männchen, die 30–50% größer und schwerer als die Weibchen sind, bestimmen wesentlich die Modalitäten der Weitergabe des genetischen Materials. Diese erfolgt aufgrund eines hierarchischen Ausleseprinzips: Die stärksten und größten Männchen besitzen ungleich größere Chancen, ihr genetisches Material weiterzugeben.

Bei der Beschreibung der Sinne der Säugetiere werden sowohl biochemische Methoden als auch jene der Verhaltensforschung angewendet: Die meisten Säugetiere besitzen ein hervorragendes Geruchsvermögen, das mit ihrer speziellen Art der Thermoregulation zusammenhängt und ihr komplexes Sozialverhalten mitbedingt. So besitzt die thermoregulierende Haut der endothermen Säuger Talg-, Schweiß- und Duftdrüsen, die komplexe Gerüche erzeugen, welche ein reiches Spektrum an Information über Geschlecht, Rang, Alter, Paarungsbereitschaft und Ernährung des Geruchsabsenders vermitteln. Die Raubtiere (*Carnivora*) etwa sind wahre Geruchskünstler: Ihr Geruchsvermögen bestimmt ihr hochentwickeltes Sozialverhalten: ihr Territorialverhalten, ihr Jagdverhalten, das Zusammenleben im Rudel mit Sozialhierarchie (z.B. Löwen, Wölfe) und ihr Fortpflanzungsverhalten. Löwen etwa leben in Rudelverbänden auf einem bestimmten Territorium, das von den führenden Männchen mit Duftmarken versehen wird. Die Mitglieder des Rudels erkennen einander individuell am Geruch. Individuen, die nicht zum Rudel gehören, können an den Duftmarken erstens ablesen, dass sie ein besetztes Territorium erreicht haben, zweitens, welches Alter der Territoriuminhaber besitzt, und drittens, wie stark dieser ist.

Der Diskurs der modernen Zoologie wird weiter von einer Fokussierung auf genaue Messdaten, Zählungen und andere auf empirischem Weg erhobene Daten gekennzeichnet. Die Körperlänge, Schulterhöhe, Schwanzlänge, das Körpergewicht der erwachsenen Tiere (mit Geschlechtsdifferenzierung), der Neugeborenen, der einjährigen, zweijährigen usw. werden genau gemessen; weiter wird die Nahrung bestimmt und gewogen, differenziert nach Geschlecht und Alter. Die Geschlechtsreife wird in Monaten, die Tragzeit in Tagen, die Lebenserwartung in Jahren (ev. differenziert nach freier Wildbahn und Menschenobhut) angegeben. Territorien werden vermessen, Verbreitungsgebiete kartographisch beschrieben, die Verbreitungsdichte durch Zählungen bestimmt. Da viele Wildtier species vom Menschen zurückgedrängt und bedroht werden, wird der jeweilige Gefährdungsgrad (IUCN-Status) angegeben.

Löwenmännchen etwa besitzen eine Körperlänge von 110–250 cm, eine Schulterhöhe von 120 cm und ein Gewicht von 150–240 kg;[23] Löwenweibchen eine Körperlänge von 160–190 cm, eine Schulterhöhe von 110 cm und ein Gewicht von 120–185 kg. Weibchen brauchen ca. 5 kg Fleisch pro Tag, Männchen ca. 7 kg. Die Lebenserwartung des Löwen

[23] Für die Daten vgl. Macdonald (Hrsg.), *Enzyklopädie der Säugetiere* 11.

beträgt 18 Jahre in freier Wildbahn, in Menschenobhut bis zu 25 Jahre. Weibchen sind mit 36–46 Monaten geschlechtsreif. Die Tragzeit beträgt zwischen 100 und 119 Tage. Die Neugeborenen bringen 1,2 – 2,4 kg auf die Waage. Nach ca. 2,5 Jahren sind sie erwachsen und selbständig.

Die Zoologie Edward Wottons (1552): Klassifizierung, Induktion und Komparatistik als Faszinationen der neuen Wissenschaft

Die Analyse der frühneuzeitlichen Zoologien soll mit dem Werk des aus Oxford stammenden Edward Wotton (1492–1555) anfangen.[24] Zwar ist sein *De differentiis animalium* (1552) ein Jahr nach Gesners *Historia animalium* erschienen, jedoch ist Wotton, der an seinem Projekt viele Jahre hindurch gearbeitet hat, eine Generation älter als Gesner und er hat sich früher als jener mit der Zoologie auseinandergesetzt. Weiter brachte Wotton sein Werk zum Abschluss, noch bevor er Gesners *Historia animalium* konsultieren konnte. Der Widmungsbrief an den englischen König Edward VI., der geraume Zeit nach der Vollendung des Werkes geschrieben wurde, trägt das Datum des 22. 10. 1551.[25] Im Widmungsbrief geht Wotton davon aus, dass sein Werk sich in Paris im Druck befand. 1551 musste jedoch noch ein Verleger gefunden werden. Wie aus dem Brief hervorgeht, hatte Wotton das Werk 1551 seinem Freund John Mason übergeben, der es auf eine europäische Reise mitnahm und in Paris einen Verleger gewann.

Wottons Zoologie zeichnet sich durch eine Herangehensweise aus, bei der die ordnende Klassifizierung des Tierreichs und die Deskription der einzelnen Arten innerhalb des Klassifizierungssystems eine herausragende Rolle spielt. Die Bedeutung dieses zoologischen Darstellungsprinzips spiegelt sich bereits im Titel wider: Das Werk soll die „Unterschiede" zwischen den einzelnen Species beschreiben. Auf den ersten Blick ähnelt Wottons systematischer zoologischer Klassifizierungsdiskurs dem taxonomie- und systematikorientierten Diskurs der modernen Zoologie. Änne Bäumer betrachtet in teleologischer Herangehensweise seine Leistung dann auch als viel fortschrittlicher als z.B. jene Gesners, der nicht auf die Klassifizierung fokussierte: „Wottons

[24] Edward Wotton, *De differentiis animalium libri decem* (Paris, Michel de Vascosan: 1552).

[25] *De differentiis animalium* fol. a ii r. „Ad Serenissimum Angliae Regem Eduardum, eius nominis sextum".

erste Systematisierungsversuche haben alle späteren Klassifikationen beeinflusst. Die von Conrad Gesner [...] vorgelegte Systematik war eher künstlich [...], sodass seine Klassifizierung insgesamt einen Rückschritt gegenüber Wotton darstellte. Ulisse Aldrovandi hingegen bemühte sich ganz bewusst um ein natürliches System [...]. Er setzte damit die von Wotton eingeschlagene Richtung fort".[26] Im Gegensatz zu Gesner habe Wotton „auf dem Gebiet der Methodik und Systematik neue Wege gefunden, die die nachfolgende Entwicklung der Zoologie entscheidend beeinflussten".[27] Sein Systematisierungsversuch stelle „einen entscheidenden Neubeginn" dar. Seine wissenschaftsgeschichtliche Leistung sei diesbezüglich mit Copernicus vergleichbar.[28]

Nach welchen Gesichtspunkten richtete Wotton sein Werk ein? Dieses zeichnet sich zunächst durch einen ordnenden, systematisierenden und einer schlüssigen Klassifizierung zustrebenden Aufbau aus. Es setzt sich aus zehn Büchern zusammen: In den Büchern 1–2 liefert Wotton zunächst eine listenartige Aufstellung der einzelnen Unterscheidungsmerkmale zwischen den Tierarten. Buch 3 ist der allgemeinen Beschreibung der Bluttiere gewidmet, Buch 4 den Unterschieden zwischen Mensch und Tier.[29] Die Bücher 5–10 stellen den Hauptteil dar, in dem Wotton die einzelnen Species aufgrund der Merkmale, welche ihnen eignen, beschreibt („per species"). Die Unterschiede werden als intentional vollständiger Katalog der relevanten äußeren („exteriora") und inneren („interiora") Körpermerkmale formuliert. In den Büchern 5–10 werden die Species in ein gültiges Klassifizierungssystem eingeschrieben. Die Bücher 5–8 sind den Species der Bluttiere gewidmet, die Bücher 9–10 den Species der „blutlosen Tiere".

Da die Klassifizierung vom physiologischen Merkmal des Blutes ausgeht und da die inneren Körpermerkmale feste, immer wiederkehrende Bestandteile von Wottons Tierbeschreibung sind, rücken sowohl die Physiologie als auch die auf die Anatomie gestützte Morphologie verstärkt ins Blickfeld. Wenn man diese Orientierung teleologisch betrachtet, könnte es scheinen, als ob Wotton auf „fortschrittliche Weise" den Weg zur Vergleichenden Anatomie und Physiologie geebnet hat. Bei näherer

---

[26] „Zoologie der Renaissance – Renaissance der Zoologie" 290.

[27] *Zoologie der Renaissance* 41.

[28] Bäumer Ä., „Das erste zoologische Kompendium in der Zeit der Renaissance: Edward Wottons Schrift ‚Über die Differenzen der Tiere'" *Berichte zur Wissenschaftsgeschichte* 13 (1990) (13–29) 28.

[29] Zu der Beschreibung des Aufbaus vgl. Bäumer, „Das erste zoologische Kompendium" 17 und *Zoologie der Renaissance* 34–35.

Betrachtung stellt sich jedoch heraus, dass Wottons Deskriptionsdiskurs weder von der Vergleichenden Morphologie noch der Vergleichenden Physiologie bestimmt wird. Vielmehr ist ihm – nach der Einordnung in eine grobmaschige Klassifizierung – die möglichst detaillierte Beschreibung der Arten in ihrer Einzigartigkeit oberstes Gebot. Seiner Klassifizierung liegt, wie sich noch zeigen wird, keine systematische empirische anatomische Forschung zugrunde (vgl. unten).

Wottons Interesse für die Zoologie leitet sich aus dem Studium des Aristoteles ab, dessen Werke er 1506–1514 während seines Philosophiestudiums am Magdalen College in Oxford kennen lernte. Für die zoologische Methodik von Wottons *De differentiis animalium* finden sich die ausschlaggebenden Ansätze in Aristoteles' zoologischen Werken, besonders in der *Historia animalium* (*Peri zoon historiai*; 10 Bücher), in *De generatione animalium* (*Von der Fortpflanzung der Tiere*; *Peri zoon geneseos*; 5 Bücher) und im ersten Buch der Schrift *De partibus animalium* (*Von den Körperteilen der Tiere; Peri zoon morion*; 4 Bücher).

Der Diskurs der aristotelischen Zoologie zeichnet sich durch eine eingehende und umfängliche Anwendung des Induktionsverfahrens auf das Tierreich aus. Während Aristoteles anfänglich von der dichotomen und privativen platonischen Diairese[30] ausging (zum Beispiel im ersten Teil der *Historia animalium*), wich er nach und nach von ihr ab, indem er sie durch eine mehr- und vielteilige Diairese ersetzte.[31] Wie Aristoteles im ersten Buch der Spätschrift *De partibus animalium*, welche in gewissem Sinn die Summe seines zoologischen Denkens bildet, auseinandersetzt, ist es die vordringlichste Aufgabe der Zoologie, die einzelnen Erscheinungsformen des Tierreichs – die einzelnen Arten, ihre Körperteile, Lebensfunktionen, Verhaltensweisen usw. – zu beschreiben. Er entwickelte dafür ein komparatistisches Beschreibungsverfahren, bei dem er verschiedenartige Information (morphologische, physiologische; Äußeres, Inneres usw.) berücksichtigte und zudem durch Anatomie gewonnene Daten heranzog. Eine Tierart definiert sich nicht durch ein einzelnes Merkmal im Sinn einer privativen Gegenüberstellung, sondern durch eine Reihe unterschiedlicher Eigenschaften.[32] Wichtig ist, dass

[30] Vgl. dazu Herter H., „Platons Naturkunde. Zum *Kritias* und anderen Dialogen", *Rheinisches Museum* 121 (1978) 103–131.

[31] Vgl. Lloyd G.E.R., "The Development of Aristotle's Theory of the Classification of Animals", *Phronesis* 6 (1961) 59–81; Bäumer A., *Die Biologie von der Antike bis zur Renaissance*, in dies., *Geschichte der Biologie*, Bd. 1 (Frankfurt a.M. – Bern – New York – Paris: 1991) 48.

[32] Vgl. Aristoteles, *De partibus animalium* I 2–3, 642b–643a.

Aristoteles das Induktionsverfahren nicht vorrangig zur Erstellung einer geschlossenen klassifizierenden Beschreibung des Tierreichs nach Species angewendet hat.[33] Sein Hauptwerk, die *Historia animalium*, zeichnet sich durch seine thematisch angeordnete Komparatistik aus. Buch 1 ist den Teilen der Tiere, Buch 3 den gleichteiligen Stoffen (feste Teile, flüssige Teile), Buch 4 ab Kapitel 11 den Funktionen der Teile, Buch 5–7 den Zeugungs- und Entwicklungsarten der Tiere, und Buch 8 und 9 dem äußeren und inneren Leben der Tiere gewidmet. In dieser thematisch orientierten Behandlung werden die zahlreichen Unterschiede zwischen den insgesamt etwa 500 behandelten Species erarbeitet.

Obwohl diese Art der Datenerhebung Aristoteles wichtiger war als die Erstellung einer Klassifizierung, hat er dennoch das Tierreich in Kategorien unterteilt. Er gliedert es in blutlose und blutführende Tiere; die blutführenden in Menschen, lebendgebärende Vierfüßer, eierlegende Vierfüßer, fußlose Tiere, Vögel, Fische und Cetaceae; die blutlosen Tiere in Caphalopoden, Crustaceen, Testaceen und Insekten.[34] Diese Einteilung baut auf unterschiedliche Informationskategorien auf. Obwohl die Erhebung der äußerlich wahrnehmbaren Information – wie sich aufgrund der geringen technischen Möglichkeiten leicht verstehen lässt – damals weitaus vielversprechender war, geht aus der Klassifizierung Aristoteles' großes Interesse für das Innere der Tiere hervor. Z.B. bildet ein physiologisches Merkmal (Blut) die Grundlage seiner Klassifizierung, das zu einer dichotomen Zweiteilung des Tierreichs in *blutführende* und *blutlose* Tiere führt (*Historia animalium*, Buch 2–4). Bei der weiteren Unterteilung spielen jedoch die äußerlich wahrnehmbaren morphologischen und physiologischen Merkmale die Hauptrolle.

Diese Diskursorientierung lässt sich z.B. anhand der Einordnung der *Mammalia* illustrieren. Aristoteles' Definition unterscheidet sich stark von der modernen. Für ihn ist weder der äußerlich nicht klar wahrnehmbare Vorgang der Thermoregulation (Warmblüter-Kaltblüter) noch die mit der Thermoregulation verbundene Physiologie (Säugen als Anpassung an die Energierechnung) noch das aus systematischer vergleichender Anatomie hervorgehende morphologische Merkmal des einteiligen

[33] Vgl. Lennox J.G., „Between Data and Demonstration: the Analytics and the *Historia Animalium*", in ders. *Aristotle's Philosophy of Biology: Studies in the Origins of Life Science* (Cambridge: 2001) 39–71; Balme D.M., "Aristotle's use of division and differentiae", in Gotthelf A. – Lennox J.G. (eds.), *Philosophical Issues in Aristotle's Biology* (Cambridge: 1987) 69–89.

[34] Bäumer, *Die Biologie von der Antike bis zur Renaissance* 48–49.

Kiefergelenks Grundlage der Klassifizierung (eine systematische anatomische Datenerhebung hat Aristoteles nicht erstrebt), sondern ein äußerlich leicht wahrnehmbares morphologisches („Vierfüßigkeit") und physiologisches Merkmal („lebendgebärend").

Wottons zoologische Methode *De differentiis animalium* geht in mehrfacher Hinsicht von Aristoteles aus. Die Klassifizierungsansätze, die er unter anderem in der *Historia animalium* antraf, regten ihn an, diese in ein *beschreibendes zoologisches Kompendium* umzusetzen. In Anbetracht der Tatsache, dass Aristoteles' umfassende Beschreibung des Tierreichs eben nicht als systematische Klassifizierung angelegt war, erschien Wotton gerade ein dergestalt eingerichtetes zoologisches Handbuch als Desideratum. Diese Diskursorganisation lässt sich in Bezug auf die *Mammalia* beobachten. Wotton übernimmt die aristotelische Kategorie der Vierfüßer" („quadrupedes"), die er wie dieser in „lebendgebärende Vierfüßer" („quadrupedes viviperi") und eierlegende Vierfüsser (vierfüssige Reptilien), die *quadrupedes viviperi* in Mehrzeher[35] („multifida"), Paarhufer („bisulca") und Unpaarhufer („solipedes") unterteilt. Andere Unterscheidungen verlieren demgegenüber ihren klassifizierenden Wert.[36]

Wottons Klassifizierungsdiskurs hat den wissensorganisatorischen und didaktischen Vorteil der eindeutigen Zuordnung, jedoch den Nachteil der Grobmaschigkeit und der einseitigen Überbewertung einzelner Merkmale. Die Kategorie, der er in seiner Zoologie den Löwen zurechnete, die Mehrzeher, ist dafür bezeichnend. Damit entstand ein Klassifikationssammelbecken, das völlig Unterschiedliches auffängt. Die Kategorie der Mehrzeher (fol. 56v–57v) vereint sowohl vom äußeren Erscheinungsbild als auch von der Anatomie, der Physiologie und vom Verhalten her – und zwar bereits aufgrund der zu Wottons Zeit

[35] Die Übersetzung von „multifida" in „Vielzeher" ist weniger zutreffend, da es bekanntlich um maximal fünf Zehen geht.

[36] Nach Bäumer, „Das erste zoologische Kompendium" 18 habe Wotton „durchgehend [...] die Unterscheidung zwischen Land- und Wassertieren" durchgeführt (Bluttiere: A. Landtiere. B. Wassertiere; Blutlose Tiere: A. Landtiere. B. Wassertiere). Es ist jedoch fraglich, ob Wotton diese Unterscheidung in der Tat „durchgehend" durchgeführt hat. Z.B. bei den „lebendgebärenden Vierfüßern", von denen Bäumer berichtet, Wotton habe sie insgesamt als „Landtiere" aufgefasst, finden sich Tiere, die Wotton selbst explizit als Wassertiere bezeichnet, u.a. die Robbe („Seekalb"), der Otter, das Flußpferd und der Biber. Vgl. Wotton, *De differentiis animalium* V, fol. 70r–v „De quadrupedis *aquaticis* ut de fibro, lutro et quibusdam aliis"; fol. 70v–71r „De vitulo marino" und fol. 81v. Für die Definition der Robbe und des Flußpferdes als Wassertier erachtet Wotton als entscheidend, dass sich diese Species die meiste Zeit im Wasser aufhalten und ihre Nahrung aus dem Wasser holen.

bekannten Fakten – eine unglaublich bunte Ansammlung von Tieren: den Elefanten (fol. 56v–57v), die Affenarten (fol. 57v–58v), die Hundearten (fol. 58v–61v), die Wolfsarten (fol. 62r–63r), die Hyäne (fol. 63r–v), den Löwen, Tiger (fol. 65r), Leoparden („panthera pardus", fol. 65r–v), Bären (fol. 65v–66r), Fuchs (fol. 66v–67v), weiter das Stachelschwein („hystrix"), den Igel („erinaceus") und den „scyurus" (fol. 66r–v), den Hasen (fol. 66v–67v), die Katze und die Wieselartigen („mustelarum genera", fol. 67v–68v), die Maus und die Mäuseartigen (fol. 68v–69v), den Biber („fiber") und den Otter („lutra", fol. 70r–v), die Robbe („vitulus marinus", fol. 70v–71r), den Maulwurf, die Fledermaus („vespertilio", fol. 71r–v) und verschiedene Fabeltiere Indiens und Äthiopiens (fol. 71v–72r).[37] Affen (Ordnung: Primates) unterscheiden sich morphologisch, physiologisch und verhaltensbiologisch stark von Raubtieren (Ordnung: Carnivora) oder Elefanten (Ordnung: Proboscidea). Ähnliches gilt mutatis mutandis für Elefanten und Nagetiere (Ordnung: Rodentia), Fledermäuse (Ordnung: Chiroptera) und Raubtiere, Igel (Ordnung: Eulipotyphla) und Affen, Raubtiere und Hasentiere (Ordnung: Lagomorpha) usw.

Diese Ausführungen haben nicht den Sinn, Wottons Klassifizierung im Hinblick auf die moderne Taxonomie als unhaltbar abzutun. Vielmehr soll die Orientierung seines zoologischen Diskurses sichtbar gemacht werden. Wotton versuchte, Aristoteles sowohl nachzuahmen als auch im Hinblick auf die klassifikatorische Präsentationssystematik zu übertreffen.[38] Von dieser Diskursorientierung ausgehend ist er von vorneherein bereit, Ungereimtheiten, Dissonanzen und Diskontinuitäten in Kauf zu nehmen. Was die Mehrzeher betrifft, fällt auf, dass er sich nicht bemüht, weitere kategorieimmanente Gemeinsamkeiten festzustellen. Daraus darf man nicht den Schluss ziehen, dass Wotton dazu nicht imstande gewesen wäre. Diese interessierten ihn einfach zu wenig. Es erscheint mir zu einfach, Wotton wissenschaftsteleologisch die ‚fortschrittliche' ‚Errungenschaft' zuzuschreiben, ein „natürliches Klassifizierungssystem" entworfen zu haben. Man müsste die Frage stellen, wie „natürlich" und empiriebezogen diese Ordnung denn eigentlich

[37] In Bäumer „Das erste zoologische Kompendium" 24 findet sich überraschenderweise der Bericht, dass Wotton in seiner Darstellung der Mehrzeher den Biber, Otter, die Katze, den Panther und das Wiesel weggelassen hätte, obwohl Aristoteles diese Species behandelte. Diese Tiere werden von Wotton sehr wohl behandelt.

[38] Bäumers Formulierung „aristotelischer als bei Aristoteles" trifft diesbezüglich zu; vgl. *Geschichte der Zoologie* 40.

sei. Wie aus dem Untenstehenden hervorgehen wird, hat Wotton zur Erstellung seines Klassifikationssystems im Grunde keine empirischen Beobachtungen angestellt. Interessant ist nicht die Frage, wie ‚fortschrittlich' Wotton gewesen sei, sondern was den Diskurs ausmacht, in dem er operierte. Die systematisierende Weiterführung des von Aristoteles autorisierten Wissensschatzes war für Wotton offensichtlich ungleich bedeutender als die Verlagerung des Wissens zur Empirie hin.

Das gilt auch überraschenderweise auch für die empirienahen Methoden der Induktion und der Komparatistik. Beide Methoden gehen von der empirischen Deskription der wahrnehmbaren Wirklichkeit aus. Wotton war von der Induktion und der Komparatistik fasziniert. Daraus folgt jedoch nicht automatisch, dass er sich verpflichtet fühlte, seine Angaben auf *eigene empirische Beobachtungen* zu gründen, obwohl ihm Bäumer solche zuzuschreiben scheint.[39] Interessanterweise hat er gleichwohl den Versuch gemacht, sein Vorbild Aristoteles im Hinblick auf die induktive und komparatistische Erfassung der Tierarten zu übertreffen. Induktion und Komparatistik sind die Methoden, mit denen Aristoteles durch das möglichst genaue Feststellen der Unterschiede zwischen den Tieren die Definition derselben in einzelne, nicht weiter teilbare Arten vorgenommen hat.[40] Dass Wotton die aristotelische Komparatistik als Grundlage verwendete, zeigt schon der Titel seiner Zoologie (*De differentiis*) an. Interessant ist, dass er sich bemühte, die einzelnen, von Aristoteles unterschiedenen Tiersorten weiter zu differenzieren bzw. in Arten oder Unterarten[41] zu gliedern.

Das Kapitel über den Löwen ist ein Musterbeispiel für diese Diskursorientierung. Das Kapitel trägt bezeichnenderweise nicht den Titel

---

[39] Bäumer, „Das erste zoologische Kompendium" 17 schreibt Wotton zu, dass er die antike Zoologie „durch eigene Beobachtungen ergänzt" habe. Es ist wahrscheinlich, dass sie damit empirische Beobachtungen meint. Letzte lassen sich jedoch schwerlich feststellen und widersprechen Wottons eigenen Angaben aus dem Vorwort (s. unten).

[40] Balme, „Aristotle's use of differentiae in zoology".

[41] Der Artenbegriff soll an dieser Stelle bewusst offen formuliert werden. Wotton kannte weder einen genau definierten Artenbegriff, wie ihn die moderne Zoologie – von genetischen Grundlagen ausgehend – hantiert, noch machte er einen systematisch fundierten Unterschied zwischen Art (Species) und Unterart (Subspecies). Die problematisierende Diskussion des Artenbegriffs (als Korrektur des Darwinschen Artenkonzepts) fand erst gegen die Mitte des 20. Jahrhunderts statt. Grundlegend dazu: Mayr E., *Systematics and the Origin of Species* (New York: 1942) und *Animal Species and Evolution* (Cambridge Mass.: 1963).

„Vom Löwen" („De leone"), sondern „Von Löwen" („De leonibus"),[42] d.h. „Von Löwenarten". Diese von der Titelei ausgehende Diskursorientierung lässt sich mehrfach beobachten, z.B. beim Hund („Von Hundearten" – „De canibus")[43] und beim Wolf („Von Wolfsarten" – „De lupis").[44] Gleich der erste, vergleichsweise sehr ausführliche Abschnitt des Kapitels „Von Löwen" (etwa ein Drittel) ist der Differenzierung des Tieres in 7 Arten gewidmet, die durch die komparatistische Feststellung verschiedener Merkmale definiert und geographisch eingeordnet werden.

Wotton unterscheidet im Gegensatz zu Aristoteles zwei europäische, eine parthische und armenische (vorderasiatische), eine arabische (Verbreitungsgebiet: arabische Halbinsel), eine afrikanische (= nordafrikanische) und zwei äthiopische Arten des Löwen. Der gemeine europäische Löwe soll stärker als der afrikanische und der vorderasiatische sein. Daneben gibt es die Species des Donaulöwen (sic), der ebenso schwach wie der vorderasiatische sein soll. Der vorderasiatische (parthische und armenische) Löwe zeichne sich durch einen großen Kopf, „funkelnde, leuchtende Augen" („oculis splendore radiantibus"), durch „lange, bis auf die Nase herabhängende Augenbrauen" („superciliis extantibus et fere ad nares usque propendentibus") und eine dichte, reiche Mähne aus. Vom vorderasiatischen Löwen unterscheide sich der arabische Löwe, der sonst ähnliche Körpermerkmale besitze, durch seine weitaus größere Aggression und einzelgängerische Lebensweise. Der afrikanische Löwe zeichne sich im Unterschied zu den europäischen, vorderasiatischen und arabischen Löwenarten durch eine schwarze bzw. schwarzblaue Körperfarbe aus: „Sie besitzen am ganzen Körper eine schwärzliche und sehr dunkle Färbung, jene Farbe, die entsteht, wenn man blau und schwarz mischt" („Totius corporis color nigricans his fuscusque, ac si nigrum coeruleo permisceas"). Außerdem hat er eine kürzere Mähne als der europäische, der vorderasiatische und der arabische Löwe. Die eine äthiopische Löwenart hat einen schwarzen Körper und ein rötlichgelbes Gesicht, die andere ist in etwa so groß wie ein Hund oder ein ägyptischer Wolf („magnitudo Aegypti luporum"), während sie ein glattes, kurzhaariges und goldbraunes Fell, jedoch *keine Mähne*

---

42 *De differentiis animalium* fol. 63v–64r.
43 Ebd. fol. 58v–61v.
44 Ebd. fol. 62r–63r. Wotton unterscheidet 10–14 Wolfsarten.

besitzt, sehr schnell ist, Luftsprünge macht und im Boden lebende Tiere ausgräbt.[45]

Wottons Streben nach einer weiteren Differenzierung und Verfeinerung der Klassifizierung in einzelne Arten und Unterarten ist umso auffälliger, als er dafür über *keine empirischen Beobachtungen* verfügte. Er hat die betreffenden Gebiete nie besucht noch frühneuzeitliche Reise- oder Forschungsberichte systematisch ausgewertet. Schon deswegen würde ein Versuch, Wottons Löwenarten mit heute bekannten Subspecies[46] des Löwen zu identifizieren, auf nichts hinauslaufen. Auch spielt die Tatsache, dass der Löwe in nahezu allen Gebieten, die Wotton einbezieht, heute ausgestorben ist, keine wesentliche Rolle.[47]

Wotton hat seine Unterteilung in Arten/Unterarten vielmehr aus einigen antiken Schilderungen des Aussehens des Löwen abgeleitet, welche weder systematischen Klassifizierungsversuchen gewidmet sind noch überhaupt als zoologische Forschungsberichte einzustufen sind, besonders aus einem Gedicht des griechischen Schriftstellers und Redners Oppian von Apameia (3. Jh. n. Chr.), *Von der Jagd* (*Kynegetika*), einem Lehrgedicht, in welchem der Löwe als Jagdobjekt dargestellt wird. In diesem Gedicht findet sich die ‚Information', dass der afrikanische Löwe zu schwärzlicher Körperfarbe neige. An derselben Stelle stellt der Dichter die Löwen Äthiopiens als schwarz mit einem rötlichgelben Gesicht[48] und die arabischen Löwen als außerordentlich angriffslustig dar. Offensichtlich hat Oppian bei seinen Schilderungen des afrikanischen Löwen und der einen äthiopischen Löwenart den Löwen mit der Species *Panthera pardus* (Leopard) verwechselt, welche auch rein schwarz auftritt.[49] Ebenso dürfte der „hundegroße", kurzhaarige und mähnenlose Löwe Äthiopiens auf eine Verwechselung mit einer anderen Species zurückzuführen sein, am ehesten wohl mit dem Karakal (*Felis*

---

[45] Ebd. fol. 64r. Diese zweite äthiopische Löwenart soll auch in den Wüstengebieten Indiens vorkommen.

[46] Katangalöwe (*Panthera leo bleyenbergii*: Simbabwe, Angola, Demokratische Republik Kongo); Asiatischer Löwe (*Panthera leo persica*: Gir-Nationalpark); Massailöwe (*Panthera leo massaicus*: Kenia, Tansania, Uganda); Senegallöwe (*Panthera leo senegalensis*: West-Afrika); Transvaallöwe (*Panthera leo krugeri*: Transvaal).

[47] Mit Afrika meint er das ans Mittelmehr grenzende Nordafrika. Seine Konzeption von „Äthiopien" ist ungenauer und weitläufiger als das heutige Äthiopien: Wottons „Äthiopien" zieht sich tief nach Mittelafrika hinab.

[48] Oppian, *Kynegetika* IV, 147.

[49] Macdonald (Hrsg.), *Enzyklopädie der Säugetiere* 28. Leoparden gibt es übrigens auch noch heute in Nordafrika, den sog. Berberleopard (Panthera pardus panthera: Marokko, Algerien, Tunesien).

caracal; Wüstenluchs). Der Karakal weist ein glattes, gelbbraunes Fell und eine Körperlänge auf, welche in etwa einem Hund oder einem kleinen Wolf[50] entspricht, macht Luftsprünge und gräbt im Boden lebende Tiere aus.

Ähnlich ist Wotton in Bezug auf andere Tierarten vorgegangen, etwa den Wolf (Species: *Canis lupus*). Auch dort bemüht er sich, das Tier in möglichst viele Arten bzw. Unterarten zu gliedern (ca. 10–14), die er wiederum ausschließlich aus der antiken Literatur bezieht.[51] Um die Differenzierung möglichst vielgliedrig zu gestalten, ist er bereit, Hybriden (Mischwesen), potentiell artfremde Tiere und anderes Ungereimtes aufzunehmen. Beispiele sind der „Thos" oder Äthiopische Wolf, ein Mischwesen, das aus der Paarung eines Wolfsmännchens und eines Leopardenweibchens („ex lupo et panthera prognatus") hervorgehen soll (mit Wolfsgesicht und Leopardenkörper),[52] der Gepard („Lycaon"),[53] zwei Luchsarten (eine große und eine kleine), der „lupus canarius", der zugleich ein Panther sein soll, oder der „Chaus Rhaphilus", ein in Gallien vorkommendes Mischwesen zwischen Wolf und Leopard.[54] Auch in Bezug auf den Wolf ist es müßig, Wottons Arten mit den heute bekannten Subspecies des Wolfes (z.B. Europäischer Wolf; Steppenwolf; Tundrawolf)[55] identifizieren zu wollen. Unter Wottens Wolfsarten befinden sich mindestens fünf Species aus der Familie der Katzenartigen (*Felidae*). Obwohl Wotton natürlich nicht über die heute bekannten Daten verfügen konnte, war ihm dennoch klar, dass hier etwas nicht stimmen könne: Denn er wälzt die Verantwortung für die Zuschreibungen auf andere Gewährsmänner ab. „Manche zählen die Luchse dieser Tiergattung (sc. Wolf) zu".[56] Es ist bezeichnend für die Klassifizierungsverliebtheit seines zoologischen Diskurses, dass er die fragwürdigen oder unglaubwürdigen Arten dennoch in seine Liste aufnimmt.

---

[50] Unter einem „Ägyptischen Wolf" verstand man eine kleine Wolfsart; vgl. Plinius, *Naturalis historia* VIII 80. Für dieses Buch der *Naturalis historia* siehe C. Plinius Secundus d. Ä., *Naturkunde. Lateinisch – deutsch. Buch VIII. Zoologie: Landtiere.* Herausgegeben und übersetzt von Roderich König in Zusammenarbeit mit Gerhard Winkler (Kempten: 1976).

[51] *De differentiis animalium* fol. 62r–v.

[52] Ebd. 62r.

[53] Ebd. 62v.

[54] Ebd.

[55] Vgl. Macdonald (Hrsg.), *Enzyklopädie der Säugetiere* 43.

[56] *De differentiis animalium* fol. 62v.

Vorstoß ins Innere der Tiere: Der Anatomiediskurs zwischen Empirie und Antiempirie

Auffällig ist, dass in Wottons Beschreibungsmethode die inneren Körpermerkmale betont werden. Immer wieder macht Wotton Angaben zu Muskeln, Knochen, Sehnen, Innereien usw. Die Annahme liegt nahe, dass Wottons Interesse am Inneren der Tiere mit seiner Ausbildung und seinem Beruf verbunden war. Wotton war Doctor in der Medizin und übte in London den Beruf eines praktischen Arztes aus. Als Mediziner war er sowohl theoretisch als praktisch mit der Anatomie vertraut. Von besonderer Bedeutung ist, dass er in Norditalien studiert hatte (Padua, 1523–1526): Die medizinischen Fakultäten Paduas und Bolognas waren Hochburgen der Anatomie und befanden sich in einem wissenschaftlichen Wettlauf um den Kompetenzprimat in diesem Wissenschaftszweig.[57] Die Leichensektion gehörte zum Studienprogramm. Außerdem publizierten die Anatomen Paduas und Bolognas lateinische Handbücher, in denen sie ihre empirischen Erkenntnisse darlegten. Als Wotton mit dem Studium anfing, waren kurz zuvor zwei Lehrgänge erschienen, welche die Anatomie durch systematische Leichensektion und Vivisektion auf eine neue, empirische Grundlage stellten: Berengario da Carpis (um 1460–um 1530) *Commentaria cum amplissimis additionibus super anatomia Mundini* [...] (Bologna: 1521), welche die herkömmliche Schulanatomie des Raimondo Mondino de' Luzzi (ca. 1275–1326) berichtigte, und die *Isagoge brevis* (Bologna: 1522) desselben Autors. Berengario da Carpi, der wie schon sein Vater ein Chirurg war und in den nämlichen Jahren diese Disziplin an der Universität Bologna unterrichtete, setzte die Empirie als unverzichtbare, absolute Grundlage anatomischen Wissens voraus.[58] Er forderte vom Anatomen, dass er möglichst häufig Sektionen betreiben sollte.

Von besonderem Interesse ist, dass Tiere als Forschungsobjekte eine sehr wichtige Rolle spielten. Das erhellt erstens daraus, dass man sie leichter beschaffen konnte und zweitens, dass in Bezug auf sie ethische Einwände, die sich der Sektion von Menschen in den Weg stellten, nicht galten. Berengario forderte den Anatomen auf, so viele Tierarten wie möglich auf den Seziertisch zu legen, und zwar sowohl männliche als

---

[57] Vgl. French R.K., „Bologna and Padua: Anatomical Rationalities", in ders., *Dissection and Vivisection in the European Renaissance* (Aldershot: 1999) 73–100.

[58] Ebd. 94.

auch weibliche, sowohl alte als auch junge Individuen.[59] Berengario hat Hunderte Tiere seziert, sowohl tote als auch – ein modernen Tierliebhabern unerträglicher Gedanke – lebende. Unter anderem inspizierte er die Beschaffenheit des Gehirns und der Herzkammer durch Vivisektion: „und ich habe lebende Tiere seziert: und sofort beim ersten Eingriff, wenn das Leben noch da ist, wollte ich die Herzkammer betrachten, und ich traf darin ein große Menge Wasser an, während sich das Herz noch stark bewegte" („et ego anatomizavi animalia viva: et statim in primo occursu, dum adhuc staret vita, volui videre pericardium, et vidi multam aquam in ipsa adhuc stante forti motu cordis").[60]

Man darf also annehmen, dass der Mediziner Wotton schon von seinem Studium her mit der Sektion von Tieren vertraut war. Es scheint daher auf der Hand zu liegen, dass er seine anatomischen Kenntnisse auf seine zoologischen Beschreibungen anwendete.

Das Interesse Wottons am inneren Bauplan der Tiere tritt im Kapitel über die Löwen klar hervor: Unter anderem behandelt er das Skelett und den Knochenbau des Löwen, das Knochenmark, die Zunge und die Zähne.[61] Zum Skelett gibt Wotton an, dass der Löwe keine Halswirbel besitze, sondern stattdessen einen einzigen Knochen („cervix osse uno rigida constat nec vertebris ullis iungitur"; „Der starre Hals setzt sich aus einem Knochen zusammen und wird nicht mit Wirbeln verbunden"). Außerdem seien die Knochen des Löwen insofern merkwürdig, als sie nur einen winzigen Hohlraum und kaum Knochenmark aufwiesen. Die Knochen seien so hart, dass man sie wie Feuersteine zum Feuermachen verwenden könne. Von seinen Zähnen soll der Löwe nur die vier Reißzähne (Eckzähne, dentes canini) wechseln, und zwar im sechsten Monat. Übrigens unternimmt Wotton keinen Versuch, diese Merkmale funktionsanalytisch zu deuten. Er stellt nicht die Frage, welchen Sinn diese anatomischen Besonderheiten haben sollen, z.B., was der Vorteil des „starren" Nackens aus einem Bein oder des mangelnden Hohlraums im Knochen sei. Die Angaben bleiben als lose morphologische Daten im Raum stehen.[62]

---

[59] *Commentaria cum amplissimis additionibus super anatomia Mundini* [...] (Bologna: 1521) fol. 3r.

[60] Ebd. fol. 439r.

[61] *De differentiis animalium* 64r.

[62] Das gilt übrigens auch für die treffende Beschreibung der Zunge des Löwen, die rau wie eine Feile sein und „ziegelartige" Erhebungen aufweisen soll.

Aus ihnen wird interessanterweise ersichtlich, dass Wotton, obwohl er als Arzt mit der Sektion von Tieren vertraut war, weder seine Daten aus empirischer Forschung bezogen hat noch sein tieranatomisches Wissen in seinem zoologischen Deskriptionsdiskurs anwendete. Wenn er jemals einen Löwen seziert oder empirische Erkenntnisse eines anderen frühneuzeitlichen Anatomen verwertet hätte, hätte er die Angaben unschwer falsifizieren können. Er hätte z.B. erkennen können, dass die Löwen sehr wohl Halswirbel besitzen (sieben, wie alle Karnivoren und die meisten Säugetiere).[63] Auch hätte er wahrscheinlich wahrgenommen, dass es andere Eigenschaften gibt, die das Skelett des Löwen auszeichnen: z.B. dass das Schlüsselbein gegenüber anderen Säugern stark reduziert ist und Elle und Speiche sowie die Handwurzelknochen verwachsen sind. Die moderne Zoologie interpretiert diese Besonderheiten evolutionsbiologisch und funktionsanalytisch als Anpassungen an das Jagdverhalten: Die Verwachsungen wirken stabilisierend und beugen Verdrehungen bei Jagdsprints vor,[64] die Reduktion des Schlüsselbeins wirkt sich günstig auf die Laufschnelligkeit aus.

Es wäre möglich, dass Wotton auf die Sektion des Löwen verzichtete, einfach weil keiner vorrätig war. Jedoch hätte er in diesem Fall auf analoge Beobachtungen über die Anatomie anderer leicht erwerbbarer Karnivoren, etwa des Hundes oder der Hauskatze, zurückgreifen können. Denn bei Aristoteles traf er einen klaren Hinweis an, dass sich eine solche Untersuchung lohnen würde: „Die Anatomie des Löwen ähnelt in jeder Beziehung der des Hundes". Obwohl Wotton diesen interessanten Satz in seinem Kapitel „Von Löwen" zitierte („interiora item omnia canibus similia continet", fol. 64r), unternimmt er bezeichnenderweise keinen Versuch, die Daten durch empirische Augenschau zu überprüfen.[65]

Aus Wottons Beschreibung des Löwen lässt sich ableiten, dass um die Mitte des 16. Jahrhunderts zwischen dem medizinischen anatomischen Wissen und der zoologischen Beschreibungspraxis auffällige Diskursunterschiede vorhanden waren. Wottons Fall ist diesbezüglich umso aussagekräftiger, als er Arzt und Zoologe in Personalunion war.

---

[63] Die Tatsache, dass die meisten Säugetiere 7 Halswirbel besitzen, war Wotton übrigens bekannt. Vgl. *De differentiis animalium* fol. 14r „In collo septem vertebrae sunt".

[64] Vgl. Macdonald (Hrsg.), *Enzyklopädie der Säugetiere* 4.

[65] Dasselbe gilt für den Wolf, der wie der Löwe keine Halswirbel besitzen soll. Wotton hat weder einen Wolf seziert noch Daten von Wolfssektionen anderer verwendet noch die anatomischen Daten über den Wolf systematisch mit denen des Hundes zusammengeführt. Vgl. *De differentiis animalium* fol. 62r.

Während der Arzt Wotton von der Bedeutung der Empirie überzeugt war, bildete für den Zoologen Wotton empirischer Datenerwerb offensichtlich keine wesentliche Grundlage.

Denn die Angaben zur Anatomie des Löwen gehen sämtlich auf Aristoteles und andere antike Autoren zurück. Es war Aristoteles, der die merkwürdige Behauptung in die Welt gesetzt hat, der Löwe besitze keine Halswirbel, sondern nur einen einzigen Knochen.[66] Andere antike Autoren – z.B. Aelian und Plinius d. Ä. – haben sie übernommen.[67] Auch die scheinbar aus der Empirie gewonnene Beobachtung, dass Löwen nur ihre Reißzähne wechseln, und zwar im sechsten Monat, geht auf Aristoteles zurück.[68] Sie ist übrigens nicht richtig: Löwen wechseln ihr gesamtes Milchgebiss. Schließlich fußt auch die Bemerkung, man könne mit Löwenknochen Feuer schlagen, nicht auf einer empirischen Beobachtung Wottons. Die Quelle der – freilich unrichtigen – Behauptung, die übrigens auch Plinius überlieferte,[69] ist abermals Aristoteles.[70]

Den Mangel an authentischen empirischen Beobachtungen zur Anatomie darf man nicht als Unfähigkeit auffassen. Wenn Wotton gewollt hätte, hätte er sehr wohl neue empirische Daten sammeln oder empirische Forschung verrichten können, zumal er ein geschulter Mediziner war und sich in der Anatomie auskannte. Er hat dies jedoch wissentlich und willentlich vermieden. Denn im Vorwort zu seiner Zoologie hielt er sich zugute, dass das Wissen, das er vermittle, nicht auf eigener, ursprünglicher und empirischer Forschung beruhe: „Du wirst in dem Werk so gut wie nichts antreffen", sagt er im Widmungsvorwort, „was aus meinem Kopf entsprungen ist" („Neque hic fere quicquam invenies, quod e meo capite natum statuatur").[71] Was er hier wiedergebe, beschränke sich auf die an sich als wertvoll eingestuften Lehrmeinungen („sententiae") der antiken Autoren.[72]

---

66 Aristoteles, *Historia animalium* II 1, 497b 16; *De partibus animalium* IV 10, 686 a 22.

67 *Naturalis historia* VIII 177; Aelian, *Historia animalium* IV 34.

68 Aristoteles, *Historia animalium* VI 31, 579b 13; *De generatione animalium* V 8, 788b 16.

69 Plinius, *Naturalis historia* IX 214. Für dieses Buch der *Naturalis historia* siehe C. Plinius Secundus d. Ä., *Naturkunde. Lateinisch – deutsch. Buch IXI. Zoologie: Insekten. Vergleichende Anatomie.* Herausgegeben und übersetzt von Roderich König in Zusammenarbeit mit Joachim Hopp (München-Zürich: 1990).

70 Aristoteles, *Historia animalium* III 7, 516b 10; *De partibus animalium* II 9, 655a 15.

71 *De differentiis animalium* fol. a iii v.

72 Ebd.

Wottons Diskurs frühneuzeitlicher zoologischer Datenerhebung: Das Verbreitungsgebiet des Löwen

Insgesamt fällt in Wottons Beschreibung der Mangel an genauen Mess- und Zähldaten sowie empirischen Daten auf. Der Leser kann der Beschreibung nicht entnehmen, wie lang, hoch und schwer der Löwe ist, wie lange er lebt, wie viele Tage die Tragzeit dauert usw. Für dieses Diskursmerkmal sind Wottons Angaben zum Verbreitungsgebiet des Löwen besonders bezeichnend. Bäumer gibt an, dass Wotton das antike zoologische Wissen aktualisiert, „den Bedürfnissen der eigenen Zeit angepasst" und „durch eigene Beobachtungen ergänzt" habe.[73] Es erscheint mir fraglich, inwiefern dies in der Tat der Fall war. Im Kapitel „De leonibus" berichtet Wotton z.B., dass sich der Löwe „eher in Europa als in Asien aufhalte" („in Europa potius quam in Asia inveniuntur leones").[74]

Diese Angaben sind überraschend weit von den realen Gegebenheiten entfernt. In Europa war der Löwe um die Mitte des 16. Jahrhunderts längst ausgestorben (jedenfalls seit der römischen Antike), während er jedoch in Asien ziemlich häufig war.[75] Wotton wertete klar ersichtlich keine Reiseberichte aus. Hätte er dies getan, hätte er ersehen können, dass Löwen in rezenter und subrezenter Zeit in Asien nachdrücklich attestiert waren (vgl. unten). Das ‚Fakt', dass es in Europa Löwen gebe, hat Wotton aus antiken Quellen erhoben, z.B. aus Aristoteles, Plinius d. Ä. oder Aelian. Plinius teilt mit, dass es nach Aristoteles „in Europa nur zwischen den Flüssen Acheloos und Mestos Löwen gebe" („Is [Aristoteles] tradit [...] in Europa [...] inter Acheloum tantum Mestumque amnes leones esse").[76] Es ist übrigens kein Zufall, dass

---

[73] „Das erste zoologische Kompendium" 17.

[74] *De differentiis animalium* fol. 63v. Zusätzlich findet sich die Angabe, dass europäische Löwen stärker seien als die afrikanischen und sie syrischen, jedoch mit Ausnahme der Löwen, die im Donaugebiet leben („Qui vero prope Istrum Europae amnem [...] procreantur, non adeo viribus pollent", ebd.). Vgl. weiter fol. 25r „Item leones in Europa potius sunt". In diesem Kapitel (28) macht Wotton theoretische Angaben zum Verbreitungsgebiet der Tiere. Daraus geht hervor, dass die europäischen Species „stärker" („fortiores"), die asiatischen wilder („efferatiores") und die afrikanischen „vielgestaltig" („multiformes") sein sollen.

[75] Der asiatische Löwe (Pantera leo Persica) ist in der Gegenwart auf ganz geringe Bestände im Gir-Nationalpark (Indien) zusammengeschmolzen und akut vom aussterben bedroht. Keller vermutet, dass der europäische Löwe möglicherweise schon in mykenischer Zeit ausgestorben war (Keller O., *Die antike Tierwelt* I (1909) 36).

[76] Plinius, *Naturalis historia* VIII 45; Aristoteles, *Historia animalium* VI 31, 579b 7; VIII 28, 606b 15.

Plinius die Verantwortung für die Angabe seinem Gewährsmann („is tradit" „er überliefert") überträgt: Plinius war sich nicht sicher, ob sie richtig sei. In der Tat war der Löwe schon zu Plinius' Zeit in Europa ausgestorben, ja nicht einmal Aristoteles (* 384) stützte seine Angabe auf Autopsie, sondern bezog sie – wie die enge intertextuelle Verknüpfung zeigt – aus dem Geschichtswerk des ca. 100 Jahre vor ihm geborenen Herodot.[77]

Der zoologische Diskurs Wottons scheint also – im Gegensatz zu Aristoteles – überhaupt nicht wesentlich darauf ausgerichtet gewesen zu sein, das Tier in der Gegenwart und aufgrund authentischer empirischer Beobachtung festzulegen. Anstatt des Verbreitungsgebiets zu seiner eigenen Zeit vermittelt er das (von ihm für wahr gehaltene) Verbreitungsgebiet der Löwenarten in der Antike, einer Periode, die sich in einem Abstand von ca. 2550–ca. 1350 Jahren zu Wottons Zeit befindet. Dies darf man nicht als Nebensächlichkeit oder Unzulänglichkeit oder Unfähigkeit zu empirischer Forschung abtun. Es liegt vielmehr eine bewusste Diskursorientierung vor.

Humanistische Wissensverwaltung und humanistisches Sammeln

Es ist von grundlegender diskursregulierender Bedeutung, dass Wotton seine Zoologie wesentlich als ein Projekt der Wissensverwaltung verstanden hat. Wie er im Widmungsvorwort an Edward VI. mitteilt, ist es sein Hauptziel, den in den Schriften der Antike vorhandenen Wissensschatz zu bergen und dem modernen Leser in der Form eines gut zugänglichen Kompendiums zu vermitteln.[78] Da sowohl der zoologische Diskurs Wottons als auch die zoologische Information vielfach auf Aristoteles zurückgeht und die zoologischen Schriften des Aristoteles vorrätig waren, stellt sich natürlich die Frage, was den Mehrwert seines

---

77 Herodot, *Historiae* VII 125; vgl. Steier A., „Löwe" 970, 19 ff. Herodots Angabe wurde auch von Aelian, *Historia animalium* XVIII 36 und Pausanias, *Periegesis* VI 5,3 übernommen. Herodot teilt mit, dass beim Durchzug des Xerxes durch Makedonien (480 v. Chr.) die Kamele des Trosses von Löwen angegriffen wurden. Daran knüpft er die Bemerkung, dass es zwar in dieser Gegend viele Löwen gebe, ihre Verbreitung in Europa sich aber auf das Gebiet zwischen den Flüssen Nestos, der das Gebiet von Abdera durchfließt, und des Acheloos in Arkanien beschränke. Durch die präzise Intertextualität zwischen der Herodot-Stelle und den Aristoteles-Stellen kann man ausschließen, dass sich Aristoteles' Angaben auf Autopsie stützen, was Steier („Löwe", 970) überraschenderweise für möglich hält.

78 *De differentiis animalium* fol. a ii r.

Werkes im Vergleich zu Aristoteles ausmache. Wotton beantwortet sie in dem Sinn, dass sein Werk eine *systematische Sammlung* darstelle, die weiter Vollständigkeit erstrebe und nach humanistischen Prinzipien angelegt sei. *Er erhebt den Anspruch, nicht nur das aristotelische, sondern das gesamte zoologische Wissen der Antike aufbereitet zu haben.* Das Problem, das er zu lösen versucht, liegt in der Tatsache, dass dieses Wissen in vielen sehr unterschiedlichen Werken verstreut war. Wotton beschränkt seine Quellen dabei keineswegs auf naturwissenschaftliche Werke: Er zitiert gleichermaßen Historiker, philosophische und belletristische Prosa und stets auch Dichter. Für sein Kapitel „Von Löwen" hat er beispielsweise Ovids epische Mythologie, die *Metamorphosen*, des griechischen Dichters Oppian von Apameias Lehrgedicht *Kynegetika* (in Hexametern), das Werk des griechischen Historikers Herodot (5. Jh. v. Chr.) und die *Geographika* des griechischen Gelehrten Strabon aus dem 1. Jh. v. Chr. herangezogen.[79] Weiter war es Wotton ein besonderes Anliegen, die antiken medizinischen Autoren zu erschließen. Diese Texte stufte er sowohl als außerordentlich wertvoll als auch als schwer zugänglich ein. Wer soll diese spezialistischen Werke heutzutage noch lesen und verstehen? In ihnen ist ein Schatz zoologischen Wissens vorhanden, jedoch in völlig verstreuter und unsystematischer Form.[80] Als gelernter Mediziner und Humanist sei er, Wotton, in der Lage, diesen Schatz zu heben. Z.B. im Kapitel „Von Löwen" zitiert er zwei medizinische Schriften, als deren Autor der griechische Arzt Galen geführt wurde.[81]

Zoologie bedeutet für Wotton also wesentlich: das eifrige und systematische Sammeln der verstreuten antiken Information. Diesbezüglich nimmt er für sich in Anspruch, unter den Intellektuellen der frühen Neuzeit der erste zu sein.[82] Das Unternehmen Gesners, das sich diesbezüglich ähnliche Aufgaben stellte, war ihm zu diesem Zeitpunkt aller Wahrscheinlichkeit nach nicht bekannt.[83]

Weiter ging Wotton davon aus, dass das Sammeln nach *humanistischen Gesichtspunkten* stattzufinden habe: Die humanistische Belesenheit in der

---

[79] Ebd. fol. 63v.

[80] Ebd. fol. a ii r.

[81] *De usu partium* und *De simplicibus medicinis*; siehe *De differentiis animalium* fol. 63v.

[82] Ebd. fol. a ii v.

[83] Jedenfalls gehörte Wotton nicht zu den Wissenschaftlern, mit denen Gesner über Zoologie korrespondierte (Vgl. Gesner, *Historia animalium* I, „Catalogus doctorum virorum [...]").

antiken Literatur („eruditio") bildet die Grundlage der zoologischen Wissenschaft. Weiter müssten die antiken Schriften „kritisch" betrachtet werden. Damit meint Wotton keineswegs, dass man ihre Aussagen durch empirische Forschung überprüfen müsse, sondern dass man die Quellenautoren als Grundlagen des Wissens einem „Urteil" („iudicium") in Bezug auf ihren Wert und ihre Autorität zu unterziehen habe. Diesbezüglich gilt ihm als Faustregel, dass die ältesten Autoren („antiquissimi") grundsätzlich „die besten" („optimi") sind. Das bedeutet, dass griechische Quellen grundsätzlich Vorrang gegenüber lateinischen erhalten. Das Kapitel „De leonibus" kann auch in dieser Hinsicht als Beispiel dienen: Wotton gibt in der Kapitelüberschrift 8 griechische Werke, jedoch nur 2 lateinische als Quellen an. Durch diese Diskursgestaltung wird der zoologische Hauptautor Plinius von Wotton bis zu einem gewissen Grad *ent-autorisiert*. Das zeigt sich sehr augenfällig im Kapitel „De leonibus": Obwohl Wotton Plinius' Ausführungen[84] mehrfach, zum Teil in extenso benutzt und einige Male sogar wörtlich übernommen hat,[85] nennt er ihn in der Kapitelüberschrift nicht als Autor.

Für Wotton war es oberstes Gebot, dass ein guter Humanist das Griechische beherrschen soll. In der Tat verfügte er über ausgezeichnete Griechischkenntnisse. 1526, nachdem er von Bologna nach Oxford zurückgekehrt war, dozierte er Griechisch am Corpus Christi College. Der humanistisch geschulte, griechischkundige Wotton macht die griechischen Quellen einem breiteren Publikum von Intellektuellen zugänglich, indem er sie entweder ins Lateinische übersetzt oder, wo bereits lateinische Übersetzungen vorhanden waren, diese kontrolliert und, wenn nötig, korrigiert. Als Beispiel von Übersetzungen aus dem Griechischen, die gründlich korrigiert werden müssen, nennt er die Werke des um die Mitte des 15. Jh. tätigen Aristoteles-, Theophrast-

---

[84] *Naturalis historia* VIII, 41–60.

[85] Einige Beispiele für wörtliche oder nahezu wörtliche Übernahmen aus Plinius' *Naturalis historia*: fol. 63v: „in Europa [...] inter Acheloum tantum et Mestum amnes [...] ii viribus longe praestantiores sunt iis, quos Africa aut Syria gignant [...]; fol. 64r: „parit [...] catulos informes [...] magnitudine mustellarum [...], ut semestres vix ingredi possint nec nisi bimestres moveri [...]" (alle: Plinius, *Naturalis historia* VIII, 45); fol. 64v: „mingit crure elato ut canes (Plinius, *Naturalis historia* VIII, 46); [...] prostratis parcunt [...] in viros potius quam in feminas saeviunt: infantes non nisi magna fame perimunt (Plinius, *Naturalis historia* VIII, 48). [...] Immota enim cauda est placido, clemens blandientique similis, quod rarum est. [...] In principio terra verberatur. Incremento terga seu quodam incitamento flagellantur [...] vis summa leoni in pectore est" (Plinius, *Naturalis historia* VIII, 49).

und Alexandros von Aphrodisias-Übersetzers Theodor Gaza.[86] Die Übersetzungen sollten grundsätzlich nach humanistischen Gesichtspunkten angefertigt werden, d.h. sie müssen *ad sententiam* und nicht *verbatim* (Wort-für-Wort-Übersetzung) angelegt und in idiomatisch einwandfreiem Latein, das nach den Spielregeln der antiken Rhetorik strukturiert ist („orationis luce"), gestaltet werden. Als sorgfältiger Gräzist registriert Wotton die Tatsache, dass man nicht alle griechischen idiomatischen Ausdrücke geradewegs ins Latein übersetzen konnte. In diesen Fällen gibt er in einer Fußnote den griechischen Wortlaut wieder. Wichtig ist weiter, dass die antiken Quellen textkritisch überprüft werden. Die Kontrolle der Überlieferung steht dabei im Vordergrund. Dem Leser soll ein Text präsentiert werden, der der Kritik standhält. Nur kontrollierte Texte können den autoritativen Wahrheitsanspruch erheben, der die Grundlage der frühneuzeitlichen Zoologie bildet. Als Beispiele von stark korrupten antiken Texten nennt Wotton Athenaios' (um 200 n. Chr.) *Deipnosophistai* und Oreibasios von Pergamons (4. Jh. n. Chr.) Auszüge aus den Werken des Galen (*Iatrikai synagogai*).[87] Zu der diesbezüglichen erstrebten wissensverwalterischen Sorgfalt gehört, dass Wotton in Fußnoten Textvarianten verzeichnet, wodurch er dem Leser ermöglicht, die Textgrundlage nochmals zu überprüfen und etwaige Eingriffe in den überlieferten Text nachzuvollziehen. Dasselbe gilt für kommentierende Aufzeichnungen zu Textemendationen. Sie dienen der nochmaligen Kontrolle.

Die zoologische Wissensautorisierung: Beleg der Schriftquellen gegenüber empirischer Wahrnehmung

Der Titel von Wottons Werk enthält ein auffälliges Element. Für den zoologischen Diskurs, der sich in ihm entfaltet, sind Quellenangaben offensichtlich so wichtig, dass sie im Titel angekündigt werden: *De differentiis animalium libri X, cum amplissimis indicibus, in quibus primum authorum nomina, unde quaeque desumpta sunt, singulis capitibus sunt notata* [...] *Von den Unterschieden zwischen den Tierarten in zehn Büchern, mit sehr reichen Verzeichnissen, in denen erstens für jedes Kapitel die Namen der Quellenautoren*

[86] *De differentiis animalium* fol. a iii r. Zu Gaza vgl. Geanakoplos D.J., „Theodore Gaza, a Byzantinian Scholar of the Palaeologan ‚Renaissance'", *Medievalia et Humanistica* 12 (1984) 61–81.

[87] *De differentiis animalium* fol. a iii v.

angeführt werden, welchen die einzelnen Angaben entnommen wurden [...]. Es ist ausgeschlossen, dass ein modernes zoologisches Werk mit einer so angelegten Titelei öffnen würde, die den Leser eher abstoßen als zum Kauf oder zur Benutzung überreden würde. Der nicht leicht verständliche Satz bezieht sich auf Folgendes: Nach dem Widmungsbrief findet sich ein Inhaltsverzeichnis. Dieses unterscheidet sich wesentlich von modernen Inhaltsverzeichnissen, für welche die Übersichtlichkeit oberstes Gebot ist. Dieses soll in möglichst kurzer Form den Inhalt des Werkes synoptisch aufschließen. Wottons Inhaltsverzeichnis weist hingegen eine Länge von 26 Foliospalten auf [Abb. 2]![88] Es bietet nicht nur den Inhalt der jeweiligen Kapitel, sondern auch – wie es die Titelei affichierte – die betreffenden Quellen dar. Als Beispiel möge die Angabe zum Kapitel 80 „Von Löwen" dienen: „De Leonibus. Ex authoribus Aristotele in historia animalium. De partibus animalium. De generatione animalium. Oppiano De venatione. Ovidio in Metamorphosibus. Strabone. Herodoto. Solino. Galeno De usu partium. De simplicibus medicinis". Am Ende des Werkes findet sich zusätzlich ein umfängliches Quellenverzeichnis, in welchem 208 Quellenautoren aufgelistet werden [Abb. 3].

Es ergibt sich die Frage, welchen Zweck diese nachdrücklich vorgetragenen Quellenangaben haben. Bäumer deutete Wottons Quellenangaben, obwohl sie die Präzision bemängelt, auf das „Bemühen des Renaissance-Humanismus um philologische Exaktheit", das einen „methodischen Fortschritt" darstelle.[89] Ich kann mich dieser wissenschaftsteleologischen Interpretation nicht anschließen. Man betrachte als Beispiel die Quellenangaben zu Kapitel 80. Können die Angaben den Sinn haben, dem Leser die philologische Überprüfung, d.h. ein Nachschlagen der Zitate in den Originaltexten zu ermöglichen? Aristoteles' *Historia animalium* weist 9 (erhaltene) Bücher, Oppians *De venatione* 4, Ovids *Metamorphosen* 15 Bücher auf. Der Leser, der die betreffenden Stellen nachschlagen wollte, müsste Hunderte von Seiten lesen oder durchsuchen. Es ist klar, dass dies ein geradezu aussichtsloses Unterfangen wäre.

Aus diesem Befund muss man schließen, dass sie eine andere Funktion haben. Sie dienen der *Autorisierung* des dargelegten Wissens. Die

---

[88] Zum Vergleich: Macdonalds viel umfangreichere *Enzyklopädie der Säugetiere* (930 Großformatseiten) weist ein Inhaltsverzeichnis von nur 2 Seiten auf.

[89] Bäumer, „Das erste zoologische Kompendium" 16.

SVMMATIM HÆC INSVNT LIBRIS SINGVLIS, ATQVE EX IIS AVTHORIBVS EXCERPTA.

In primo libro generatim explicatur, quibus modis constituuntur animalium Differentiæ: & præterea animalium partes recensentur.

ANIMAL quid sit, & quo differat ab iis quæ non sunt animalia. Ex authoribus. Platone in Timæo. Aristotele in historia animalium. De partibus animaliũ. De generatione animalium. De Anima. De sensu & sens. De Iuuent. & senect. De somno & uigil. De Respiratione. Et Plinio. Cap.1. Fol.1

Quibus modis constitui debeant animaliũ Differẽtiæ. Ex Aristotele in histor. animaliũ. De partib. animaliũ. De sensu & sens. ca.ii. fo.1.b

Animalium genera amplissima. Ex Aristotele in historia animaliũ. cap.iii. fol.2.b

De partibus animalium. Authores. Aristoteles in historia animalium. De partb. animaliũ. Galenus in li. De placitis Hippocratis & Platonis. In Methodo medẽdi. De inæquali intemperie. De facultatibus naturalibus. De usu partium. cap.iiii. fol.3

De partibus exterioribus dissimilaribus. Author Aristoteles in historia animalium. De partibus animalium. cap.v. ibid.

De capite & eius partibus. Ex authoribus Aristotele in historia animaliũ. De partibus animaliũ. Cicerone De natura Deorũ. Cornel. Celso. Plinio. Galeno De ossibus. De Alimẽtorũ facultatibus. De compositione medicamẽtorũ per loca. In Cõmẽtariis in Aphorismos. Ex libro qui Introductorius inscribitur. Ex Iulio Polluce. cap.vi. fol.3.b

De Collo & eius partibus. Ex Aristotele in animalium historia. De partibus animaliũ. Ex Galeno De Sanitate tuenda. In Anatomicis. De usu partium. De locis affectis. De uerarum sectione. Ex Cornel. Celso. Plinio. & Iulio Polluce. cap.vii. fol.5

De Thorace & eius partibus. Authores. Aristoteles in animaliũ historia. De partibus animalium. De Generatione animalium. Galenus in lib. Anatomicis. De usu partium. In Methodo medendi. De Motu musculorum. De ossibus. In Commentariis de fracturis. Et iis qui inscribũtur κατ᾽ ἰητρεῖον. Hippocrates in Aphorismis. Et Galenus in eundem librum. cap. viii. fol.5.b

De Brachiis & eorum partibus. Ex authoribus. Aristotele in historia animalium. De partibus animalium. Galeno de motu musculorum. De ossibus. De usu partium. In Commentario κατ᾽ ἰητρεῖον. Ex lib. qui Introductorius inscribitur. Cornel. Celso. Plinio. Festo Põpeio. Iulio Polluce. cap. ix. fol.6.b

De Cruribus & eorum partibus. Authores. Aristoteles in historia animalium. De partibus animaliũ. Galenus De usu partium. De ossibus. In Commentariis κατ᾽ ἰητρεῖον. Lib. qui Introductorius inscribitur. Cornel. Celsus. Iulius Pollux. Fest. Pompeius. cap. x. fol.7

De Situ partium uario. Authores. Aristoteles in historia animaliũ. De partibus animaliũ. De Animalium incessu. De Cælo & mũdo. Plato in Timæo. Corn. Celsus. cap.xi. fo.7.b

De Capitis partibus interioribus. s. De Cerebro & eius partibus. Authores. Aristoteles in animalium historia. De partibus animaliũ. De Somno & Vigil. Galenus in Anatomicis. De neruorum dissectione. De olfactus instrumẽto. Lib. Introductorius. Iul. Pollux. ca.xii. fo.8

De Oculi partibus interioribus. Authores. Galenus De neruorum dissectione. De usu partium. De Placitis Hippocratis & Platonis. In Methodo medẽdi. Author libelli qui De Anatomia oculorum inscribitur, Galeno adscripti. Iul. Pollux. cap. xiii. fol.8.b

De Colli & Pectoris partibus interioribus. Ex Authoribus. Aristotele in historia animaliũ. De partibus animalium. Galeno De usu partium. Iulio Polluce. cap. xiiii. fol.9.b

De uisceribus & Septo transuerso. Ex authoribus Aristotele in historia animalium. De partibus animaliũ. Galeno in Anatomicis. De usu partium. De Temperamẽtis. De pla-

ã

Fig. 2. Edward Wotton, *De differentiis animalium libri decem* (Paris, Michel de Vascosan: 1552), Inhaltsangabe.

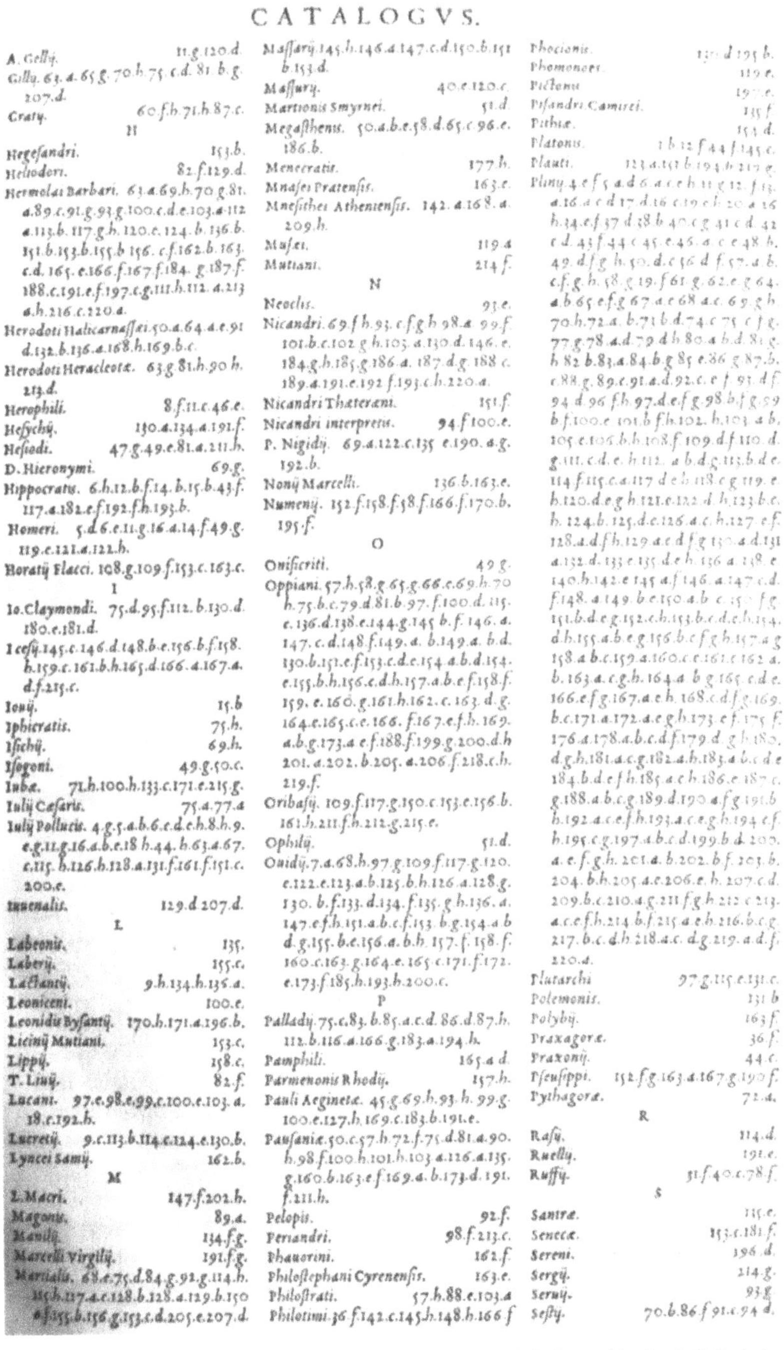

CATALOGVS.

A. Gellij. 11.g.120.d.
Gellij. 63.a.65.g.70.h.75.c.d.81.b.g.
207.d.
Cratij. 60.f.h.71.h.87.c.

H

Hegesandri. 153.b.
Heliodori. 82.f.129.d.
Hermolai Barbari. 63.a.69.h.70.g.81.
a.89.c.91.g.93.g.100.c.d.e.103.a.112
a.113.b.117.g.h.120.c.124.b.136.b.
151.b.153.b.155.b.156.c.f.162.b.163.
c.d.165.e.166.f.167.f.184.g.187.f.
188.c.191.e.f.197.c.g.111.h.112.a.213
a.h.216.c.220.a.
Herodoti Halicarnassei.50.a.64.a.e.91
d.132.b.136.a.168.h.169.b.c.
Herodoti Heracleotæ. 63.g.81.h.90.h.
213.d.
Herophili. 8.f.11.c.46.e.
Hesychij. 130.a.134.a.191.f.
Hesiodi. 47.g.49.e.81.a.211.h.
D. Hieronymi. 69.g.
Hippocratis. 6.h.12.b.f.14.b.15.b.43.f.
117.a.182.e.f.192.f.h.193.b.
Homeri. 5.d.6.e.11.g.16.a.14.f.49.g.
119.e.121.a.122.h.
Horatij Flacci. 108.g.109.f.153.c.163.c.

I

Io. Claymondi. 75.d.95.f.112.b.130.d.
180.e.181.d.
Iesij.145.c.146.d.148.b.e.156.b.f.158.
b.159.c.161.b.h.165.d.166.a.167.a.
d.f.215.c.
Ionij. 15.b
Iphicratis. 75.h.
Isichij. 69.h.
Isogoni. 49.g.50.c.
Iubæ. 71.h.100.h.133.c.171.e.215.g.
Iulij Cæsaris. 75.a.77.a
Iulij Pollucis. 4.g.5.a.b.6.c.d.e.h.8.h.9.
e.g.11.g.16.a.b.e.18.h.44.h.63.a.67.
c.115.b.126.h.128.a.131.f.161.f.151.c.
200.e.
Iuuenalis. 129.d.207.d.

L

Labeonis. 135.
Laberij. 155.c.
Lactantij. 9.h.134.h.135.a.
Leoniceni. 100.e.
Leonidis Bysantij. 170.h.171.a.196.b.
Licinij Mutiani. 153.c.
Lippij. 158.c.
T. Liuij. 82.f.
Lucani. 97.e.98.e.99.c.100.e.103.a.
18.e.192.h.
Lucretij. 9.c.113.b.114.c.124.e.130.b.
Lyncei Samij. 162.b.

M

L. Macri. 147.f.202.h.
Magonis. 89.a.
Manilij. 134.f.g.
Marcelli Virgilij. 191.f.g.
Martialis. 68.e.75.d.84.g.92.g.114.h.
115.b.117.a.c.128.b.128.a.129.b.150
a.f.155.b.156.g.153.c.d.205.e.207.d.
Massarij.145.h.146.a.147.c.d.150.b.151
b.153.d.
Massurij. 40.e.120.c.
Martionis Smyrnei. 51.d.
Megasthenis. 50.a.b.e.58.d.65.c.96.e.
186.b.
Menecratis. 177.h.
Mnasei Pratensis. 163.e.
Mnesithei Athenienfis. 142.a.168.a.
209.h.
Musei. 119.a
Mutiani. 214.f.

N

Neoclis. 93.e.
Nicandri. 69.f.h.93.c.f.g.h.98.a.99.f.
101.b.c.102.g.h.103.a.130.d.146.e.
184.g.h.185.g.186.a.187.d.g.188.c.
189.a.191.e.192.f.193.c.h.220.a.
Nicandri Thaterani. 151.f.
Nicandri interpretis. 94.f.100.e.
P. Nigidij. 69.a.122.c.135.e.190.a.g.
192.b.
Nonij Marcelli. 136.b.163.e.
Numenij. 152.f.158.f.58.f.166.f.170.b.
195.f.

O

Onisicriti. 49.g.
Oppiani. 57.h.58.g.65.g.66.c.69.h.70
h.75.b.c.79.d.81.b.97.f.100.d.115.
e.136.d.138.e.144.g.145.b.f.146.a.
147.c.d.148.f.149.a.b.149.a.b.d.
130.b.151.e.f.153.c.d.e.154.a.b.d.154.
e.155.b.h.156.c.d.h.157.a.b.e.f.158.f.
159.e.160.g.161.h.162.c.163.d.g.
164.e.165.c.e.166.f.167.e.f.h.169.
a.b.g.173.a.e.f.188.f.199.g.200.d.h
201.a.202.b.205.a.206.f.218.c.h.
219.f.
Oribasij. 109.f.117.g.150.c.153.e.156.b.
161.h.211.f.h.212.g.215.e.
Ophilij. 51.d.
Ouidij.7.a.68.h.97.g.109.f.117.g.120.
e.122.e.123.a.b.125.b.h.126.a.128.g.
130.b.f.133.d.134.f.135.g.h.136.a.
147.e.f.h.151.a.b.c.f.153.b.g.154.a.b
d.g.155.b.e.156.a.b.h.157.f.158.f.
160.c.163.g.164.e.165.c.171.f.172.
e.173.f.185.h.193.h.200.c.

P

Palladij. 75.c.83.b.85.a.c.d.86.d.87.h.
112.b.116.a.166.g.183.a.194.h.
Pamphili. 165.a.d.
Parmenonis Rhodij. 157.h.
Pauli Aeginetæ. 45.g.69.h.93.h.99.g.
100.e.127.h.169.c.183.b.191.e.
Pausaniæ.50.c.57.h.72.f.75.d.81.a.90.
h.98.f.100.h.101.h.103.a.126.a.135.
g.160.b.163.e.f.169.a.b.173.d.191.
f.211.h.
Pelopis. 92.f.
Periandri. 98.f.213.c.
Phauorini. 162.f.
Philostephani Cyrenensis. 163.e.
Philostrati. 57.h.88.e.103.a
Philotimi.36.f.142.c.145.h.148.h.166.f
Phocionis. 130.d.195.b.
Phomonees. 119.e.
Pictonis. 197.e.
Pisandri Camirei. 135.f.
Pithiæ. 153.d.
Platonis. 1.b.12.f.44.f.145.c.
Plauti. 123.a.151.b.194.h.217.g.
Plinij.4.e.f.5.a.d.6.a.c.e.h.11.g.12.f.13.
a.16.a.c.d.17.d.16.c.19.e.h.20.a.16
h.34.e.f.37.d.38.b.40.c.g.41.c.d.42
c.d.43.f.44.c.45.e.46.a.c.e.48.b.
49.d.f.g.h.50.d.c.56.d.f.57.a.b.
c.f.g.h.58.g.19.f.61.g.62.e.g.64.
a.b.65.e.f.g.67.a.e.68.a.c.69.g.h.
70.h.72.a.b.73.b.d.74.c.75.c.f.g.
77.g.78.a.d.79.d.h.80.a.b.d.81.g.
h.82.b.83.a.84.b.g.85.e.86.g.87.b.
c.88.g.89.e.91.a.d.92.c.e.f.93.d.f.
94.d.96.f.h.97.d.e.f.g.98.b.f.g.99
b.f.100.e.101.b.f.h.102.h.103.a.b.
105.e.106.b.h.108.f.109.d.f.110.d.
g.111.c.d.e.h.112.a.b.d.g.113.b.d.e.
114.f.115.c.a.117.d.e.h.118.c.g.119.e.
h.120.d.e.g.h.121.e.122.d.h.123.b.c.
h.124.b.125.d.e.126.a.c.h.127.e.f.
128.a.d.f.h.129.a.c.d.f.g.130.a.d.131
a.132.d.133.e.135.d.e.h.136.a.138.e
140.h.142.e.145.a.f.146.a.147.c.d.
f.148.a.149.b.e.150.a.b.c.150.f.g.
151.b.d.e.g.152.c.h.153.b.c.d.e.h.154.
d.h.155.a.b.e.g.156.b.c.f.g.h.157.a.g
158.a.b.c.159.a.160.c.e.161.e.162.a.
b.163.a.c.g.h.164.a.b.g.165.c.d.e.
166.e.f.g.167.a.c.h.168.c.d.f.g.169.
b.c.171.a.172.a.e.g.h.173.e.f.175.f.
176.a.178.a.b.c.d.f.179.d.g.h.180.
d.g.h.181.a.c.g.182.a.h.183.a.b.c.d.e
184.b.d.e.f.h.185.a.c.h.186.e.187.c.
g.188.a.b.c.g.189.d.190.a.f.g.191.b
h.192.a.c.e.f.h.193.a.c.e.g.h.194.e.f.
h.195.c.g.197.a.b.c.d.199.b.d.200.
a.e.f.g.h.201.a.b.202.b.f.203.b.
204.b.h.205.a.e.206.e.h.207.c.d.
209.b.c.210.a.g.211.f.g.h.212.c.213.
a.c.e.f.h.214.b.f.215.a.e.h.216.b.c.g.
217.b.c.d.h.218.a.c.d.g.219.a.d.f.
220.a.
Plutarchi 97.g.115.e.131.c.
Polemonis. 131.b
Polybij. 163.f.
Praxagoræ. 36.f.
Praxonij. 44.c.
Pseusippi. 152.f.g.163.a.167.g.190.f.
Pythagoræ. 72.a.

R

Rasij. 114.d.
Ruellij. 191.e.
Ruffij. 31.f.40.c.78.f.

S

Santræ. 115.e.
Senecæ. 153.c.181.f.
Sereni. 196.d.
Sergij. 214.g.
Seruij. 93.g.
Sestij. 70.b.86.f.91.c.94.d.

Fig. 3. Edward Wotton, *De differentiis animalium libri decem* (Paris, Michel de Vascosan: 1552), Quellenverzeichnis.

formelhafte Diktion „Ex authoribus" ist dafür ausschlaggebend. Wottons zoologischer Diskurs wird dadurch gekennzeichnet, dass die Zuordnung und Rückführung der Information auf antike „Gewährsmänner" (die eigentliche Bedeutung von „author") den ultimativen Beleg ihrer Gültigkeit liefern soll. Die Grundlage der Zoologie bildet für Wotton damit klar ersichtlich nicht der empirische Nachweis, sondern der antike Autoritätsbeleg.

Der zoologische Diskurs Wottons unterscheidet sich diesbezüglich wesentlich von dem moderner zoologischer Übersichtswerke. In diesen betrachtet man Quellenangaben eher als Ballast. Antike Autoren oder nicht-wissenschaftliche Quellen werden grundsätzlich nicht zitiert. Der Diskurs ist so angelegt, dass die Befindungen prinzipiell auf empirisch Nachvollziehbares und Wiederholbares rückführbar sind. Auch wird zumeist nicht unterschieden, ob die jeweiligen Beobachtungen vom Autor selbst oder von anderen gemacht wurden. Dies wird sowohl von der Forschungspraxis (zoologische Teamforschung) als auch von der Forschungstheorie eingegeben. Letzterer liegt der Ausgangspunkt zugrunde, dass niemals die Autorität der Quelle, sondern die empirische Rückführbarkeit den Wahrheitsgehalt der Darstellung ausmache. Dabei wird die Subjektivität der Warte des Betrachters grundsätzlich neutralisiert oder am liebsten ganz ausgeschaltet. Die Diskursanlage impliziert, dass die Subjektivität des Beobachters die Beschaffenheit der Beobachtung nicht nennenswert beeinflusse.

Folgende Art, empirische Beobachtungen wiederzugeben, ist bezeichnend: „Häufig kommt es vor, dass nur ein oder zwei Mitglieder einer größeren Gruppe (nml. der Species *Panthera leo*) einen Jagdversuch unternehmen, während der Rest des Rudels aus sicherer Entfernung zusieht".[90] Aus den Angaben geht nicht hervor, wer die Beobachtungen gemacht hat (der Autor oder ein anderer); jedoch wird durch die Betonung des iterativen Charakters der Beobachtung („häufig kommt es vor [...]") ihre empirische Glaubwürdigkeit attestiert: Die Beobachtung wurde schon oft gemacht und sie kann auch in Zukunft beliebig oft gemacht werden – also ist sie wahr. Dass die Subjektivität zugeschüttet wird, ist umso auffälliger, als die zitierte Beobachtung subjektive Elemente enthält wie „aus sicherer Entfernung zusieht". Diese Angabe interpretiert das Verhalten der nicht-jagenden Löwen in dem Sinn, dass sie sich vor den Beutetieren oder der Gefahr des Jagens fürchten

[90] Macdonald (Hrsg.), *Enzyklopädie der Säugetiere* 10.

würden. Dafür liegen allerdings keine schlüssigen Beweise vor. Die bibliographische Nachvollziehbarkeit der Information wird in modernen Übersichtswerken insgesamt stark reduziert, bei der Beschreibung der einzelnen Species oft ganz ausgeklammert.[91] Das hat nicht einfach mit der objektiven Kürze des Lemmas zu tun. Wottons Abschnitt „Von Löwen" etwa ist nicht einmal halb so lang wie das betreffende Lemma in Macdonalds *Enzyklopädie der Säugetiere*, weist jedoch eine beträchtliche Anzahl von Quellenangeben auf, während sich in jener keine einzige findet.

Moderne zoologische Werke weisen in Bezug auf den Wahrheitsbeleg eine andere Diskursorientierung auf. Statt auf Literaturangaben rekurrieren sie auf Tierphotographien, die sozusagen die Funktion des Wahrheitsbeweises übernehmen. Sie sollen nicht nur die äußere Gestalt des Tieres sichtbar machen, sondern die verhaltensbiologischen (z.B. Sozialverhalten; Jagd) und äußerlich wahrnehmbaren physiologischen Vorgänge (z.B. Fortpflanzung) vorführen. Im Abschnitt über den Löwen in Macdonalds *Enzyklopädie der Säugetiere* (8 Seiten) werden nicht weniger als 10 Photographien dargeboten, die insgesamt *fast die Hälfte des verfügbaren Platzes* beanspruchen. Obwohl Photographien natürlich auch schlicht die Attraktivität eines Buches erhöhen, zeigt ihr massierter Einsatz in modernen zoologischen Übersichtswerken, dass sie die Rolle einer Autorisierungsinstanz übernehmen. Was der Leser mit eigenen Augen sehen kann, muss wahr sein.

Der Glaube an den Wahrheitsbeleg des Bildes ist so stark ausgeprägt, dass Bilder nicht selten als Belege für Vorgänge oder Sachverhalte eingesetzt werden, die auf dem Bild nicht wahrgenommen werden können. Deswegen werden sie mit Begleittexten versehen, die den Betrachter instruieren, was er auf dem Photo „sehen" soll. Immer wieder geht aus Text-Bild-Kombinationen hervor, wie problematisch der augenscheinlich glasklare Erkenntnisvorgang dieser zoologischen Empirie ist. Z.B. findet sich in Macdonalds *Enzyklopädie der Säugetiere* das Bild eines männlichen Löwen [Abb. 4] mit der Beischrift: „Links: Das Brüllen des Löwen, das weit über die ostafrikanische Savanne trägt, lässt potentielle Eindringlinge unmissverständlich wissen, das dieses Revier

[91] In Goulds und McKays *Enzyklopädie der Säugetiere* beschränken sich die Angaben auf eine kurze Übersichtsbibliographie am Ende des Buches, die nicht mehr als eine halbe Seite beansprucht; in Macdonalds *Enzyklopädie der Säugetiere* gibt es auf ca. 800 Seiten keine Literaturangabe, am Ende des Buches findet sich eine Übersichtsbibliographie von 4 Seiten.

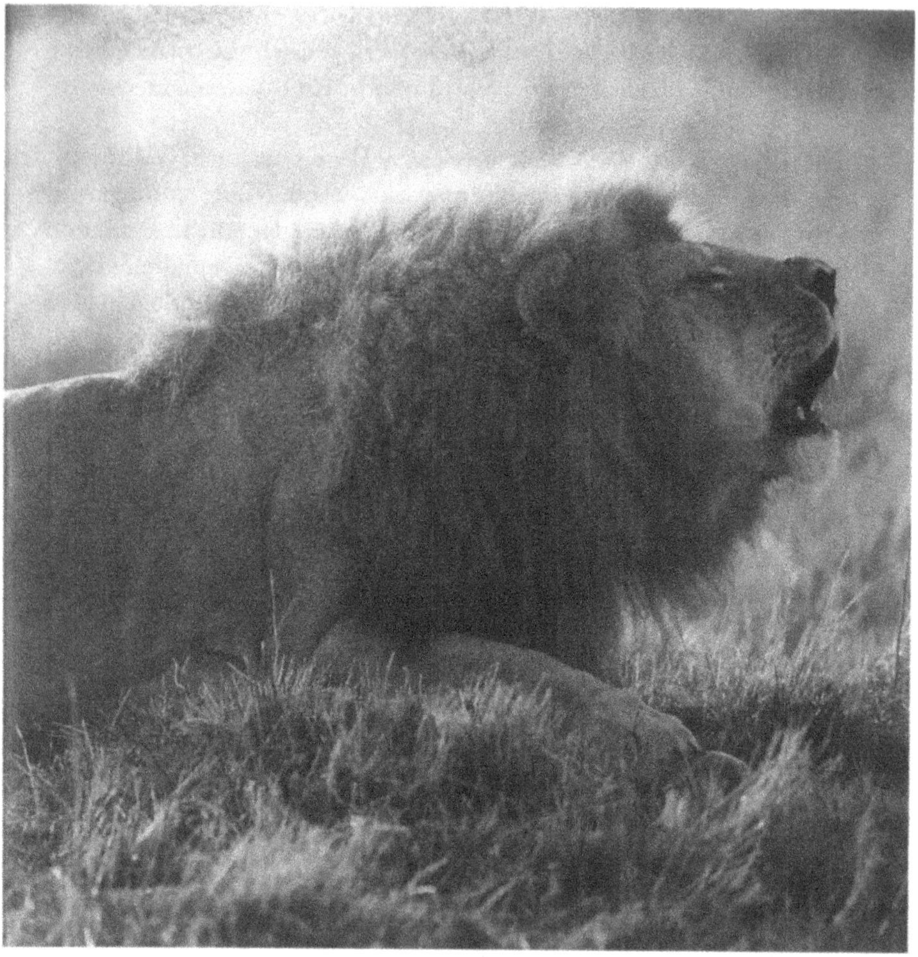

Fig. 4. ‚Das Brüllen des Löwen'. Tierfotografie aus Macdonald D. (Hrsg.), *Enzyklopädie der Säugetiere* (Königswinter: 2003; urspr. English Oxford: 2001) 16–17.

besetzt ist. Meist hört man das Brüllen nach Sonnenuntergang oder im Anschluss an ein sättigendes Mahl" (17). Was belegt das Bild eigentlich? Das Brüllen kann man zunächst nur auditiv wahrnehmen, zu sehen ist es nicht. An sich könnte der abgebildete Löwe auch andere Laute von sich geben. Weiter geht aus dem Bild in keiner Weise hervor, dass sich der Löwe in Ostafrika befindet. Das Photo könnte genauso gut im südlichen Afrika, in einem Reservat oder in irgendeiner künstlich geschaffenen Graslandschaft genommen worden sein. Das Bild liefert schließlich auch keinen Beleg für die abschließende Bemerkung: Es wurde weder nach Sonnenuntergang gemacht noch ist auf ihm in irgendeiner Form wahrnehmbar, dass dieser Löwe gerade gefressen hat. Dass man das Bild dennoch bemüht, zeigt die Bildgläubigkeit des empirischen Diskurses an. Wottons Werk hingegen weist keine Illustrationen auf. Wahrheitsgrundlage ist in Wottons Zoologie – im Gegensatz etwa zu Gesners Publikationen – ausschließlich das Wort der antiken Autoren.

Die Konstituierung der Zoologie als Universalwissenschaft: Gesners Historia quadrupedum viviparorum (1551):

Universalwissenschaft durch Ent-Klassifizierung: Diskursverlagerung von systematischer Taxonomie zu lexikalischer Wissensorganisation, am Beispiel der Species Panthera leo

Der Schweizer Conrad Gesner (1516–1565) brachte Voraussetzungen mit, die denen Edward Wottons auffällig ähnlich sind. Wie Wotton hatte er eine gründliche Ausbildung in den Altsprachen, besonders im Griechischen, erhalten. Wie Wotton bekleidete er eine Professur in der griechischen Sprache (an der Akademie in Lausanne, 1537–1540). Wie Wotton hatte er eine besondere Zuwendung für die Schriften der antiken Mediziner, welche er seit 1533 intensiv rezipierte. Wie Wotton studierte er Medizin (Basel und Montpellier, 1536–37; 1540–41). Wie Wotton übte er den Beruf des Arztes aus (seit 1541 Praxis in Zürich; 1552 Unterstadtarzt; 1553 Stadtarzt in Zürich). Wie Wotton hatte er zoologische Studien viele Jahre hindurch betrieben, bevor er die erste Publikation fertigstellte. Wie Wotton hatte er die Werke des Aristoteles sehr eingehend studiert. Sein Handexemplar, das erhalten ist (Zürich, Zentralbibliothek, B 83), zeigt die Spuren dieser intensiven Lektüre – es ist mit marginalen Aufzeichnungen übersät.

Von daher scheint vorprogrammiert zu sein, dass die Anlage des zoologischen Wissenschaftsdiskurses, in dem die beiden Autoren operieren, identisch oder quasi-identisch ist. Bis zu einem gewissen Grad mag dies auch aufgehen und es gibt – wie unten gezeigt werden wird – eine Reihe von Ähnlichkeiten. Dennoch erscheinen mir – besonders in Anbetracht der fast identischen Ausgangslage – die Diskursunterschiede bemerkenswerter. Gesners Zoologie hinterlässt schon bei einem ersten Durchblättern einen anderen Gesamteindruck. In der Tat ist seine Zoologie in einem anderen Diskurs situiert. Statt einer systematischen Klassifizierung – ein Hauptmerkmal von Wottons Konstituierung der zoologischen Wissenschaft – erstellt Gesner eine diametral entgegengesetzte Wissensorganisation. Er ist keineswegs gewillt, die *Mammalia* in Kategorien zu gliedern. Stattdessen wendet er die klassifizierungsfeindlichste und hierarchiefremdeste Anordnung an, die sich denken lässt: die *alphabethische Reihenfolge*. Das mag von einer teleologischen Warte aus betrachtet als beträchtlicher Rückschritt gegenüber Wotton erscheinen. Bäumer bemerkte dann auch: „Die von Conrad Gesner [...] vorgelegte Systematik war eher künstlich [...], sodass seine Klassifizierung insgesamt einen Rückschritt gegenüber Wotton darstellte.“.[92] Jedoch verstellt uns diese Betrachtungsweise den Blick auf die Eigenheiten und Parameter von Gesners Wissenschaftsorientierung.

Für Gesner war die Zoologie nicht eine für wenige Spezialisten bestimmte Fachwissenschaft, sondern ein Bildungsfundament, das sich auf diverse Bereiche auswirkte und für unterschiedliche Gruppen von Intellektuellen relevant war. Auf der Titelseite nennt er Philosophen, Ärzte, Grammatiklehrer, Philologen sowie Dichter in allen Sprachen. Über die Philosophie bietet die Erkenntnis der Natur auch für frühneuzeitliche Theologen eine wichtige Grundlage der Gotteserkenntnis. Das galt nicht zuletzt für das protestantische Zürich, Gesners Wirkungsstätte. Man vergleiche hierzu z.B. die Ausführungen des Reformators Zwingli, der an der Zürcher Prophezei unterrichtete, in seinem Bildungstraktat *Quo pacto ingenui adulescentes informandi sint* (1523).[93] Gesner hat die Naturerkenntnis als Fundament der protestantischen Bildung in die Praxis umgesetzt, als er 1541 an der nämlichen Bildungsanstalt Natur-

---

[92] „Zoologie der Renaissance – Renaissance der Zoologie“ 290.

[93] Vgl. Leu U.B., *Conrad Gesner als Theologe. Ein Beitrag zur Zürcher Geistesgeschichte des 16. Jahrhunderts* (Bern – Frankfurt a.M. – New York – Paris: 1990; Zürcher Beiträge zur Reformationsgeschichte 14) 47–48.

philosophie und Ethik unterrichtete. Zoologie war für ihn ein wesentlicher Teil dieses Wissens von der Natur, das für die Allgemeinbildung der protestantischen Oberschicht bestimmt war. Der protestantische Gläubige hatte „a positive duty to acknowledge the divine artistry of the natural world in which he was placed, and indeed to delight in it".[94] Dieser Gedanke findet sich nicht von ungefähr im Vorwort des ersten Buchs von Gesners *Historia animalium.*[95] Auch von daher lässt sich verstehen, dass Gesner den Nutzen betonte, den die Zoologie bringe.

Man sollte also nicht die Frage stellen, wie fortschrittlich oder rückschrittlich Gesners Zoologie von einer modernen Warte her betrachtet war, sondern wie die Diskurse, in denen er sich bewegte, beschaffen sind und inwiefern sie mit den Zielen, die er verfolgte, zusammenhängen. Man muss also fragen, inwiefern es seinem Ziel, eine Bildungsgrundlage der (protestanischen) Oberschicht zu erstellen, gedient hätte, seine Zoologie auf die Klassifizierung zu fokussieren bzw. diese zur Grundlage der Wissenschaft zu machen. So betrachtet lässt sich nicht leicht erkennen, welchen Nutzen der Leser aus der spezialistischen Taxonomie, z.B aus der taxonomischen Aufgliederung der „Vierfüßer" in Einzeher, Zweizeher und Mehrzeher, beziehen hätte sollen. Die Nachteile liegen auf der Hand: Nehmen wir einmal an, der Leser hätte sich für das „Seekalb" (vitulus marinus), d.h. eine Robbenart, interessiert. Inwiefern hätte ihm die Klassifikation geholfen, das Tier zu finden? Hat das Tier überhaupt Füße? Wenn dem so wäre, gehört es den Paarhufern oder den Unpaarhufern oder den Mehrzehern zu? Man kann sich vorstellen, dass Gesner angesichts solcher Schwierigkeiten das alphabethische Ordnungssystem, welches eindeutig ist und an Einfachheit nicht übertroffen werden kann, für zielführender hielt.

Diese im Vergleich zu Wotton diametral entgegengesetzte Diskursorientierung lässt sich anhand des Kapitels über den Löwen[96] besonders anschaulich zeigen. Zunächst verzichtete Gesner in seiner Beschreibung auf eine nähere Feststellung des systematischen Ortes des Löwen. Z.B. fehlt jeder Hinweis darauf, dass das Tier der Kategorie der Mehrzeher zugehöre.[97] Weiter ist Gesner nicht geneigt, durch die Anwendung der

[94] Rudwick M.J.S., *The Meaning of Fossils* (London – New York: 1972) 16.
[95] Fol. aa 3v.
[96] Ebd. 642–681.
[97] Lediglich am Ende der morphologischen Beschreibung macht Gesner eine kurze Angabe zur Anzahl der Zehen (647), jedoch ohne daraus klassifizierende Rückschlüsse zu ziehen.

Induktion und der Komparatistik zu einer weiteren Klassifikation der Species in einzelne Arten bzw. Unterarten zu gelangen. Im Gegenteil: Z.B. war bei Aristoteles, Plinius und anderen antiken Autoren überliefert, dass sich die Species in zwei Unterarten gliedere: in eine kleinere, schwächere, kurz- und kraushaarige bzw. kurz- und krausmähnige und in eine größere, längere, stärkere, aggressivere mit langer, glatter Mähne.[98] Gesner ‚*ent-klassifizierte*' diese durch die antike Zoologie autorisierte Unterscheidung mit harter Hand. Er versuchte sie als Irrtum zu entlarven, indem er sie als Geschlechtsunterschied deutete. Bei der ersten „Gattung" handle es sich recht eigentlich um Weibchen, bei der zweiten um Männchen: „Ich freilich meine, dass alle männlichen Löwen lange Mähnen haben".[99] Dementsprechend schickt er dem Kapitel das artbestimmende Bild eines männlichen Löwen mit langer Mähne voraus [Abb. 5].

Es ist auffällig, wie selbstbewusst der protestantische Zoologe Gesner seine individuelle Einsicht und Wahrnehmung über die sanktionierte Autorität der Antike stellt. Die Entlarvung des ‚Irrtums' erscheint durch ihre Einfachheit und ihren Common-Sense-Charakter plausibel, ist es jedoch nicht ganz, da Löwenweibchen nicht eine kurze, sondern gar keine Mähne haben.[100] Gesner scheint sich dieser Schwierigkeit bewusst gewesen zu sein. Deshalb zieht er Theodor Gazas Aristoteles-Übersetzung (kurz- und kraushaarig), die Wotton als wenig vertrauenswürdig angemerkt hatte, der des Plinius (kurz- und krausmähnig) vor. Entscheidend ist übrigens nicht, inwiefern Gesners Bemerkung einen wissenschaftlichen Fortschritt darstellt, sondern welche Diskursanlage ihr zugrunde liegt.

Diese zielt in der Tat auf eine ‚ent-klassifizierende' Deskription ab. Sie bedingt auch, dass Gesner die Unterscheidung der Löwen in „Erzlöwen" („Archoleones") und aggressionsarme („mansueti"), also gewöhnliche

---

[98] Z.B. Plinius, *Naturalis historia* VIII 46 „Leonum duo genera, compactile et breve crispioribus iubis. Hos pavidiores esse quam longos simplicique villo; eos contemptores vulnerum". Vgl. Aristoteles, *Historia animalium* VIII 5, 594 b.

[99] Gesner, *Historiae animalium lib. 1* 643 „Atqui ego leones mares omnes iubatos esse puto".

[100] Worauf die antike Unterscheidung in langmähnige und kurzmähnige Löwen zurückgeht, ist unklar. H. Leitner hat sie in seiner Studie *Zoologische Terminologie beim älteren Plinius* (Hildesheim: 1972) 152 als Subspecies-Unterscheidung zwischen dem asiatischen Löwen (Leo persicus) und dem Berber-Löwen gedeutet. Das ist jedoch keineswegs schlüssig, ja es ist unsicher, ob der Berberlöwe in der Tat eine Subspecies darstellt. Vgl. Macdonald, *Enzyklopädie der Säugetiere* 12. Der Unterscheidung können genauso gut unrichtige, lückenhafte, übertriebene und simplifizierende Beobachtungen oder altersbedingte Differenzen (junge Löwen haben kurze Mähnen) zugrunde liegen.

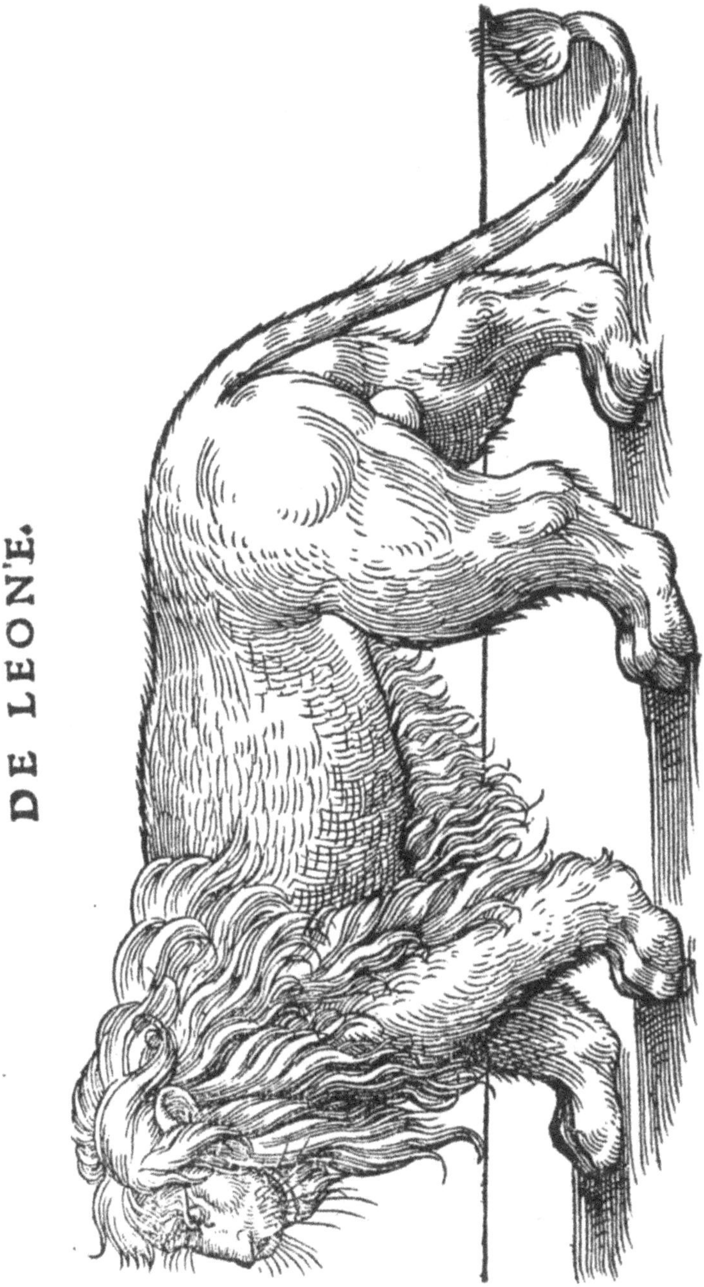

Fig. 5. Abbildung des Löwen. Aus Conrad Gesner, *Historiae animalium I de quadrupedis viviparis* (Zürich, Christoffel Froschauer: 1551) 642. Universiteitsbibliotheek Leiden.

Löwen, welche er bei dem römischen Historiker Julius Capitolinus antraf,[101] als Irrtum ‚entlarvt'. Der Name Archoleo(n) bezeichne keine Art oder Unterart, sondern beziehe sich auf individualspezifische Merkmale (besonders große, starke Individuen).[102]

Gesners Diskursorientierung zeigt sich noch klarer im Vergleich mit dem taxonomieverliebten Wotton, der die Löwen in 7 Arten oder Unterarten gliederte und dabei auch mähnenlose Löwenarten, z.B. die lybische (Leopard) und die hundegroße (wohl ein Karakal) und die schwarze äthiopische (Leopard) konstituierte. Schon der Titel des Kapitels („De leone", Einzahl) zeigt, dass Gesner geneigt war, den Löwen als eine einzige, *homogene* Species aufzufassen.[103] Oppians Ausführungen wertete er, anders als Wotton, nicht als Klassifizierungsgrundlage, sondern lediglich als Materialien zur geographischen Verbreitung der Species Leo aus.[104] Die merkwürdige hundegroße und mähnenlose äthiopische Löwenart, die Wotton aus Oppian konstruierte (s. oben), akzeptiert Gesner nicht. Er vermeldet die betreffende Information überhaupt nicht. Im Abschnitt über das Aussehen des Löwen schreibt er der Art insgesamt eine große Körperlänge zu („Leo corpore longo est"), indem er sich auf den Arzt, Mathematiker und Universalgelehrten Gerolamo Cardano (1501–1576), dessen damals gerade erschienenes Werk *De subtilitate* (1550) eine seiner Hauptquellen ist, beruft.[105] Es war unter anderem diese Hauptquelle, die Gesner dazu brachte, die von der antiken *auctoritas* Oppian überlieferte Unterart des hundegroßen äthiopischen Löwen abzuschreiben. Gesner hat also die Species *Panthera leo* mit Hilfe moderner Quellen homogenisiert.

Für dieses Diskursmerkmal ist die Art bezeichnend, mit der Gesner eine andere aus der Antike überlieferte Angabe verarbeitete. Bei Oppian traf er – wie Wotton – den Bericht an, dass sich die Löwen Libyens durch ein blauschwarzes Fell auszeichnen, und eine äthiopische Löwenart durch ein schwarzes Fell.[106] Wenn man von der zoologischen Diskursorientierung Wottons ausgeht, gibt es zwei Möglichkeiten. Entweder

[101] Julius Capitolinus, *Gordianus* III 33.

[102] Gesner, *Historia animalium* I 643: „Quid si archoleontes genus aliquod leonum praegrande intelligas? Ut qui magnitudinis ratione caeteris praestare et quodammodo imperare videantur".

[103] Ebd. 642.

[104] Ebd. 643–644.

[105] Ebd. 644. Girolamo Cardano, *De subtilitate libri XXI* [...] (Nürnberg, Joannes Petreius: 1550).

[106] Vgl. oben.

man akzeptiert die Angabe nicht, indem man auf den Autoritätsbeleg des Aristoteles verweist, der allen Löwenarten ein rötlichgelbes Fell zugeschrieben hatte. Wenn man diese Angabe als solche akzeptiert und wie Wotton die Klassifizierung als Hauptaufgabe der Zoologie betrachtet, liegt es nahe, aus ihr Arten bzw. Unterarten des Löwen zu konstruieren. Wie oben gezeigt wurde, hat sich Wotton für die letzte Möglichkeit entschieden.

Gesner fand jedoch einen dritten Weg. Dazu zog er die Angaben heran, die er sich in Bezug auf asiatische Löwen erarbeitet hatte. Daraus ging hervor, dass es schwarze Löwen auch in Syrien, Indien und in der Mongolei gab.[107] Gesner, der sowohl bestrebt war, das Tierreich zu entklassifizieren als auch seine Quellen zu homogenisieren, zog aus der Vielzahl seiner Belege die Schlussfolgerung, dass ein an so vielen unterschiedlichen Orten vorkommendes Merkmal kein Artspezifikum sein könne. Anders gesagt: Die schwarze Farbe sei eine Spielart derselben Species.

Auf diese Weise kam Gesner dem Melanismus auf die Spur, den es bei den Katzenartigen in der Tat gibt, zu dem ihm allerdings keine systematischen empirischen Untersuchungen vorlagen. Sein Homogenisierungsversuch ist hochinteressant, obwohl oder gerade weil er nicht mit den heute bekannten zoologischen Fakten übereinstimmt. Der Melanismus stellt kein artimmanentes Merkmal des Löwen dar, sondern lässt sich für den Leoparden (*Panthera pardus*, Afrika und Asien), den Jaguar (*Panthera onca*, Südamerika) und den Serval (*Felis serval*, Afrika) nachweisen.[108] Es ist interessant, mit welchem Selbstvertrauen Gesner seine neue Erkenntnis des Melanismus bei Löwen gegenüber der im 16. Jahrhundert akzeptierten Autorität par excellence, Aristoteles, durchsetzt. Sie führt zu einer regelrechten Berichtigung: „Alle Löwen sind rötlichgelb (gelblich, „fulvus"), sagt Aristoteles. [...] Ich habe schon oben gezeigt, dass es in einigen Regionen schwarze, in anderen blauschwarze Löwen gibt".[109]

---

[107] Gesner, *Historia animalium* I 644.

[108] Dem Leoparden schrieb Gesner jedoch nicht den Melanismus als Artspezifikum zu. Vgl. Ebd. I 935–948. Gleichwohl vermelden Oppian, *Kynegetika* III 74–75, Solin 17,8 und Isidor von Sevilla, *Etymologiae* XII 2,8 schwarze Exemplare.

[109] Gesner, *Historia animalium* I 644.

Zum Verbreitungsgebiet der Species Panthera leo: Diskursverlagerung der zoologischen Wissensautorisierung vom antiken Autoritätsdiktat zum ‚modernen' Wissensaustausch

Gesners Angaben zum Verbreitungsgebiet des Löwen unterscheiden sich grundlegend von denen Wottons. Der Wissenschaftsdiskurs, in dem er seine Zoologie situierte, ist, obwohl er ungefähr zur gleichen Zeit arbeitete, ein grundsätzlich anderer als der des Oxforder Gelehrten, der einseitig auf antike Autoritätsbelege rekurrierte. Wottons Diskursautorisierung hatte dazu geführt, dass er Europa als das Hauptverbreitungsgebiet des Löwen präsentierte, während die Species dort schon seit der Antike ausgestorben war.[110] Gesners gelangt zu einer diametral entgegengesetzten Darstellung: „Es ist sicher, dass es in unserer Zeit in Europa nirgends Löwen in freier Wildbahn gibt" („Leones nostro tempore nusquam in Europa nasci certum est").[111] Als Verbreitungsgebiet des Löwens konstituiert er die südlich und östlich von Europa liegenden Erdteile, also Afrika und Asien, eine Angabe, die in etwa den Tatsachen zu Gesners Zeit entspricht. Im Gegensatz zu Wotton hebt Gesner hervor, dass der Löwe in Nordafrika und weiten Teilen Asiens *häufig* vorkomme. Besonders auffällig ist, dass Gesner überhaupt *Asien als Hauptverbreitungsgebiet des Löwen* präsentiert. Der längste und ausführlichste Teil des Abschnitts ist dem Löwen in Asien gewidmet. Während der asiatische Löwe heute unmittelbar vom Aussterben bedroht ist und auf eine geringe Anzahl im indischen Gir-Nationalpark zusammengeschmolzen ist, war er zu Gesners Zeit in der Tat noch über weite Teile des Vorderen Orients bis nach Indien hin verbreitet. Wie gelangte Gesner zu seiner auf den ersten Blick realistischeren Darstellung?

Eine auf der Hand liegende Erklärung wäre, dass man sie als Berichtigung der Schrifttradition durch Empirie auffasst. Im Fall der Species *Panthera leo* geht sie jedoch nicht auf. Gesner hat ebenso wenig wie Wotton Nordafrika oder Asien bereist, wenngleich er, wie er selbst im Vorwort seiner *Historia animalium* vol. I angibt, gerne weite Reisen unternommen hätte. Gesner gelangte zu seiner abweichenden Darstellung durch eine andersartige Orientierung der Wissensautorisierung. Er gab den einseitigen Rekurs auf die Antike, wie er sich bei Wotton findet, auf und *zog systematisch rezente Informationsquellen heran.*

[110] Vgl. oben.
[111] *Historia animalium* I 643.

In seinem Quellenverzeichnis finden sich neben 120 antiken (griechischen und römischen) auch 105 rezente oder subrezente Autoren (Nr. 146–251). Werke, die er als Hauptinformationsquellen benutzte, versah er mit einem Asterix. Aus diesen Markierungen geht noch klarer die Verlagerung des Wissenschaftsdiskurses zur ‚Moderne' hin hervor: Von den „recentiores" sind 80 Werke mit einem Asterix versehen, von den Antiken nur 48. Gesners zoologische Wissensgrundlage stammt also zu einem überwiegenden Teil aus der ‚Moderne'.[112]

Besonders wichtig ist, dass Gesner rezente und subrezente Reiseberichte systematisch ausgewertet hat. Aus dem Verzeichnis der „libri recentiorum" geht hervor, dass er die Beschreibung der Reisen Amerigo Vespuccis (Nr. 151), des Burchardus' de Monte Sion Beschreibung der Reise ins Heilige Land (*Descriptio terrae Sanctae*, 1283; Nr. 162: „Brocardus monacus de terra sancta"),[113] Columbus' Bericht *De insulis nuper inventis* (Nr. 169), Ludovico Varthemas (ca. 1472–ca. 1517) Tagebuch über seine Reisen in den Orient 1502–1508 (*Itinerario de Ludovico Varthema Bolognese nello Egypto, nella Surria, nella Arabia deserta et felice; nella Persia, nella India et nella Ethiopia* [...] [Rom: 1510]; Nr. 207 Ludovici *Vartomanni Romani patritii Navigationum* libri VII), Marco Polos (1254–1324) Berichte über seine Reisen ins Innere Asiens (Nr. 209), Paulo Giovios (1486–1552) Bericht über eine Gesandtschaftsreise nach Moskau (*De Moscovitarum legatione*, Nr. 219),[114] Pietro Martyres *Decades de orbe novo*[115] (Alcalà: 1530; Nr. 223), Vicente Yanez Pinzons (ca. 1460–1524) Reisen ins Amazonasgebiet und ins nördliche Brasilien und Fernao Magalhaes (1480–1521) Reisen zu den Molukken (1505–1512) (Nr. 225; *Pinzoni navigationes* et *Magellani ad insulas Moluchas*) benutzt hat. Interessant ist, dass er die meisten der hier genannten Reiseberichte mit einem Asterix als Hauptquellen markiert hat.

Es sind in der Tat gerade die Reiseberichte Lodovico Varthemas und Marco Polos, mit welchen Gesner das häufige Vorkommen des Löwens in Asien in der Gegenwart (ca. 1300–1510) nachweist. Varthema reiste 1502 nach Ägypten, wo er sich als Muslimpilger ausgab und bis nach Medina vorstieß; von dort zog er nach Yemen und Äthiopien und setzte

---

112 Vgl. Leu, *Conrad Gesner als Theologe* 45.

113 Vgl. die Textausgabe von W.A. Neumann (1880).

114 Zu Giovio vgl. Zimmermann T.C. Price, *Paolo Giovio. The Historian and the Crisis of Sixteenth Century Italy* (Princeton: 1995).

115 Vgl. Peter Martyr von Anghiera, *Acht Dekaden über die Neue Welt*, übersetzt von H. Klingelhöfer, 2 Bde. (Darmstadt: 1972–1973).

nach Indien über, wo er sich jahrelang aufhielt. Erst 1508 kehrte er nach Italien zurück. In seinem Itinerar[116] zeichnet er auf, dass es in der Umgebung Adens zahlreiche (menschenfressende) Löwen gebe. Weiter leitet Gesner aus Varthemas Itinerar ab, dass der Löwe in Indien besonders häufig sei („Leones in regno Narsingae plurimi [...]").[117] Aus der Lektüre von Marco Polos („Paulus Venetus") Reiseberichten destillierte Gesner den Nachweis des Löwen in der Mongolei.

Gesners Abweichen vom Rekurs auf die Antike ist nicht einfach als progressiver Schub zur Empirie hin zu deuten.[118] Die Mehrzahl der ‚neuen' Angaben Gesners beruhen nicht auf authentischen empirischen Wahrnehmungen, die er selbst angestellt hätte. Jedoch ist die Diskursverlagerung der Faktenerhebung zu ‚modernen' Informationsquellen hin von entscheidender Bedeutung. Die zur Abgeschlossenheit tendierende Ausrichtung des Wissenserwerbs auf antike *auctoritates* wird zu ‚modernen' Gewährsmännern hin geöffnet, wodurch neues, rezentes Wissen aufbereitet werden kann. Die neue Wissenschaft der Zoologie wird durch eine rezente Diskussion unter Gelehrten gespeist. Sie ist, wie Gesner in seinem Korrespondentenkatalog, in dem er zumeist die Herkunft der insgesamt 52 Gelehrten angibt, betont, international vernetzt. Sein zoologischer Wissenschaftsdiskurs ist auf den dialogischen Wissenschaftsaustausch mit Franzosen (Galli), Engländern (Angli), Schweizern (Rhaeti), Deutschen (Germani), Italienern (Itali) und Polen (Poloni) ausgerichtet [Abb. 6].

Die Konstituierung der Zoologie als Universalwissenschaft durch Systematisierung und Erweiterung der Beschreibungsmethode

Statt auf die klassifizierende Systematik verwendete Gesner seine Schaffenskraft darauf, die Reichweite der zoologischen Deskription zu vergrößern, die Beschreibungsmethode zu verbessern und den Umfang der zoologischen Information zu erweitern. Dafür entwickelte er ein eigenes Deskriptionssystem, das dazu dient, die Beschreibung jedes lexikalisch

---

[116] Siehe dazu Barozzi P., *Ludovico de Varthema e il suo itinerario* (Rom: 1996).

[117] *Historiae animalium lib. 1* 644.

[118] Dies scheint Änne Bäumer, *Zoologie der Renaissance* 42, zu suggerieren: „Er hatte zwar großen Respekt vor dem überkommenen tradierten Wissen, verband seine Studien der antiken Autoren aber mit den Erfahrungen, die er aus der unmittelbaren Beobachtung der Natur gewonnen hatte".

CATALOGVS DOCTORVM VIRORVM, QVI VT OPVS hoc nostrum & rempub. literariam illustrarent, uel aliunde imagines animalium, aut nomina & descriptiones miserunt: uel præsentes communicarunt. Horũ nonnulli superius quoq; nominati sunt, quòd insuper scriptis eorum publicatis adiutus sim.

Achilles P. Gassarus medicus Germanus.
Alexander Peijer Scaphusianus.
Aloisius Mundella Brixiensis medicus.
Andreas Martinus Rostochiensis.
Antonius Eparchus Corcyræus, Græcæ linguæ professor Venetijs.
Antonius Musa Brasauolus illustrissimi Ferrariæ ducis Herculis Estensis archiatros.
Antonius Stuppa Rhætus.
Arnoldus Peraxylus Arlenius Germanus.
Bartolemæus à Castromuro canonicus Curiensis in Rhætia.
Cælius Secundus Curio Italus.
Cælius Sozinus Senensis.
Caspar Hedio ecclesiastes Argentinensis.
Christophorus Clauserus Tigurinus archiatros.
Cornelius Sittardus medicus Germanus.
Dominicus Monthesaurus medicus Veronẽsis.
Florianus Susz Rolitz à Varshauia, Polonus.
Franciscus Belinchettus mercator Bergomensis.
Ge. Agricola consul Kempnicij.
Ge. Fabricius poëta, Scholæ rector Misenæ.
Gisbertus Horstius Amsterodam9 medicus Romæ.
Gregorius Mangolt Constantiensis.
Gulielmus Gratarolus Bergomensis, medicus.
Guilielmus Turnerus Anglus medicus.
Henricus Stephanus Roberti filius, Parisiensis.
Hieronymus Fracastorius Veronensis medicus.
Hieronymus Frobenius Basiliẽsis typographus.
Hieronymus Tragus Germanus.
Io. Altus Hessus.
Io. Culmannus Goppingensis.
Io. Dernschwam Germanus.
Io. Estwycus Anglus.
Io. Falconerus medicus Anglus.
Io. Kentmannus Dresdensis medicus.
Io. Oporinus Basiliensis typographus.
Io. Ribittus sacrarum literarum interpres Lausannæ.
Iustinus Goblerus I. C. & principi Nassauiensi à consilijs.
Lucas Gynus medicus Italus.
Michaël Alysius Gallus Trecensis medicus.
Nicolaus Gerbelius Phorcensis I. C.
Petrus Dasypodius Græcarum literarum professor Argentorati, præceptor meus.
Petrus Gyllius Gallus.
Petrus Merbelius Germanus, Carolo v. à consilijs Mediolani.
Petrus de Mesnil Gallus.
Petrus Paulus Vergerius, olim episcopus Iustinopolitanus.

γ

Fig. 6. Gesners Korrespondentenliste. Am Conrad Gesner, *Historiae animalium I de quadrupedis viviparis* (Zürich, Christoffel Froschauer: 1551). Universiteitsbibliotheek Leiden.

erfassten Tieres auf genau dieselbe Weise zu gestalten. Gesner hat acht Kategorien (A–H) entworfen, nach denen jedes Lemma gegliedert ist. In Kategorie A werden die Namen des betreffenden Tieres in möglichst vielen Sprachen erfasst. Es geht dabei weder um eine klassifizierende Maßnahme noch um eine Vereinheitlichung der Nomenklatur, wie manchmal angenommen wird,[119] sondern um die Anlage eines Thesaurus, einer Schatzkammer oder Fundgrube, die einerseits ein möglichst reiches Sprachmaterial darbieten (aus dem unter anderem sprachkomparatistische Schlussfolgerungen gezogen werden können), andererseits durch Miteinbeziehung der Volkssprachen den Zusammenhang mit der sprachlichen Gegenwart herstellen soll. Ein mehrfacher Index, in dem die Tiernamen in verschiedenen Sprachen aufgegliedert werden, bietet den zeitgenössischen Benutzern einen weiteren nützlichen Zugang. Missverständnisse und Namensverwechslungen, die sich aus der Inkongruenz zwischen der lateinischen Sprache und diversen Volkssprachen ergeben, sollen dadurch ausgeschlossen werden.

Kategorie B ist der Tierbeschreibung im engeren Sinn gewidmet. In dieser Kategorie behandelt Gesner: das Verbreitungsgebiet und die Morphologie, einschließlich der anatomischen Daten. Die Behandlung der Morphologie erfolgt systematisch, wobei Gesner die einzelnen Teile des Tiers, vom Kopf bis zum Schwanz von ‚oben' nach ‚unten' bzw. von ‚vorne' nach ‚hinten' behandelt. Gesners Kategorie C ist der Physiologie und der mit den Körperfunktionen zusammenhängenden Verhaltensbiologie (Nahrungsaufnahme, Fortpflanzung etc.) sowie den Krankheiten der Tiere gewidmet. Kategorie D setzt sich mit der Tierpsychologie auseinander. Die Kategorien E–G sind der Nutzung des betreffenden Tiers durch den Menschen gewidmet. In Kategorie E werden Tierfang, Zähmung, Tierhaltung, Veterinärmedizin, Hilfsmittel zur Tiernutzung (Pflug, Sattel etc.), die Verwendung des betreffenden Tiers in Sport und Spiel, zur Kleidung (Felle), Düngung, Wettervorhersage usw. behandelt, in Kategorie F die Nutzung des betreffenden Tiers als Nahrungsmittel, in Kategorie G als Heilmittel. In Kategorie

---

[119] Fischer H. et alii, *Conrad Gessner 1516–1565. Universalgelehrter, Naturforscher, Arzt* (Zürich: 1967) 131 „Eine der vordringlichsten Aufgaben, welche die damalige [...] Naturwissenschaft zu lösen hatte, bestand darin, *Klarheit in die Nomenklatur zu bringen.* [...] *Erst wenn Ordnung in die Nomenklatur gebracht*, die synonymen Bezeichnungen in den alten und modernen Sprachen ermittelt waren, konnten die antiken und neueren Nachrichten über ein Tier zusammengestellt werden" (Kursivierungen von K.A.E.E.). Gmelig-Nijboer, *Conrad Gesner's „Historia animalium"* 49–50 scheint dieser Darstellung zuzustimmen.

H erläutert Gesner die Verwendung des Tiers in Literatur, Kunst und Religion. In dieser Kategorie erschließt er unter anderen Tierembleme, Tierfabeln und Sprichwörter.

Diese differenzierende Systematisierung der Beschreibung in 8 Kategorien hat den wichtigen Vorteil, dass sie dem Benutzer der Zoologie einen effizienteren und schnelleren Zugang ermöglicht. Der Benutzer, der auf der Suche nach Tiervergleichen ist – z.B. weil er gerade eine Predigt zusammenstellt – kann sich sofort der Kategorie H zuwenden, ohne eine seitenlange Lektüre der vorangehenden Teile des Lemmas auf sich nehmen zu müssen. Der Benutzer, der sich für die Haltung eines bestimmten Tieres interessiert, findet die gewünschte Information sofort unter der Kategorie E. Der Apotheker, der ein Heilmittel anfertigen will, kann sich sofort der Kategorie G zuwenden usw.

Die Beschreibungssystematik Gesners ist so angelegt, dass sie die Differenzierung des frühneuzeitlichen Leserpublikums optimal berücksichtigt. Ein entscheidender Vorteil der Systematisierung ist, dass die Quantität der zoologischen Information stark erweitert werden kann, ohne dass der Benutzer einem undurchdringlichen Dickicht diverser Angaben gegenübersteht. Der Wissensschatz kann also angereichert werden, während die Zugänglichkeit erhalten bleibt. Dadurch wird es weiter möglich, verstärkt ins Detail zu treten. Die zoologische Beschreibung kann dadurch im Hinblick auf Detailangaben in eine andere Dimension vorstoßen. Gesners Kapitel „Vom Löwen" ist auch in dieser Hinsicht illustrativ: Es ist wohlgemerkt etwa zwanzigmal so lang wie Wottons Kapitel „Von Löwen".

Ein interessanter Punkt betrifft die Trennung von Morphologie (Kategorie B) und Physiologie (Kategorie C). Wenn man sie wissenschaftsteleologisch betrachtet, könnte man sie als Herausdifferenzierung „zoologischer Spezialgebiete" wie „Physiologie, Vergleichende Anatomie" verstehen, in der Bäumer einen entscheidenden Fortschritt der frühneuzeitlichen Zoologie erkennt.[120] Obwohl dies bis zu einem gewissen Grade zutreffen könnte, ist zuerst eine nähere Spezifizierung des Diskurses erforderlich. Es erscheint mir fragwürdig, ob Gesner mit seiner Beschreibungsmethode 1. Spezialdisziplinen schaffen, 2. sich in die Richtung der Vergleichenden Anatomie vorwagen wollte. Interessant ist zunächst, dass er sich bei der Aufgliederung der morphologischen Deskription von Quellen anregen ließ, die den Schritt zur Vergleichenden Anatomie durchaus nahe legen, z.B. vom 11. Buch von Plinius'

---

[120] „Zoologie der Renaissance – Renaissance der Zoologie" 279.

Naturalis historia (121–284). Plinius behandelt in diesem Abschnitt seiner Zoologie die einzelnen Körperteile der Tiere[121] – den Kopf (§ 121ff.) mit Hörnern (§ 123–128), Knochbau (§ 132), Gehirn (§ 133–135), Ohren (§ 136–137), Gesicht (§ 138), Augen (§ 139–157), Wangen (§ 158), Backen (§ 158), Lippen (§ 159), Kinn (§ 159), Zähne (§ 160–170) und Zunge (§ 171–174); Kehle und Nacken mit Mandeln (§ 175) sowie Luft- und Speiseröhre (§ 176–180); Herz (§ 181–187); Lunge (§ 188); Leber und Galle (§ 189–195); Magen und Darm (§ 197–204); Nieren (§ 206); Brustkorb (§ 207ff.); Knochen (§ 212–216); Nerven und Sehnen (§ 217–218); Blutgefässe und Blut (§ 219–225); Haut (§ 226–227); Haare (§ 228–231); Zitzen (§ 232–235); Milch (§ 236–242); Gliedmassen (§ 243–246); Klauen und Nägel (§ 247–248); Gelenke (§ 249–251) usw., indem er eine komparatistische Zusammenschau entwirft.

Im Kapitel „Vom Löwen" wird zunächst ersichtlich, dass Gesner den Schritt zur Vergleichenden Anatomie *nicht* gemacht hat. Zwar behandelt er die einzelnen Körperteile des Löwen, jedoch nicht in der Art eines komparatistischen Vergleichs, wie er ihn etwa bei Plinius antraf. Vielmehr sammelt er aus dem vergleichenden Abschnitt der Plinius-Zoologie die Angaben zum Löwen lediglich, um Einzeldaten, die nur den Löwen betreffen, zu erheben.

Es gibt jedoch Ausnahmen. Diese sind allerdings nicht so beschaffen, dass man sie ohne weiteres als wissenschaftlichen Fortschritt bewerten könnte. Eine solche Ausnahme bilden die Ausführungen zum Knochenbau des Löwen. Bei Aristoteles fand Gesner einen interessanten Schlüssel zur Vergleichenden Anatomie: Aristoteles hatte bemerkt, dass der innere Bauplan des Löwen „in allem" dem Hunde ähnle.[122] Dies lässt sich geradezu als Auftrag verstehen, die aus der Literatur übermittelten Daten durch empirische Autopsie zu untersuchen. Hunde waren ja, im Gegensatz zu Löwen, jederzeit vorrätig, wenn man Sektionen vornehmen wollte. Wie aus obigem hervorging, wurden bei Sektionen von Medizinern leicht zugängliche Tiere wie Hunde und Katzen verwendet. Nun fand Gesner bei Aristoteles die Angabe, dass der Löwe keine Halswirbel besitze, sondern stattdessen nur einen langen Knochen. Was hätte näher gelegen als einmal einen Hund zu sezieren und nachzusehen, ob dieses Tier Halswirbel besitze?

---

[121] *Naturalis historia* XI 121: „Nunc per singulas corporum partes [...] membratim tractetur historia" („Nun soll die Beschreibung nach den einzelnen Körperteilen [...] Glied für Glied erfolgen").

[122] Gesner, *Historia animalium* I, 644 „Interiora omnia canibus similia continentur".

Es ist bezeichnend, dass Gesner diesen Schritt nicht gemacht hat. Stattdessen verstärkt er die Gültigkeit der aristotelischen Angabe, indem er sie mit weiteren Autoritäten (Aelian, Plinius, Manuel Philes von Ephesos [1275–1345], Ibn Ruschd [Averroes, 1126–1198], Albertus Magnus, Ambrosius) und Texten untermauert. So erfährt der Benutzer Gesners zusätzlich, dass der Löwe seinen Hals nicht beugen könne (aus Manuel Philes), dass er ihn nicht nach hinten wenden könne, etwa um umzublicken (aus Ibn Ruschd über Albertus Magnus), ja dass Löwen ohnehin nur einen kurzen Hals hätten, weil sie nicht zu grasen bräuchten wie das Weidevieh (aus Ambrosius).[123]

Daraus muss man den Schluss ziehen, dass die Vergleichende Anatomie kein Anliegen von Gesners Zoologie war. Es geht ihm zuvorderst um die Vervollständigung der Beschreibung. Die Behandlung der Morphologie als separate Kategorie führt allerdings zu einer detaillierteren und vollständigeren Wiedergabe der morphologischen Merkmale. In der Tat ist Gesners Beschreibung der Morphologie des Löwen vollständiger und präziser als die Wottons. Klar ist weiter, dass die Zusammenführung der Angaben den Vergleich des aus diversen Quellen stammenden Informationsmaterials und damit eine kritische Betrachtung desselben fördert. Widersprüche können auf diese Weise leichter ans Licht gebracht werden. Freilich impliziert der Vergleich nicht, dass dieser auf empirischer Basis stattzufinden habe. Er konzentriert sich zuvorderst auf die Ebene der Textüberlieferung. Wenn die Textüberlieferung nicht widersprüchlich ist, zementiert der Vergleich die überlieferte Information als *communis opinio*. Wenn die Texttradition falsche Information enthält, führt die Separierung der Morphologie zur Zementierung von Irrtümern. Insofern bildet Gesners Differenzierung der Beschreibungsmethode an sich keinen Indikator für wissenschaftlichen Fortschritt.

Für die separate Darstellung der Physiologie gilt mutatis mutandis Ähnliches wie für die Morphologie. Die Konzentration ermöglicht eine detailliertere Beschreibung und die systematische Zusammenführung der Quellen kann ihre kritische Betrachtung fördern. Diese Vorteile treten im Kapitel „Vom Löwen“ u.a. in Bezug auf die Nahrungsaufnahme hervor. Diesbezüglich liegen Quellenangaben vor, die die außerordentliche Jagdfreudigkeit und ‚edle' Tapferkeit des Löwen betonen und diese mit der Angabe verbinden, dass Löwen ausschließlich frische und selbst

[123] Ebd. 646.

erlegte Beute verspeisen. Aus Ibn Sina (Avicenna, 980–1037) bezog Gesner über Albertus Magnus die Information: „Löwen ernähren sich ausschließlich von der Jagd, und sie jagen mindestens einmal pro Tag".[124] In Joannes Tzetzes' (ca. 1110–ca. 1180) Homerkommentar traf er die Angabe an, dass sich Löwen nur von frisch erlegter Beute ernähren. Was sie nicht unmittelbar fressen können, lassen sie liegen und kehren später nicht mehr zu ihrer Beute zurück. Bei Manuel Philes, Aelian und Plinius fand er jedoch den Bericht, dass die Löwen ihre Beute mit ihrem stinkenden Atem anhauchen, um andere Tiere von ihr abzuhalten.[125] Dieses merkwürdige Verhalten macht natürlich nur dann Sinn, wenn der Löwe an den Folgetagen zu der verwesenden Beute zurückkehren will, also auch Aas frisst. Aus der Zusammenführung der Quellen ergibt sich somit ein widersprüchliches Bild, das eine kritische Betrachtung und eine Auflösung des Widerspruchs erfordert.

Es war für Gesner keineswegs leicht, den Widerspruch aufzulösen, da er über keine empirischen Beobachtungen bezüglich der Ernährungsweise des Löwen in freier Wildbahn verfügte. Systematische Beobachtungen der modernen Zoologie haben ergeben, dass es mit dem vermeintlichen ‚Edelmut' des Löwen nicht weit her ist. Löwen sind sowohl bereit, die Beute anderer Karnivoren, wie Geparden, Leoparden, Wildhunden und Hyänen, zu rauben als auch Aas zu fressen. Interessant ist zu beobachten, auf welche Weise sich Gesner einen Weg durch das schwer durchdringbare Dickicht bahnte. Er schloss von der ihm zugänglichen Empirie auf die unbekannten Daten. Er hatte Löwen in Gefangenschaft beobachtet. Dabei war ihm aufgefallen, dass Löwen in Gefangenschaft nicht nur Aas fressen, sondern sogar Honigkekse und andere Süssigkeiten" („tamen [captivi inclusique leones] non mortua solum corpora [...] devorare, sed etiam ex melle placentas et huiusmodi"). Daraus leitete er richtig ab, dass die Annahme verfehlt sein müsse, dass Löwen Aas prinzipiell verschmähen würden, z.B. weil sie vom Gestank abgestoßen werden würden. Zugleich war er sich der Beschränkung seiner empirischen Beobachtung bewusst. Es konnte sein, dass gefangene Wildtiere atypisches Verhalten an den Tag legen. Löwen in Gefangenschaft hatten ja keine Wahl, sie durften in der Regel nicht jagen, d.h. sie mussten, wenn sie überleben wollten, totes Fleisch zu sich nehmen. Er wollte daher nicht ausschließen, dass Löwen in freier

---

[124] Ebd. 648.
[125] Ebd.

Wildbahn sich im Prinzip auf frisches Fleisch zulegten. Dies verband er interessanterweise mit einer aus der Textüberlieferung tradierten Auffassung über die psychischen Eigenschaften des Löwen, seinem vermeintlichen „Hochmut" („superbia"). Es sei diesem zuzuschreiben, dass wilde Löwen frisches und selbsterlegtes Fleisch vorzögen. Aus dem Abschnitt über die Tierpsychologie (D) geht hervor, dass Gesner die aus der literarischen Überlieferung vermittelten Eigenschaften ohne Einschränkungen unterschrieb: Tapferkeit („fortitudo") und Edelmut („generositas") seien die wichtigsten Eigenschaften des Löwen,[126] mit denen er übrigens die idealen Eigenschaften des Mannes am besten verkörpere („Leo ideam maris maxime omnium refert").[127] Da Gesner entsprechende empirische Beobachtungen fehlten, konnte er nicht wissen, dass die meiste Beute nicht von den tapferen, kraftstrotzenden, edelmütigen Männchen, sondern gerade von den als weniger tapfer eingestuften Weibchen erlegt wird, da diese aufgrund ihres geringeren Körpergewichtes bei Jagdsprints gegen die normalen Beutetiere, Zebras, Gnus und Antilopen, bessere Chancen haben als die schweren Männchen.

Zoologie als utilitaristische Wissenschaft: Die universale Ausschlachtung des Tieres

Eine wesentliche Erweiterung von Gesners Deskriptionsdiskurs findet auf Gebieten statt, die in modernen zoologischen Übersichtswerken ausgeklammert bleiben: Sie betreffen die Nutzung des Tieres durch den Menschen. In modernen Zoologien leitet man die Berechtigung der zoologischen Deskription nicht aus dem Maß ab, in dem das betreffende Tier dem Menschen nützt. Vielmehr ist man sich der Tatsache bewusst, dass zahlreiche Tierarten von der Existenz des Menschen bedroht werden. Moderne Zoologie ist bis zu einem gewissen Grad die Wissenschaft von gefährdeten Lebewesen. Ein fester Bestandteil von Übersichtswerken ist deshalb eine normierte Angabe des Gefährdungsgrades der beschriebenen Species (IUNC-Status). Zahlreiche Säugetierarten zählen zu den gefährdeten Species, besonders unter den Karnivoren, die als Spitzen der Nahrungskette sowohl mit dem

[126] Ebd. 652–660.
[127] Ebd. 656.

Menschen konkurrieren als auch aufgrund der Lebensweise ihrer Beutetiere oft weitläufige Jagdterritorien benötigen, welche von der modernen Zivilisation eingeschränkt oder beseitigt werden. Zum Beispiel wird die Species *Panthera leo* mit dem Prädikat „gefährdet" versehen, ebenso wie der Nebelparder (*Neofelis nebulosa*), die Chilenische Waldkatze (*Felis guigna*), die Asiatische Goldkatze (*Felis temmincki*), die Fischkatze (*Felis viverrinus*), die Rostkatze (*Felis rubiginosus*), der ostasiatische Rothund (*Cuon alpinus alpinus*), der westasiatische Rothund (*Cuon alpinus hesperius*) und der Waldhund (*Speothos venaticus*); der Tiger (*Panthera tigris*) sogar mit dem Prädikat „stark gefährdet" ebenso wie der Gepard (*Acinonyx jubatus*), der Schneeleopard (*Panthera uncia*), der Pardelluchs (*Felis pardinus*), die Iriomotekatze (*Felis iriomotensis*), die Bergkatze (*Felis jacobitea*) und die Borneo-Goldkatze.[128] Die Löwen-Subspecies *Leo persica* ist unmittelbar vom Aussterben bedroht, ebenso wie der Afrikanische Wildhund (*Lycaon pictus*), der Abessinische Fuchs (*Canis simensis*) und drei Subspecies des Leoparden. Drei der insgesamt 8 Tigerunterarten sind ausgestorben.

Diese Liste, die keineswegs exhaustiv ist und nur als Beispiel dienen soll, zeigt die Ausrichtung des Diskurses an. In diesem wird von Artenschutzmassnahmen, Artenrettungs- und Rückzüchtungsprogrammen berichtet (z.B. in Bezug auf den Europäischen Wolf, den Berberlöwen, den Abessinischen Fuchs, den Afrikanischen Wildhund, den Rothund usw.). Der Diskurs wird durch ein verstärktes Verantwortungsbewusstsein des Menschen charakterisiert, der den Artenschutz als moralischen Imperativ betrachtet. Ein Beispiel für die diesbezügliche zoologische Ökorhetorik: „Die Erhaltung und Wiedereinführung überlebensfähiger Wolfspopulationen als Bestandteil europäischer Ökosysteme ist eine der größten Herausforderungen für den europäischen Artenschutz im nächsten Jahrtausend".[129]

Wie anders Gesner! Dass Gesner gleich vier Kategorien (E–H) der Tiernutzung widmete, zeigt an, dass dies ein diskursbestimmendes Merkmal ist. Daraus geht hervor, dass die Tierdeskription nicht als Selbstzweck gedacht ist. Die Ausführungen zu den Namen der Tiere, zu ihrem Verbreitungsgebiet, zu Morphologie, Physiologie und Psychologie führen organisch zur Auseinandersetzung des Nutzens hin. Der Nutzen erscheint damit gewissermaßen von übergeordneter Bedeutung. Das Tier soll dem Menschen zur Mobilität, zum Zeitvertreib,

---

[128] Macdonald, *Enzyklopädie der Säugetiere* 11, 36–37, 66.
[129] Ebd. 45.

zur Kleidung, zur Düngung, zum Krieg, weiter als Schutz, Nahrung und Heilmittel dienen. Es ist kein Zufall, dass die Tiere, welche dem Menschen den meisten Nutzen bringen, am ausführlichsten behandelt werden. Das „nützlichste" Tier, das Pferd, beansprucht bei Gesner das längste Lemma: 187 Folioseiten (442–619)!

In welchem Masse das Dogma der Tiernutzung Gesners zoologischen Diskurs bestimmte, lässt sich vielleicht noch signifikanter anhand einer Species wie *Panthera leo* zeigen. Denn die katzenartigen Raubtiere zählen zu den am wenigsten nützlichen Tieren. Der Löwe ist kein Reittier, lässt sich kaum zähmen und nur schwer halten, sein Fleisch ist ungenießbar (während er im Gegenteil viel Fleisch benötigt) und sein Kot eignet sich nicht zur Düngung. Außerdem eignet sich der Löwe nicht als Milchtier, schon weil er sich aufgrund seiner Aggression nicht melken lässt. Überhaupt wird sein Nutzen von vorneherein schon dadurch praktisch verunmöglicht, dass er zu Gesners Zeit in Europa nicht mehr vorkam.

Wenn man dies berücksichtigt, ist umso erstaunlicher, dass Gesner die betreffenden Kategorien dennoch behandelt. In Kategorie E behandelt er den Löwen auf ca. 2 Folioseiten als Jagdobjekt. Das mutet seltsam an, da der Löwe nicht nur in Gesners Schweiz, sondern auch im übrigen Europa nicht als Jagdtier vorrätig war. Einige seltene Exemplare wurden in Menagerien zu Schauzwecken gehalten. Man hütete sich jedoch, die wertvollen Schauobjekte aufs Spiel zu setzen. Wenn Gesner den Löwen als Jagdobjekt darstellen wollte, musste er auf die Antike zurückgreifen.[130] Besonders Oppians Jagdgedicht *Kynegetica* stellt diesbezüglich eine ergiebige Quelle dar. Aus Oppian übernimmt Gesner verschiedene Arten der Löwenjagd: Mit Pferden, Hunden, Netzen, Fackeln, Lärm,[131] mit Schildern, Peitschen und Stricken[132] sowie mit Ködern und Fallgruben.[133] Noch merkwürdiger sind Gesners Angaben zum Löwen als Jagdgefährten. Man soll ihn zur Jagd abrichten können wie den Hund. Wie ein Jagdhund sei der Löwe imstande, Beute aufzuspüren. Weiter lasse er sich wie ein Pferd vor den Wagen spannen. Diese unrealistischen Daten hat Gesner aus antiken Quellen und aus den subrezenten Reiseberichten des Marco Polo bezogen.

---

[130] Vgl. Steier, Art. „Löwe", Sp. 978–980.
[131] *Kynegetica* IV 112 ff.
[132] Ebd. IV 147–212. Auf diese Weise soll man in Äthiopien Löwen gejagt haben.
[133] Ebd. IV 77–111.

Die kulinarische Abteilung musste notgedrungen kurz bleiben, da Löwenfleisch, wie das der meisten Karnivoren, ungenießbar ist. Bezeichnend ist jedoch, dass sich Gesner gezwungen fühlte, die Gründe zu erarbeiten, weshalb man Löwenfleisch lieber nicht essen sollte. Das Löwenfleisch besitze eine außerordentlich trockene Komplexion, weshalb sein Verzehr viel „schwarze Galle" erzeuge, die ihrerseits zu Leibschmerzen, Bauchkrämpfen und Flatulenz führe („dolorem et torsionem et inflationem ventris generat"). Die kuriose Sachlage – es war in Europa kein Löwenfleisch vorrätig, Löwenfleisch ist überhaupt ungenießbar – zeigt das zugrundeliegende Diskursdogma – Tiere sind zum Verzehr durch den Menschen bestimmt – umso klarer auf.

Dasselbe gilt mutatis mutandis vom Löwen als Lieferanten von Heilmitteln: Obwohl Löwen nicht in ausreichendem Maß vorrätig waren, um Löwenarzneien zu produzieren und obwohl dem praktizierenden Arzt Gesner die geringe medizinische Relevanz derselben bewusst gewesen sein muss, widmet er dem Thema eine Folioseite. Löwenfleisch eigne sich als Medizin für Paralytiker. Es soll dazu dienen, das Aufkommen von Trugbildern („phantasmata") zu verhindern. Ein Streifen Löwenfell soll gegen Fußschmerzen helfen; Löwenblut und Löwenfett soll vor Angriffen wilder Tiere schützen. Schmerzen, die durch Hämorrhoiden verursacht werden, können durch das Sitzen auf einem Löwenfell gelindert werden. Diese wenig hilfreichen Angaben zeigen a fortiori das Dogma, das Gesners zoologischen Diskurs bestimmt. *Das Tier wird bis zum letzten Rest genützt – ausgeschlachtet.* Dabei geht es nicht immer um den tatsächlichen praktischen Wert, sondern ums Prinzip.

Dieses Diskursmerkmal zeigt einmal mehr auf, dass vieles ausgeblendet wird, wenn man die frühneuzeitlichen Zoologien wissenschaftsteleologisch als Vorläufer oder Wegbereiter der modernen Zoologie betrachtet. Das Bild, das sich aus der Diskursanalyse der Großprojekte Wottons und Gesners ergibt, ist durchaus vielgestaltiger, reicher, komplexer und voll von Gegensätzen und Diskontinuitäten. Es liegt keine eindeutige, ‚fortschrittsbezogene', geradlinige oder schrittweise Entwicklung vor. Der Vergleich der beiden, ungefähr gleichzeitig ausgeführten Projekte führt zudem vor, wie die Wissenschaft der Zoologie zur selben Zeit auf ganz unterschiedliche Weise konstituiert wurde. Es lohnt sich, die Vielgestaltigkeit, die Gegensätze und Diskontinuitäten zu registrieren. Dadurch wird es möglich, die Wissenschaftsorganisation der frühen Neuzeit in ihrer schillernden Vielfarbigkeit und anregenden Alterität zu verstehen.

Auswahlbibliographie

ACKERMANN J.M., „Early Renaissance ‚naturalism' and scientific illustration", in Ellenius A. (Hrsg.), *The Natural Sciences and the Arts* (Uppsala: 1985) 1–17.

ASHWORTH W.B. Jr., „Natural history and the emblematic world view", in Lindberg D.C. – Westmann R.S. (Hrsg.), *Reappraisals of the Scientific Revolution. Journal of the History of Ideas* 52 (1991) 373–398.

——, „Emblematic natural history of the Renaissance" in Jardine N. – Secord J.A. – Spary E.C. (Hrsg.), *Cultures of Natural History* (Cambridge: 1996) 17–37.

BALME D.M., "Aristotle's use of division and *differentiae*", in Gotthelf A. – Lennox J.G. (eds.), *Philosophical Issues in Aristotle's Biology* (Cambridge: 1987) 69–89.

BALSS H., *Albertus Magnus als Biologe* (Stuttgart: 1947).

BARANTAY E. – HARDOUIN-FUGIER E., *Zoos. Histoire des jardins zoologiques en l'Occident XVIe–XXe siècle* (Paris: 1998).

BÄUMER A., *Die Biologie von der Antike bis zur Renaissance*, in Dies., *Geschichte der Biologie*, Bd. 1 (Frankfurt a.M. – Bern – New York – Paris: 1991).

——, *Zoologie der Renaissance – Renaissance der Zoologie*, in Dies., *Geschichte der Biologie*, Bd. 2 (Frankfurt a.M. – Bern – New York – Paris: 1991).

——, „Das erste zoologische Kompendium in der Zeit der Renaissance: Edward Wottons Schrift ‚Über die Differenzen der Tiere'", *Berichte zur Wissenschaftsgeschichte* 13 (1990) 13–29.

BÄUMER-SCHLEINKOFER A., „Zoologie der Renaissance – Renaissance der Zoologie", in Döring K. – Wöhrle G. (Hrsg.), *Antike Naturwissenschaft und ihre Rezeption*, Bd. I und II (Bamberg: 1992) 275–295.

——, „Biblische Zoologie (Hermann Frey, Wolfgang Franz, Heinrich von Hoevel)", in Kattenstedt H. (Hrsg.), *‚Grenzüberschreitung'. Festschrift zum 70. Geburtstag von Manfred Büttner* (Bochum: 1993) 3–17.

BAY J.CH., „Conrad Gesner (1516–1565), the Father of Bibliography. An Appreciation", *Papers of the Bibliographical Society of America* 10 (1958) 53–88.

BEAGON M., *Roman Nature: the Thought of Pliny the Elder* (Oxford: 1992).

BERENGARIO DA CARPI, *Commentaria cum amplissimis additionibus super anatomia Mundini* [...] (Bologna: 1521).

——, *Isagoge brevis* (Bologna: 1522). Engl. Übers. *A Short Introduction to Anatomy*, trans. Lind L.P. (Chicago: 1959).

BOAS M., *The Scientific Renaissance 1450–1630* (New York: 1962).

BYLEBYL J., "The School of Padua : Humanistic Medicine in the 16th Century", in Webster Ch. (Hrsg.), *Health, Medicine and Mortality in the 16th Century* (Cambridge: 1979) 335–370.

CASTIGLIONI A., „The Origin and Development of the Anatomical Theatre to the End of the Renaissance", *Ciba Symposium* 3 (1941) 826–844.

CÉARD J., „Analogie et zoologie chez naturalistes de la Renaissance", in Christian L.C. (Hrsg.), *Theatrum mundi. The History of an Idea* (New York: 1987).

COOK H.J., „Physick and natural history in the 17th century", in Ariew R. – Barker P. (Hrsg.), *Revolution and Continuity: Essays in the History of Philosophy of the Early Modern Science* (Washington: 1991) 63–80.

——, „Physicians and natural history" in Jardine N. – Secord J.A. – Spary E.C. (Hrsg.), *Cultures of Natural History* (Cambridge: 1996) 91–105.

CUNNINGHAM A., *The Anatomical Renaissance* (Aldershot: 1997).

DAHL F., „Zur Geschichte der Zoologie. Von Aristoteles bis Plinius", *Sitzungsberichte der Gesellschaft Naturforschender Freunde zu Berlin* 50 (1924) 62–104.

DELAUNY P., *La Zoologie au XVIe siècle*, Histoire de la pensée 7 (Paris: 1962; reprint Paris: 1997).

DRIESCH A. VON DEN, *Geschichte der Tiermedizin: 5000 Jahre Tierheilkunde* (München: 1989).

DURLING R.J., „Conrad Gesner's Liber amicorum, 1555–1565", *Gesnerus* 22 (1965) 134–159.

FALOPPIO GABRIELE, *Observationes anatomicae* (1561), in Ders., *Opera genuina omnia*, 3 Bde. (Venedig: 1606).

FINDLEN, P., *Possessing Nature. Museums, Collecting and Scientific Culture in Early Modern Italy* (Berkely: 1999).

FISCHER H., „Conrad Gessner"; „Conrad Gessner als Bibliograph" und „Conrad Gessner als Arzt", in Ders. et alii, *Conrad Gessner 1516–1565. Universalgelehrter, Naturforscher, Arzt* (Zürich: 1967) 11–22; 31–37; 39–47.

——, *Conrad Gesner. Leben und Werk* (Zürich: 1966).

——, „Conrad Gesner als Arzt", *Gesnerus* 22 (1965) 205–215.

FRANZIUS WOLFGANG, *Historia animalium sacra, in qua plerorumque animalium praecipuae proprietates in gratiam studiosorum Theologiae et ministrorum verbi ad usum eikonologikon breviter accommodantur, in Academia Witenbergensi* [...] dictata (Wittenberg, Zarcharias Schurer und Johannes Gormann: 1612).

FRENCH R.K., *Ancient Natural History: Histories of Nature* (London: 1994).

——, *Dissection and Vivisection in the European Renaissance* (Aldershot: 1999).

FRENCH R.K. – WEAR A. (Hrsg.), *Medical Revolution in the 17th century* (Cambridge: 1989).

GESNER CONRAD, *Historiae animalium I de quadrupedis viviparis* [...] (Zürich, Christoffel Froschauer: 1551).

——, *Icones animalium quadrupedum vivparorum et oviparorum quae in Historiae animalium Conradi Gesneri libro I et II describuntur* [...] (Zürich, Christoffel Froschauer: 1553).

——, *Thierbuch, das ist ein kurze beschreybung aller vierfüssigen Thieren, so auff der erden und in wassern wonend, sampt irer waren conterfactur: alles zu nutz und gutem allen liebhabern der künsten, Artzeten, Malern, Bildschnitzern, Weydleuten und Köchen gestelt. Erstlich durch Cunrat Gesner in Latin beschriben, yetzunder aber durch Cunrat Forer zu mererem nutz aller mengklichen in das Teutsch gebracht, und in ein kurtze komliche ordnung gezogen* (Zürich, Christoffel Froschauer: 1563).

GILLES PIERRE, *Ex Aeliani historia per Petrum Gyllium latini facti itemque ex Porphyrio, Heliodoro, Oppiano, tum eodem Gyllio luculentis accessionibus aucti libri XVI de vi et natura animalium. Eiusdem Gyllii liber unus de Gallicis et Latinis nominibus piscium* (Lyon, Sebastian Gryphius: 1535).

GMELIG-NIJBOER, C.A., *Conrad Gesner's „Historia Animalium": An Inventory of Renaissance Zoology* (Diss. Utrecht: 1977).

——, „Conrad Gesner (1516–1565) Considered as a Modern Naturalist", *Janus* 60 (1973) 41–51.

GOULD E., MCKAY G., *Enzyklopädie der Tierwelt. Säugetiere* (München: 2002; urspr. englisch, Sidney: 1990).

GUERRINI A., *Experimenting with Humans and Animals. From Galen to Animal Rights* (Baltimore – London: 2003).

HARMS W., „On natural history and emblematics in the 16th century", in Ellenius A. (Hrsg.), *The Natural Sciences and the Arts* (Uppsala: 1985) 67–83.

HELMCKE J.G., „Der Humanist Conrad Gesner auf der Wende von mittelalterlicher Tierkunde zur neuzeitlichen Zoologie", *Physis* 12 (1970) 329–346.

HERR MICHAEL, *Gründtlicher Underricht, wahrhaffte und eygentliche beschreibung wunderbarlicher seltsamer art, natur, krafft und eygenschafft aller vierfüssigen thier, wild und zahm, so auff und in der erden oder wassern wonen* [...] (Straßburg, Balthasar Beck: 1546).

HÜNEMÖRDER Ch., „Aims and Intentions of Botanical and Zoological Classification in the Middle Ages and in the Renaissance", *History and Philosophy of the Life Science* 5 (1983) 53–67.

JAHN I. (Hrsg.), *Geschichte der Biologie* (Hamburg: 2004; sonderausgabe nach dritter Auflage Jena: 1998).
——, *Grundzüge der Biologiegeschichte* (Jena: 1990).
JARDINE N. – SECORD J.A. – SPARY E.C. (Hrsg.), *Cultures of Natural History* (Cambridge: 1996).
KAUFMANN T.D., *The Mastery of Nature. Aspects of Art, Science and Humanism in the Renaissance* (Princeton: 1993).
KELLER O., *Die antike Tierwelt*, 2 Bde., I (Leipzig: 1909), II (Leipzig: 1913), Register (1920). Nachdruck (Hildesheim: 1963).
——, *Thiere des classischen Alterthums in culturgeschichtlicher Beziehung* (Innsbruck: 1887; Nachdruck Hildesheim: 1983).
KNIGHT D., *Zoological Illustration. An Essay towards a History of Printed Zoological Pictures* (Folkstone: 1977).
LEITNER H., *Zoologische Terminologie beim älteren Plinius* (Hildesheim: 1972).
LENNOX J.G., „Between Data and Demonstration: the Analytics and the *Historia Animalium*", in: Idem, *Aristotle's Philosophy of Biology: Studies in the Origins of Life Science* (Cambridge: 2001) 39–71.
LEU U.B., „Theologia naturalis", in Ders., *Conrad Gesner als Theologe. Ein Beitrag zur Zürcher Geistesgeschichte des 16. Jahrhunderts* (Bern – Frankfurt a.M. – New York – Paris: 1990; Zürcher Beiträge zur Reformationsgeschichte 14) 31–100.
LEY W., Conrad Gesner. *Leben und Werk* (München: 1926).
LLOYD G.E.R., *Science, Folklore, and Ideology. Studies in the Life Sciences in Ancient Greece* (Cambridge: 1983).
——, "The Development of Aristotle's Theory of the Classification of Animals", *Phronesis* 6 (1961) 59–81.
LLOYD J.E. Barclay, *African Animals in Renaissance Literature and Art* (Oxford: 1971; New York: 1972).
LONITZER ADAM, *Naturalis historiae opus novum, in quo tractatur de natura et viribus arborum, fructicum, herbarum animantiumque terrestrium, volatilium et aquatilium, item gemmarum, metallorum succorumque concretorum* (Frankfurt a.M., Christoph Egenolph: 1551–1555).
MACDONALD D. (Hrsg.), *Enzyklopädie der Säugetiere* (Königswinter: 2003; urspr. englisch, Oxford: 2001).
MAYERHÖFER J., "Conrad Gesner als Bibliograph und Enzyklopädist", *Gesnerus* 22 (1965) 176–195.
MAYR E., *Systematics and the Origin of Species* (New York: 1942).
——, *Animal Species and Evolution* (Cambridge Mass.: 1963).
MESNARD P., *L'horizon zoologique de la Renaissance. Les animaux anciens et les animaux modernes*, in Sciences de la Renaissance. VIIe congrès international de Tours (Paris: 1973).
MEYER H., „Frühe Neuzeit", in Dinzelbacher P. (Hrsg.), *Mensch und Tier in der Geschichte Europas* (Stuttgart: 2000) 293–403.
MILT B., „Zürichs Vergangenheit in Naturwissenschaft und Medizin (Mittelalter und 16. Jahrhundert", *Gesnerus* 4 (1947) 19–43.
NISSEN C., *Die naturwissenschaftliche Buchillustration*, 2 Bde. (Bad Münster am Stein: 1950).
——, *Die zoologische Buchillustration, ihre Bibliographie und Geschichte* (Stuttgart: 1971).
OGILVIE B., *The Science of Describing. Natural History in Renaissance Europe* (Chicago – London: 2006).
PARK K., „The Criminal and Sainly Body: Autopsy and Dissection in Renaissance Europe", *Renaissance Quarterly* 47 (1994) 1–33.
PELLEGRIN P., *La classification des animaux chez Aristote* (Paris: 1982).
PETIT G., „Conrad Gessner, zoologiste" in Fischer H. et alii, *Conrad Gessner 1516–1565. Universalgelehrter, Naturforscher, Arzt* (Zürich: 1967) 49–56.
——, „Conrad Gesner zoologiste", *Gesnerus* 22 (1965) 195–205.
PETIT G. – THEODORIDES J., *Histoire de la zoologie des origines à Linné* (Paris: 1962).

PLINIUS, GAIUS PLINIUS SECUNDUS DER ÄLTERE, *Naturkunde. Lateinisch – deutsch. Buch VIII. Zoologie: Landtiere.* Herausgegeben und übersetzt von Roderich König in Zusammenarbeit mit Gerhard Winkler (Kempten: 1976).

——, *Naturkunde. Lateinisch – deutsch. Buch XI. Zoologie: Insekten. Vergleichende Anatomie.* Herausgegeben und übersetzt von Roderich König in Zusammenarbeit mit Joachim Hopp (München – Zürich: 1990).

RAVEN CH. E., *English Naturalists from Neckham to Ray: A Study of the Making of the Modern World* (Cambridge: 1947).

SARTON G., *The Appreciation of Ancient and Medieval Science during the Renaissance (1450–1600)* (Philadelphia Penn.: 1953).

STEIER A., Art. "Löwe", in Wissowa G. – Kroll W., *Paulys Encyclopädie der classischen Alterthumswissenschaft* XIII,1 (1926) Sp. 968–990.

——, *Aristoteles und Plinius. Studien zur Geschichte der Zoologie* (Würzburg: 1913).

SUUTALA M., *Tier und Mensch im Denken der Deutschen Renaissance* (Helsinki: 1990).

THOMAS K., *Man and the Natural World: Changing Attitudes in England 1500–1800* (London: 1983).

VELDE A.J.J. VAN DE, „Rond Gesner's *Historiae animalium liber I* van 1551", *Mededelingen van de Koninklijke Vlaamse Academie voor wetenschappen letteren en schone kunsten van België. Klasse der Wetenschappen* XIII,17 (Brüssel: 1951).

WEAR A. – FRENCH R.K. – LONIE I.M. (Hrsg.), *The Medical Renaissance of the 16th Century* (Cambridge: 1985).

WOTTON EDWARD, *De differentiis animalium libri decem* (Paris, Michel de Vascosan: 1552).

ON TOUCANS AND HORNBILLS: READINGS IN EARLY MODERN ORNITHOLOGY FROM BELON TO BUFFON*

Paul J. Smith

In *Les Delices de Leide*,[1] an anonymous travel guide from 1712, French tourists and 'autres curieux' are invited to take a walk along the curiosities of Leiden. The walk is made up of concentric circles from the outer canals to the town centre. Guidebook in hand, the tourist arrives at the well-known Anatomical Theatre. At that moment – in the centre of the town and in the middle of the book – the very readable style changes dramatically into a catalogue: seventy pages listing and numbering every single object of the Anatomical Theatre. Some curiosities are singled out and described in much detail: an Egyptian mummy, a Grecian vase and – the lengthiest description of all – the bill of a hornbill. Paradoxically, Leiden's main curiosities, stored and exhibited in the town centre, turn out to be exotic objects that transport the visitor away from the centre to remote times and far away countries. It thus announces the image that Holland will have in the coming centuries until at least Baudelaire's 'pays de Cocagne': a little, neat country furnishing an entrance to the outer world.

In the description of the hornbill (see appendix), some strange things happen. From the very opening of the text the bird is confused with another species – the toucan –, and authorities are quoted from times long before both birds were known to the learned world: Pliny, Hesychius of Alexandria (5th century AD), and Varinus Phavorinus (16th century). Another sixteenth-century authority, André Thevet, in fact never spoke of hornbills, only of toucans. There is also a strange anecdote about a long-billed bird shot at the battle of Lepanto (1571) in the Corinthian Sea – a place where neither toucans nor hornbills are found.

This curious text will serve us as a point of reference, not only for its own sake, but also because its broad intertext will provide a more

---

* I thank Peter Mason (Rome) for having corrected my English.

[1] Anonymus, *Les Delices de Leide, Une des célèbres villes de l'Europe* [...] (Leiden: 1712). See Smith P.J., "Wandelen in de *Délices de Leide* (1712)", *De zeventiende eeuw* 22 (2006) 185–208.

general insight in early modern ornithology and into the sometimes confusing way in which unknown, exotic birds – from both the West and the East Indies – are seen, described, classified and thus introduced into Western scientific discourse.

For the sake of clarity, it is necessary to give some brief relevant ornithological information on the birds under discussion. Toucans are birds with enormous, brightly coloured bills, sometimes almost as long as their bodies. They are found wild only in American tropical forests. Within the suborder of *Piciformes* (woodpeckers and picarian allies), they form a family of thirty-seven species (from the largest crow-sized species of the genus *Ramphastos* to the lesser toucanets and aracaris). Like woodpeckers, toucans have two toes in front and two behind.

Hornbills only superficially resemble toucans in the size of their bills, but these are less fancifully coloured and are usually surmounted by a horny casque. The hornbill family consists of forty-five species, all from the Old World. Except for the two species of African ground hornbills, which forage in open country, they live in the tropical forests of Africa and Asia. According to current ornithological taxonomy, the hornbills (*Bucerotidae*) belong to the suborder of the *Coraciiformes*, to which kingfishers, bee-eaters, rollers and hoopoes also belong, characterised, as can be read in modern manuals, 'by having their three front toes joined for part of their length, a condition known as "syndactyly"'.[2] Compared to the toucans, the size of some hornbills is enormous, especially the Great Hornbill (*Buceros bicornis*, 120 cm), the Rhinoceros Hornbill (*Buceros rhinoceros*, 110 cm), and the Wreathed Hornbill (*Aceros undulatus*, 100 cm). A bird in flight is impressive, not only for its format but also because of its noise: each wing stroke 'makes a loud "whoosh" that sounds in the distance exactly like the chuffing of an old steam locomotive'.[3] This characteristic can explain why, in the heat of the battle of Lepanto, the hornbill mentioned in the *Delices* was spotted and shot.

Although to modern eyes the two families are very distinct from each other, when they were first introduced to Europe toucans and hornbills were confused by early modern zoologists, as can be deduced from *Les Delices de Leide*. The birds have this in common with other non-related species that more or less resemble one other by way of 'convergent

[2] Austin O.L. Jr., *Birds of the World. A survey of the twenty-seven orders and one hundred and fifty-five families* (London: 1962) 175.
[3] Austin, *Birds of the World* 184.

evolution', such as the Antarctic penguins and the Arctic auks (especially the extinct Great Auk), the New World armadillos and the Old World pangolins, and the Old and New World warblers. Their origins are also unknown because the trade routes from the East and the West mostly went via West Africa, where cargos were transshipped, so that it was often difficult to tell from which continent the exotic material originated. This confusion is still clear from the current names of some well-known animals: the English name *guinea pig* ignores the animal's South American origin, and the word *turkey* suggests that the bird comes from Turkey and not from North America, – just as the bird's French name *dindon* or *poule d'Inde* suggests India, and the Dutch name *kalkoen* the town Calcutta as the bird's place of origin.

The following is a chronological discussion of the principal texts on both families, beginning in the sixteenth-century – the age of their discovery – and ending with Buffon, who wrote the most influential ornithology of the Enlightenment. This corpus enables us to outline the history of the scientific knowledge of both bird families. But as a case study it also hopes to show more: because the newly discovered birds were incorporated in the prevailing ornithological discourse and taxonomy, thus making continuous revisions necessary, this teaches us a lot about early modern zoology. As we shall see, the rare, fragmentary and transitory material and the scattered oral, hand-written and printed information about the birds underline the need for every zoologist to maintain a wide and well-functioning social and scientific network and to have access to collections of non-professionals.

The first toucans

The toucan was known in Europe long before the hornbill, although its entry into European ornithology was hesitant. The first description of the bird comes from the Spanish naturalist Gonzalo Fernández de Oviedo y Valdés (1478–1557) in his *Sumario de la natural y general historia de las Indias* (1526), quoted here in a modern English translation:

> There is in Tierra Firme a bird that the Christians call *picudo* ['big bill,' toucan] because it has a small body and large beak. The beak weighs more than the whole body. This bird is only slightly larger than a quail, but it appears much larger because of its beautiful thick varicoloured plumage. Its turned-down beak is as much as six inches long. At the head the beak

> is about three finger-breadths wide. The bird's tongue is a feather,[4] and it can whistle loudly. With its beak it makes holes in trees, where it nests and raises its young. It is a very strange bird, and different from any other that I have seen because of its feather tongue, its appearance, and its disproportionately large bill. No bird can so surely and safely protect its young from cats. Because of the location of the nests, cats cannot enter the hole in the tree to take the eggs or young birds. When the birds hear a cat they get into the nest and turn their bills out, and give the cat such pecks that the cat deems it wise not to molest the young birds.[5]

Oviedo's very interesting description of the toucan remained unnoticed until at least the work of the German scholar Johannes Faber (1628),[6] extensively quoted and commented upon in the *Ornithology* (1678) by Francis Willughby and John Ray (see below). The influential ornithological work by Conrad Gesner (1516–1565), *Historiae animalium liber III, qui est de avium natura* (1555), does not mention the toucan (or the hornbill). We shall see that his first mention of a *Rhampast* only occurs in his *Icones avium* (1560). The bird became more generally known to zoologists through the description and illustration of its bill given by Pierre Belon du Mans (ca. 1517–1565) in his influential *Histoire de la nature des oyseaux* (1555) [Fig. 1].[7] Although Belon only had the bird's bill at his disposal, while the rest of the bird was unknown to him – the illustrated bill clearly belongs to a Toco Toucan (*Ramphastos toco*), the biggest and most common of the toucan family –, his brief description was to remain, as we shall see, highly authoritative until the eighteenth century. He begins by indicating that the bird is unknown to the Ancients – a detail which is not without importance, because early ornithology defines itself in regard to the Ancients, often affirming or amending the texts of Aristotle, Pliny and others. He continues

[4] On the bird's tongue, whose strangeness will be developed by Buffon, see Austin, *Birds of the World* 189: 'The tongue, which extends almost to [the bill's] tip, is very thin and narrow, rather stiff and hard, and fringed on both sides with bristles. The function of these bristles is uncertain, and perhaps as ambiguous as the bill itself'.

[5] Gonzalo Fernández de Oviedo y Valdez, *Natural History of the West Indies*, transl. and ed. Stoudemire S.A. (Chapel Hill: 1959) 68.

[6] Before Faber, the Italian scientist Gerolamo Cardano proved to have read Oviedo, although wrongly, because he amalgamated Oviedo's description of the tongue ('Illud in eo mirum, quod pennam habent loco linguae') with Oviedo's chapter on the *alcatrez* (the Brown Pelican [*Pelecanus occidentalis*]). Gerolamo Cardano, *De subtilitate* (Basle: 1554) 337.

[7] Quotations are from Pierre Belon du Mans, *L'Histoire de la nature des oyseaux. Facsimilé de l'édition de 1555*, ed. Glardon P. (Geneva: 1997).

184 LIVRE III. DE LA NATVRE

Du bec d'vn oyſeau des terres neufues,incognu aux anciens.

CHAP. XXVIII.

CEVX qui nauigent aux terres neufues, faiſants leur profit de toutes choſes,apportent ce qu'ils trouuent de bon,pour vendre aux marchands.Or eſt-ce qu'il y à vn oyſeau en ce païs lá,ayát le bec long de demy pied,gros cóme le bras d'vn enfant, poinctu & noir par le bout, mais blanc en touts autres endroits, & quelque peu coché par les bords.Il eſt creux par dedens : eſtant ſi finement delié, qu'il en eſt tranſparent & tenue comme parchemin, & par ce eſt moult legier.Ceſt ſa beauté qui fait qu'on en voit ia pluſieurs par les cabinets des hommes curieux de choſes nouuelles: Car au demeurant,lon ne s'en ſert à aucune choſe.Et nous n'ayants veu l'oyſeau qui l'à produit, n'en pouuons dire autre choſe,ſinon que par ſoupçon le penſons eſtre de pied plat. Et par ce l'auõs mis en ceſt endroit auec les oyſeaux de riuiere. Mais pour faire voir quel eſt ce bec, en auons cy mis le portraict.Il eſt ſeul entre touts ceux qu'auons obſeruez, à qui n'ayons veu conduicts pour odorer.

Portraict d'vn bec d'oyſeau apporté des terres neufues.

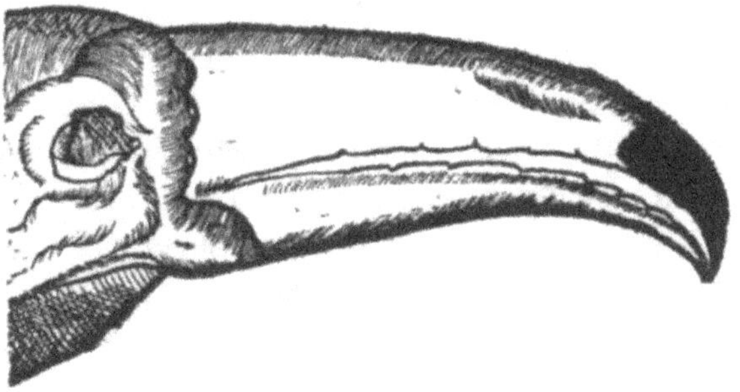

FIN DV TROISIESME LIVRE.

Fig. 1. Skull of a Toco Toucan from Brazil. From Pierre Belon du Mans, *L'Histoire de la nature des oyseaux* (Paris: 1555).

by remarking that the bird is a popular merchandise, which is another detail expanded upon by Belon's first readers: Thevet and Léry.

Belon's brief but rather accurate description of the bill is followed and developed by Aldrovandi and many others. Thus the bill's slightly crenate edges and its amazing lightness due to its parchment structure ('Il est creux par dedans: estant si finement delié, qu'il en est transparent & tenue comme parchemin, & par ce est moult legier') are details often repeated after Belon. Furthermore, Belon did not discover any nostrils ('Il est le seul [oiseau] entre touts ceux qu'avons observez, à qui n'ayons veu conduicts pour odorer') – the bird's nostrils, at the base of its bill, are indeed difficult to find. From Aldrovandi on until the description in the *Delices de Leide*, this anomalous lack of nostrils is explained by the two other characteristics mentioned by Belon: the bird does not need any nostrils to breathe because the bill's material is transparent; and its edges make it impossible for the bird to close the bill completely. Even the hypothesis that the unknown bill belongs to a webbed-footed river bird – a hypothesis already refuted by Thevet – will leave its traces: until Buffon an aquatic species of toucan or hornbill, a *hydrocorax*, often emerges.

What Belon left implicit is the question of why the bird is placed in the second ('waterfowl') of the six ornithological groups that he distinguished.[8] The reason for this placement is probably due to the lengthy, saw-edged mandibles – characteristics Belon also noted for webbed-footed fish-eaters such as pelicans, mergansers, and some other species of ducks. And Belon's statement that the bill can be seen in many *cabinets de curiosités* is particularly important for our argument, since it shows the interdependency of private collections and zoological treatises. As we shall see in more detail, the 'cabinets des hommes curieux de choses nouvelles' often furnished the material for zoology, just as, inversely, zoological treatises were often at the basis of collections.

Thevet, Léry, Gesner and Aldrovandi on toucans

André Thevet is the first to give a description of the toucan, complete with illustration. Thevet (1516–1592) was a Franciscan and author of

[8] These groups are: raptors, waterfowl, shore birds, terrestrial birds, large arboreal birds, small brushwood birds.

some influential 'cosmographic' works, partly based on his own travel experiences. After travelling to the Middle East he wrote a *Cosmographie de Levant* (1554). In 1555–1556 he briefly sojourned in 'Antarctic France', *i.e.* Brazil. Upon falling ill at his arrival at Rio de Janeiro, he had to return to France after only ten weeks. In spite of his brief stay, he managed to acquire enough material to write his *Singularitez de la France Antarctique* (1557). In this book he gives a description of the toucan, clearly a response to the recently published description of Belon, though without naming him. Like Belon, Thevet begins by mentioning the bird's quality as merchandise, not for its bill, but for its plumage. He goes on to give the name of the bird and a brief description of the bird's feathers:

> Sur la coste de la mer, la plus frequente marchandise est le plumage d'un oiseau, qu'ils appellent en leur langue *Toucan*, lequel descrirons sommairement, puis qu'il vient à propos. Cest oyseau est de la grandeur d'un pigeon. Il y en a une autre espece de la forme d'une pie, de mesme plumage que l'autre: c'est à sçavoir noirs tous deux, hors-mis autour de la queuë, ou il y a quelques plumes rouges, entrelacées parmy les noires, soubs la poitrine plume iaune, environ quatre doigts, tant en longueur que largeur: & n'est possible trouver iaune plus excellent que celuy de cest oiseau: au bout de la queuë il a petites plumes rouges comme sang. Les Sauvages en prennent la peau à l'endroit qui est jaune, & l'accommodent à faire des garnitures d'espées à leur mode, & quelques robes, chapeaux, et autres choses.[9]

In order to establish himself as a reliable and highly esteemed authority, Thevet continues: 'J'ai apporté un chapeau fait de ce plumage, fort beau et riche, lequel a été présenté au roi comme chose singulière'. Thevet's superiority over Belon is suggested in the following observation: 'Certains pourraient penser qu'il fût aquatique, ce qui n'est pas vraisemblable comme je l'ai vu par expérience'.

That Thevet is more interested in the bird as a commodity than as a zoological species is clear from the fact that he only mentions *en passant* the bird's bill, without further details, in which, once again, his superiority to Belon is suggested: 'Au reste, cet oiseau est merveilleusement difforme est monstrueux, ayant le bec plus gros et plus long quasi que le reste du corps. J'en ai apporté un qui me fut donné par-delà avec les peaux de plusieurs de diverses couleurs [...]'. This competitive intertextual relationship with Belon's description is confirmed by the

---

[9] André Thevet, *Les Singularitez de la France antarctique* (Paris: 1558) 90–91.

illustration [Fig. 2]: the bird's bill clearly presents the nostrils that Belon could not find. But the nostrils of Thevet's toucan are depicted at a very unlikely place in the bill (in the middle of the upper mandible), and the bird's toes are wrongly represented (three front toes and one back toe) – a mistake that Belon, known as the 'father of comparative anatomy', would not have committed had he known the bird.

It is probable that Thevet received some criticism from his readers. This is clear from his description of the bird in his *Cosmographie universelle* (1575), printed by André Wechel. Indeed, in this prestigious publication, Thevet, recently appointed King's cosmographer, expands on his former description by giving more detailed information on the bird (for instance, as we saw in the *Delices*, the bird's preference for peppers will be a much quoted – and much debated – detail until the eighteenth century). The illustration [Fig. 3] is a remake of the 1557 one, but the nostrils have disappeared, and its toes are represented in such a way that their correct position is more or less visible.

Thevet's 1575 text and illustration, recognizable by the unnatural, triangle-like posture of the whole of bill, head and body, and the bird's large staring eye, is copied again and again by Ambroise Paré, Ulisse Aldrovandi, Jan Jonston, and others. This is remarkable because, at the end of the sixteenth century, the bird was regularly to be seen in *cabinets de curiosités*. For instance, in his book *Des monstres et prodiges* (1573), also printed by Wechel, Ambroise Paré (ca. 1509–1590) first alludes to Thevet's *Singularitez*, and then describes a bird that he had dissected himself:

> Un gentilhomme Provençal en feit present d'un au feu Roy Charles neufiesme, ce qu'il ne peut faire vif, car en l'apportant mourut, neantmoins le presenta au Roy: lequel apres l'avoir veu, commanda à Monseigneur le Mareschal de Rets me le bailler, pour l'anatomiser et embaumer, à fin de le mieux conserver; toutesfois bien tost apres se putrefia. [...] Je le garde comme une chose quasi monstrueuse. La figure duquel t'est icy representee.[10]

But, curiously, Paré's illustration represents not the bird that he dissected, but Thevet's bird. As far as I can see, André Wechel and later Gabriel Buon, Paré's printers, continued to use the same woodcut from Thevet's *Cosmographie*.

[10] Ambroise Paré, *Des monstres et prodiges*, ed. Céard J. (Geneva: 1971) 128–129.

DE LA FRANCE ANTARCTIQVE. 91

Deſcription du Toucan, oyſeau de l'Amerique.

pellent en leur langue *Toucan*, lequel deſcrirons ſommairement, puis qu'il vient à propos. Ceſt oyſeau eſt de la grandeur d'vn pigeon. Il y en á vne autre eſpece de la forme d'vne pie, de meſme plumage que l'autre: c'eſt à ſçauoir noirs tous deux, hors-mis autour de la queuë, ou il y à quelques plumes rouges, entrelacées parmy les noires, ſoubs la poitrine plume iaune, enuirõ quatre doigts, tant en longueur que largeur: & n'eſt poſsible trouuer iaune plus excellét que celuy de ceſt oiſeau: au bout de la queuë il à petites plumes rouges cõme ſang. Les Sauuages en prénent la peau, à l'endroit qui eſt iaune, & l'accommodét à faire garnitures d'eſpées à leur mode, & quelques robes, chapeaux, & autres choſes. I'ay apporté vn chapeau fait de ce plumage, fort beau & riche, lequel a eſté preſenté au

Chapeau eſtrange cõpoſé de plumages.

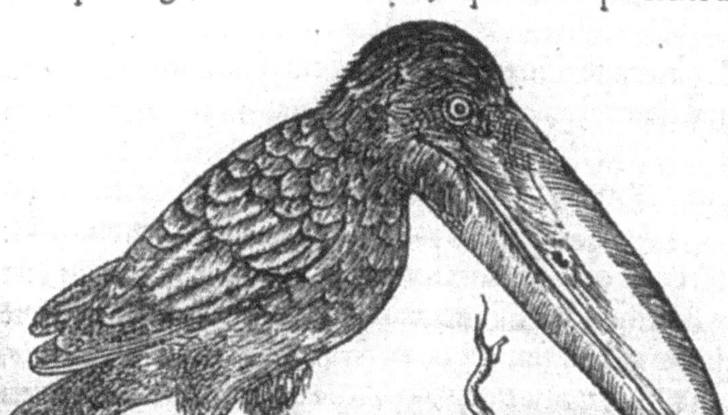

Roy, comme choſe ſinguliere. Et de ces oyſeaux ne ſ'en trouue ſinon en noſtre Amerique, prenant depuis la ri-

z iij

Fig. 2. Toucan. From André Thevet, *Les Singularitez de la France antarctique* (Paris: 1558) fol. 91r.

Coſmographie Vniuerſelle

ſinguliere, digne d'eſtre admiree, veu la gentileſſe de l'œuure, où ces Sauuages font le tiſſu du plumage ſi mignonement auec leur filet d'eſcorce d'arbre, que à grand peine le ſçauroit on faire plus proprement par deça à tout le fil de ſoye. Au reſte, il ne ſe trouue de ces oyſeaux, ſinon vers le Cap de Frie, & où nous eſtions ſur la riuiere de *Ianaire*, prenant neantmoins le cours du païs, depuis la riuiere de *Plate*, iuſques à celle de *Darlane*. Il eſt vray, qu'il ſ'en trouue quelques vns au Peru, mais beaucoup plus petits que les autres. D'en voir en Mexique, Terre neufue, ou à la Floride, n'en faut point parler, à cauſe que le pays n'y eſt de telle temperature, &

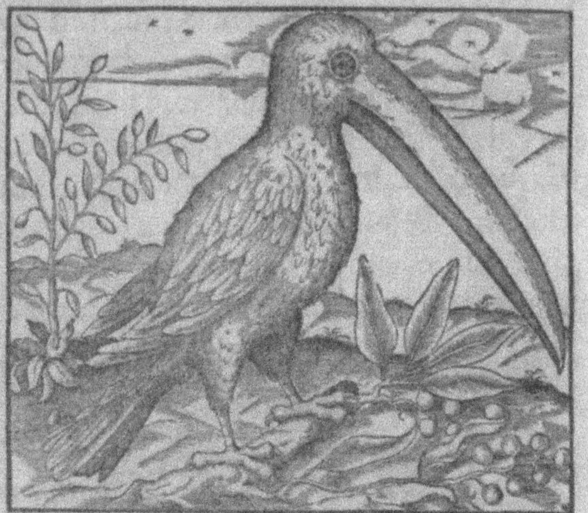

que ceſt oyſeau ne ſçauroit viure parmy la froidure, qu'il craint merueilleuſement. Ce *Toucan* eſt treſ-monſtreux & difforme, entant qu'il a le bec plus gros & long preſque que tout le reſte du corps: & n'eſt point aquatique, comme pluſieurs l'ont penſé, car i'en ay veu l'experience au contraire, d'autant qu'il ſe recule touſiours le plus qu'il peult des riuieres. Ceſt oyſeau ne vit que de certains fruicts parmy les boys, où il fait ordinairement ſa reſidence, & mange auſſi de certain poiure long & rouge, duquel ſe trouuent deux eſpeces, l'vn plus long que l'autre, & le plus petit eſt fait tout ainſi qu'vne fraiſe, vn peu toutesfois plus pointu, & ſe nomme *Quin Apoua*: le plus grand ſ'appelle en leur patoys *Quein Boucoup*. De ce poiure ſe nourrit non ſeulemēt le *Toucan*, ains encor vn autre oyſeau, que les Sauuages appellent *Sumarb*, lequel eſt de la grandeur d'vn Merle, duquel ſ'en voyent deux eſpeces, l'vn tout noir, & l'autre auſſi finement rouge que Eſcarlatte, tel que encor i'en ay dans mon Cabinet diuerſes peaux que ce peuple eſcorche. Quand ces oyſeaux ont mangé de ce poiure, en quelque lieu qu'ils fientent, ſoit ſur vn rocher ou ailleurs, ceſte matiere, bien digeree & cuitte, ou non, ne faudra de prendre en terre, & ſe conuertir en herbe, tout ainſi que ſi lon y auoit ſemé de ce meſme

poiure

Fig. 3. Toucan. From André Thevet, *Cosmographie universelle* (Paris: 1575) vol. II fol. 938v. Leiden, University Library.

Paré's description shows us the difficulties of keeping animals alive during and after their transportation over sea, and therefore the importance of taxidermy. It was difficult to prepare good skins and to protect them against insects. It has been rightly stated that the development of ornithology in the eighteenth century was particularly due to the advanced techniques of taxidermy.[11] Paré's description is also interesting because it gives him the opportunity of promoting himself by naming his highly placed connections and patrons: the King and the Maréchal de Retz.

One of the most interesting descriptions of the toucan is given by another French traveller, Jean de Léry (1534–1613), who belonged to the group of Calvinist colonists in Brazil. In 1557 they had a conflict with their Catholic compatriots and were obliged to flee and take refuge with the Tupinamba, who welcomed them heartily. Léry stayed for two months with the Indians. The Catholics, including the Franciscan Thevet in his *Cosmographie*, blamed the failure of the French colonial enterprise in Brazil on the Huguenots. Years later, Léry published his *Histoire d'un voyage faict en la terre du Bresil, autrement dite Amerique* [...] (1585), as an answer to the Catholic calumnies. Because of its accurate descriptions of the Tupinamba, this book is now considered as a forerunner of modern ethnology – it was even the *livre de chevet* of Claude Lévi-Strauss during his stay among the Tupinamba in the 1930s.

Léry's emulation of Thevet is mostly implicit, but the reader should bear in mind that Léry wrote from a direct, personal experience with the Indians of several months, whereas Thevet only spent some weeks in Brazil. This difference is also evident in Léry's description of the toucan: he notices the same features as Thevet did, but with more direct knowledge. Thus Léry provides much more detailed information on the bird and its feathers, on the ornamental and commercial use of the feathers by the Indians, as well as the etymology of the word 'toucan' – meaning 'feather'. His treatment of Belon's description of the bill is much more sympathetic, and he remains silent regarding Thevet's information on the pepper-eating habits of the bird:

> Et au surplus, ayans en leur pays un oyseau qu'ils nomment Toucan, lequel (comme je le descriray plus amplement en son lieu) a entierement le plumage aussi noir qu'un corbeau, excepté sous le col, qu'il a environ

[11] See Schulze-Hagen K. et al., "Avian taxidermy in Europe from the Middle Ages to the Renaissance", *Journal für Ornithologie* 144(4) (2003) 459–478.

> quatre doigts de long et trois de large, tout couvert de petites et subtiles plumes jaunes, bordé de rouge par le bas, escorchans ses poitrals (lesquels ils appellent aussi Toucan du nom de l'oyseau qui les porte) dont ils ont grande quantité, apres qu'ils sont secs, ils en attachent avec de la cire qu'ils nomment Yra-yetic, un de chacun costé de leurs visages au dessus des oreilles: tellement qu'ayans ainsi ces placards jaunes sur les jouës, il semble presques advis que ce soyent deux bossettes de cuivre doré aux deux bouts du mord ou frain de la bride d'un cheval.[12]

This other place ('en son lieu') gives some other detailed information:

> Les autres oyseaux du pays de nos Ameriquains sont, en premier lieu, celuy qu'ils appellent Toucan, (dont à autre propos j'ay fait mention ci-dessus) lequel est de la grosseur d'un Ramier, et a tout le plumage, excepté le poictral, aussi noir qu'une Corneille. Mais ce poictral (comme j'ay aussi dit ailleurs) estant l'environ quatre doigts de longueur et trois de largeur, plus jaune que saffran, et bordé de rouge par le bas: escorché qu'il est par les sauvages, outre qu'il leur sert, tant pour s'en couvrir et parer les jouës qu'autres parties du corps, encores parce qu'ils en portent ordinairement quand ils dansent, et pour ceste cause le nomment Toucan-tabouracé, c'est à dire plume pour danser, ils en font plus d'estime. Toutesfois, en ayans grande quantité, ils ne font point de difficulté d'en bailler et changer à la marchandise que les François et Portugais, qui traffiquent par delà leur portent.
>
> Outre plus, cest oyseau Toucan, ayant le bec plus long que tout le corps, et gros en proportion, sans luy parangonner ni opposer celuy de grue, qui n'est rien en comparaison, il le faut tenir non seulement pour le bec des becs, mais aussi pour le plus prodigieux et monstrueux qui se puisse trouver entre tous les oyseaux de l'univers. Tellement que ce n'est point sans raison que Belon en ayant recouvré un, l'a par singularité fait pourtraire à la fin de son troisiesme livre des oyseaux: car combien qu'il ne le nomme point, si est-ce sans doute que ce qui est là representé, se doit entendre du bec de nostre Toucan.

In reaction to Léry, Thevet, in one of his last writings on his Brazilian experience, returns once again to subject of the toucan. In this book, which was never published during his life, Thevet, in retrospective, gives us the following information about some specimens of flying fish and toucans he acquired:

> J'en retins quelques uns [i.e. flying fishes] que je sallay, et emportay en France. Depuis j'en envoyay en Allemagne a Gesnerus, le plus docte, medecin en nostre siecle, pour le mettre en son livre des Poissons, comme

[12] I quote Jean de Léry's *Histoire d'un Voyage Faict en la Terre du Brésil* from http://www.mafua.ufsc.br/BT1730002.html (last consultation April 10, 2007).

> estant chose fort rare. Ce poisson est fort bon, horsmis qu'il a beaucoup d'arrestes fort deliées, et n'est non plus gros qu'un barbeau. Il a quatre aisles, dont deux sont aussy longues que le poisson, les autres deux petites. Je luy envoyay par mesme moyen un bec de l'oiseau *Tocan*, long d'un pied, et gros comme le bras d'un homme, encore que cét oyseau ne soit non plus gros qu'un pigeon, qu'il a mis dans son livre des bestes, qu'il a fait cinq ans devant de mourir. J'en ay deux de reste dans mon Cabinet à Paris.[13]

The difference from his earlier texts is that Thevet appears to become increasingly aware of the importance of his network, not only for the exchange of *singularitez*, but especially for reaffirming his own authority by explicitly referring to the authority of his correspondent.

This famous correspondent, Gesner, did include the toucan in his *Icones avium* of 1560, completed with ample allusions to Thevet's book and with an illustration of the bird.[14] In his commentary Gesner explains how he has created this illustration. It is a sort of collage: the bird's bill is based on the bill that was sent to him by one of his correspondents, the historian and humanist writer Giovanni Ferrerio (Joannes Ferrerius), while the rest of the bird is based on the description given by Thevet. The result is a non-existent bird: whereas this bill evidently belongs to a Toco Toucan (such as the one depicted in Belon), the colouring of the bird is not that of a Toco,[15] and the position of the toes (three in front, one in back) is wrong too.

This collage turns out to be a fruitful mistake. For his *Ornithologiae hoc est de Avibus historiae libri* (1599–1603),[16] Ulisse Aldrovandi (1522–1605) commissioned a beautiful but incorrect watercolour,[17] based on Gesner's

[13] Ambroise Thevet, *Histoire d'André Thevet, de deux voyages par luy faits aux Indes Australes et Occidentales*, ed. Lestringant F. et Laborie J.-C. (Geneva: 2005) 376. On the toucan, see also 232, 235–236.

[14] Conrad Gesner, *Icones avium omnium, quae in historia avium Conradi Generi describuntur* (Zurich: 1560) 130.

[15] For details, see my note 17.

[16] Ulisse Aldrovandi, *Ornithologiae, hoc est de avibus historiae libri XII* (Bologna: 1599–1603), vol. I 802.

[17] This watercolour is now in the Museum Aldrovandi, and reproduced in Biancastella A. (ed.), *Les animaux et les créatures monstrueuses d'Ulisse Aldrovandi* (Arles: 2005) 136–137. The editors noticed the peculiarity of the bird: 'L'oiseau que nous voyons ici a certes le bec du *Ramphastos toco*, mais la couleur de son plumage ne correspond absolument pas; le sujet présente en effet une couverture de plumage blanc sur la poitrine, mais la coloration brun-roux est limitée aux sous-caudales. Absentes également la caractéristique caroncule jaune autour de l'oeil' (230), but they were not aware of Aldrovandi's models: Thevet and Gesner. There is a hand-coloured illustration of the toucan in one of the copies held by the University of Bologna which corresponds perfectly to the

composite toucan, which served as a model for the engraving in the book. This engraving, or Gesner's, has a long life, for it reappears in Jan Jonston's *Historia naturalis de avibus libri VI* (1650)[18] [Fig. 4], exhibiting Thevet's toucan in the company of some other birds taken from Clusius and Nieremberg (see below). It is even found in J.H. Lochner's famous *Rariora Musei Besleriani* (1716) among other ornithological curiosities (two species of birds of paradise, a hornbill, a spoonbill, a hummingbird, and an unidentifiable bird).[19]

Aldrovandi's hornbill

Aldrovandi is the first to give a description and illustration of a hornbill [Fig. 5].[20] This bird is introduced directly after the toucan. Aldrovandi begins by saying that, except for the illustrated bill, he has no certitude about the bird. He continues by motivating the name 'aves rhinocerote' with references to classical authorities: Claudianus (4th century AD), Hesychius of Alexandria (5th century AD), Varinus (according to whom the name 'rhinocerote' was used in Ethiopia to indicate both a mammal and a bird), and Pliny, whose description of the bird 'Tragopan' was imitated by Solinus and Pomponius Mela, and commented upon by Cardano and J.C. Scaliger. These authorities are the same ones adduced in the text of the *Delices de Leide*: Aldrovandi indeed appears to be the main source of this text, but in a very truncated form, as we shall see below.

After these introductory lines, Aldrovandi comments upon his illustration of the hornbill's head, which can be identified as the Rhinoceros Hornbill (*Buceros rhinoceros*) in a subspecies that originates from the isle of Borneo. He informs the reader that this illustration is based upon

---

watercolour. See http://diglib.cib.unibo.it/diglib.php?inv=26&int_ptnum=&term_ptnum=826&format=jpg&comment=0&zoom=&x=4&y=6 (last consulted 11 April 2007). These pictures are in accordance with the toucan depicted in the Museum of Rudolph II: Haupt H. – Vignau-Wilberg T. – Irblich E. – Staudinger M., *Le Bestiaire de Rodolphe II. Cod. Min. 129 et 130 de la Bibliothèque nationale d'Autriche* (Paris: 1990), figure 169.

[18] Jan Jonston, *Historiae naturalis de avibus libri VI* (Amsterdam: 1650) 46 and tab. 56.

[19] See the illustration in George W., "Alive or Dead: zoological collections in the seventeenth century", in Impey O. – MacGregor A. (eds.), *The Origins of Museums. The cabinet of curiosities in sixteenth- and seventeenth-century Europe* (Oxford: 1985) 179–187, figure 69. It is not clear whether these figures represent paintings or stuffed birds.

[20] Aldrovandi, *Ornithologiae*, 1 804–805.

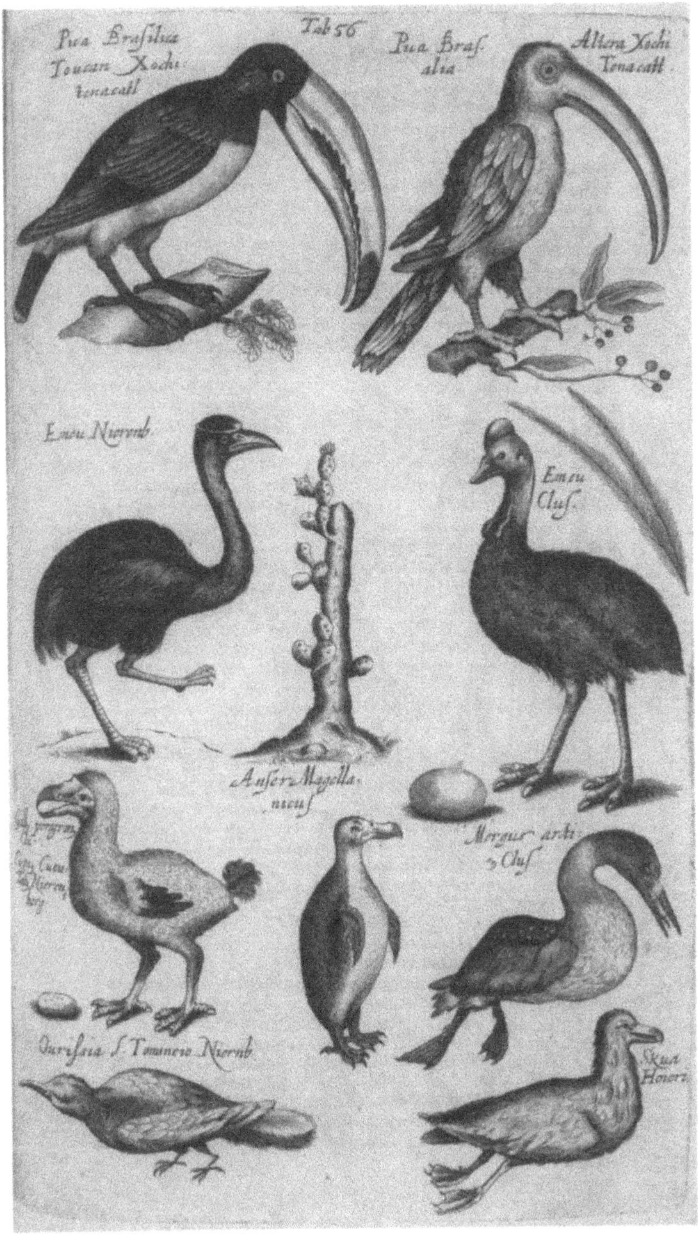

Fig. 4. The left toucan belongs to the Gesner-Aldrovandi type, the right one to the Thevet type. Other birds represented: an emu (?), a dodo, a hummingbird (all from Nieremberg), a cassowary, a penguin, a merganser and a skua (all from Clusius). From Jan Jonston, *Historiae naturalis de avibus libri VI* (Amsterdam: 1650) Table 56. Leiden, University Library.

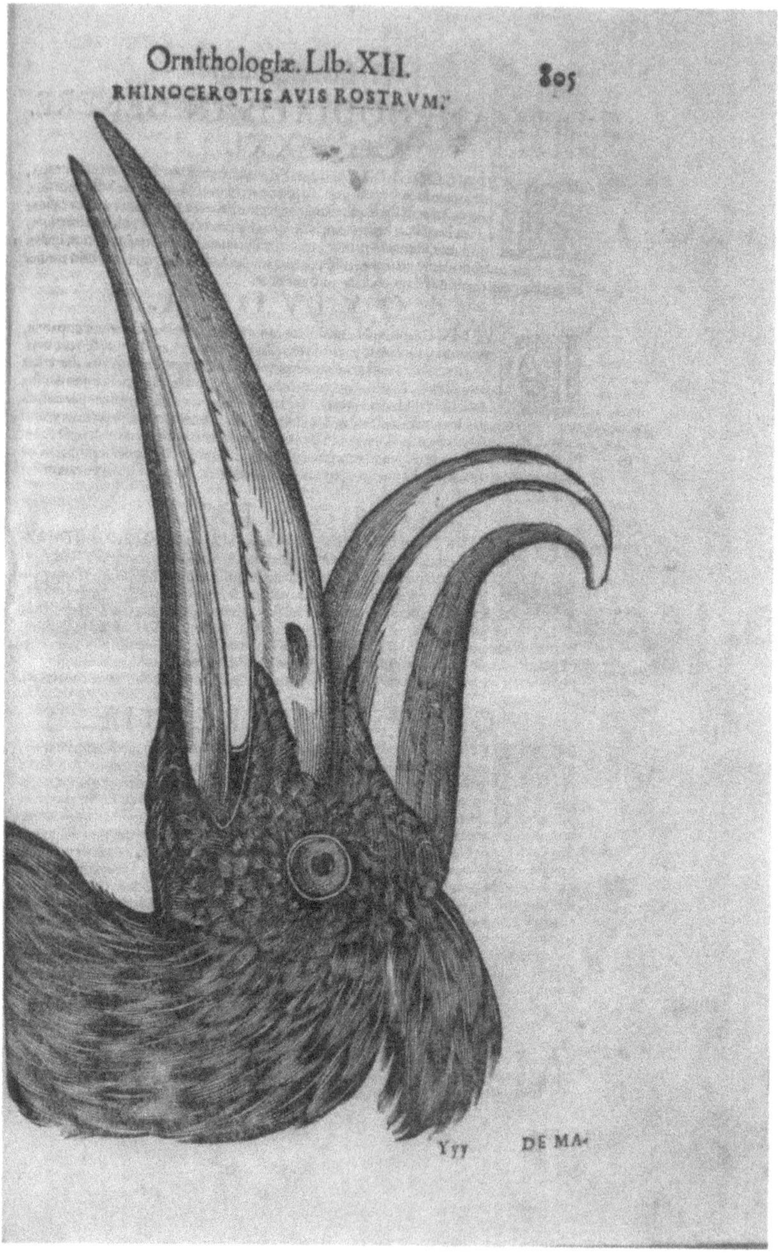

Fig. 5. Head of a Rhinoceros Hornbill from Borneo. From Ulisse Aldrovandi, *Ornithologiae, hoc est de avibus historiae libri XII* (Bologna: 1599) vol. I 805. Leiden, University Library.

two watercolours which he received independently of each other by two persons who had been present at the Battle of Lepanto, both claiming that this bird was shot during the battle. Because the bird is not resident in the Middle East, the only explanation I can think of is that the hornbill at Lepanto had escaped from the Turkish fleet that was carrying a lot of precious items and *exotica* from the East.

Like those of the toucan, the two watercolours of the hornbill are kept in the Museum Aldrovandi in Bologna and can be consulted on internet.[21] These watercolours remind us once again that Renaissance ornithology was based on transitory material, due to the lack of good taxidermal methods: whereas one watercolour represents the bird's original colours and feathering, the other shows the same specimen, but at a later state of decay, reduced to a mere skull. The watercolours and Aldrovandi's precise indications of how he obtained them – both correspondents receive some honorary lines – also underline the growing importance of the zoologist's networks to obtain the necessary materials, as we shall see in the section on Carolus Clusius's hornbill.

Like Belon's chapter on the toucan's bill, Aldrovandi's chapter on the hornbill ends with some prudent hypotheses regarding the rest of the bird, concluding: 'Verum alij, qui avem aliquando nacturi sunt, exactius omnia tradere poterunt, & de coniectura nostra vere iudicare' [In fact, once they have found this bird, others will be able to treat all this more precisely, and to verify our conjecture]. Thus, in presenting unknown *naturalia* together with hypotheses, Belon, Aldrovandi and several other zoologists hoped to open up a debate that would further the progress of zoology.

The hornbill and Clusius's network

The second mention of a hornbill is found in Carolus Clusius's *Exoticorum libri decem* (1605), a book in which the famous botanist, at the end of his long life (1526–1609), published the exotic naturalia he received

---

[21] See Tavole vol. 006–2, c. 92–93, reproduced in http://www.filosofia.unibo.it/aldrovandi/pinakesweb/main.asp?language=it (last consult: April 11, 2007). The first of these watercolours corresponds perfectly with the hand coloured illustration of the Bologna copy of the *Ornithologia*.
See http://diglib.cib.unibo.it/diglib.php?inv=26&int_ptnum=&term_ptnum=829&format=jpg&comment=0&zoom=&x=2&y=8 (last consult: April 11, 2007).

during his position as prefect of the hortus botanicus in Leiden. The illustration of the bill [Fig. 6] indicates that this bird is a species different from the one described by Aldrovandi, namely one of the two African ground hornbills of the genus *Bucorvus*.[22] Convinced, however, that the bird was of South American origin, Clusius referred to the work of Oviedo, though not to Oviedo's chapter on the toucan, but to his description of the *alcatraz*, the Brown Pelican. This explains Clusius's title *An Alcatraz Oviedi? sive verius Corvi Marini genus?* (Oviedo's Alcatraz, or rather a kind of Sea-Crow?). The whole chapter expands upon the hypothesis announced in the title.[23]

The relevance of Clusius's chapter for our argument lies not only in its hypotheses about the bird, but also in the information it gives on the importance of scientific networks for the development of zoological knowledge. In comparing the descriptions given by Clusius and Aldrovandi with the texts of Belon, Thevet and Paré, an important difference emerges. Whereas in the French authors the details on the provenance of the specimen primarily serve to highlight the scientific credibility of the author (Belon) or his high social position (near the King, in the cases of Thevet and Paré), the detailed information furnished by Gesner, Aldrovandi and Clusius on their informants and donors clearly serves another purpose: it keeps the network going. This is particularly evident from Clusius's description of the hornbill and his correspondence on the subject.

Chapters XII and after, all on unknown birds, begin by mentioning their donor: Jacques Plateau, one of Clusius's correspondents, in the service of Charles de Croy, Archduke of Aerschot, and himself the owner of a *cabinet de curiosités*, a botanical garden and some living animals, including a tame seal (as he mentions in his correspondence with Clusius). Plateau is also mentioned explicitly in the opening sentence of Clusius's chapter on the hornbill: 'In earum avium numero, quarum picturas Iacobus Plateau mittebat [...]' [Among these birds, of which Jacques Plateau sent me the illustrations [...]]. The accompanying letter by Plateau has been kept in Leiden University Library[24] and is worth

[22] Most probably *Bucorvus leadbeateri*, the other species *Bucorvus abyssinicus* having a casque open in front.

[23] The confusion seems to begin with Cardano who amalgated Oviedo's descriptions of the toucan and the Brown Pelican (Cardano, *De Subtilitate* 336–337; see my note 6).

[24] University Library Leiden ms. Vul. 101.

An Alcatraz Oviedi? ſive verius Corvi Marini genus?

CAP. XII.

IN earum avium numero, quarum picturas Iacobus Plateau mittebat, cujuſdam etiam avis caput erat cum collo pectore tenus, valde ingenti & craſſo roſtro præditum, cui ſanè pauca ſimilia videre memini: Iſtam picturam, licet integræ avis non eſſet, quia tamen non minùs forſitan peregrina, quàm quæ ſuperiore capite exhibita, dignam exiſtimabam cujus exemplar in tabella expreſſum Lectoris oculis proponeretur.

Avis peregrinæ caput.

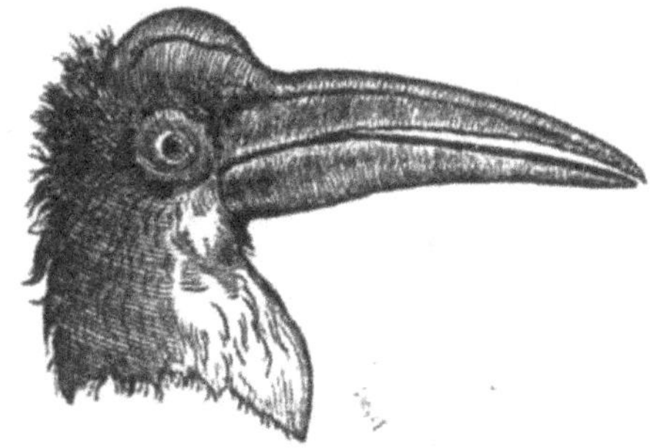

MAGNAM porrò fuiſſe eam avem oportet, quæ ſimile caput geſtaverit, ſi pictura fideliter eſt expreſſa: nam à ſummo capite ad extremam colli partem, dodrantalem habebat longitudinem, ſive novem uncias erat longa, colli latitudo paullò infra roſtrum, ſeptem unciarum; roſtrũ cum capite pedalem longitudinem ſuperabat, ipſum verò roſtrum directum, & minimè curvum in dodrantalem mucronem deſinebat, quà autem parte capiti jungebatur, palmum erat latum: capitis colliq́ue prona pars cum roſtro nigri erant coloris, ſupina verò & guttur cineracei, cujus etiam coloris maculà caput ſub oculis ad roſtri ſupinæ partis radicem, inſignitum erat: pennas in vertice pænè, nonnihil reflexas criſtæ inſtar habebat: nullam autem ingluviem à roſtro propendentem perſpiciebam, ſed guttur & infimum collum in anteriorem partem valdè prominula & protuberantia. Ejuſmodi erant notæ quas in miſſa pictura obſervare poteram.

Fig. 6. Head of a Southern Ground Hornbill from Africa. From Carolus Clusius, *Exoticorum libri decem, quibus animalium, plantarum, aromatum aliorumque peregrinorum fructuum historiae describuntur* (Antwerp: 1605) 106.

quoting at some length because it gives us insight into the functioning of the network to which Clusius belongs. As is usual, the opening of the letter re-establishes the practical contact between the correspondents, informing the addressee of the modes of reception of previous letters and of the writing and sending of the present one. Moreover, these opening lines confirm what we saw already in Aldrovandi's text: the reliance on good painters for the communication of scientific information – the objects themselves being too fragile and expensive to send, and their mere description without illustrations being judged inadequate:

> Monssr la cause principale que n'ay donné responce aux vostres du 8e Juillet et 2 Aoust, pensant tousiours acompaigner mes lettres d'aucunes pourtrais, combien que eusions plusieurs paintres en Tournay, il n'en y a pas un quy sauroit pourtraire une seule plante a la plume. Il n'en y a qu'un lequel sait aucunement paindre sur papier avec couleur, chose;

> grossier. Il est advenu que Dieu auroit appelé sa femme de ce monde et n'a gaire voulu besogner de puis, iusque a ce qu'il a recouvert une nouvelle femme. Il m'a falu par force avoir pasience de luy, iusques a ce que la volunté luy est venue de besogner, combien touttesfois que ie l'ay prié par plusieurs fois.

Although Plateau does not send any watercolours of plants, he manages to send some illustrations of animals (probably in black and white), several of which are reproduced in Clusius's *Exoticorum*: a hitherto unknown species of an armadillo, three mergansers, a 'pa[o]n marin' (a 'sea-peacock', which turns out to be an African Crowned Crane (*Balearica regulorum*)) and a hornbill:

> Je croy de vous avoir encoire envoyé de deux sortes de grand bec d'oyseau. J'en ay trouvé en mon cabinet un plus grand que le precedens, duquel ie vous envoye aussi le pourtraict sous lequel i'ay mis quelque escriture. Vous en porez dire votre opnion, vous avez meilleur loisir de feulleter les livres que moy. Je ne scay si Aldrovandus en auroit fait mention en ses livres de Avibus, car ie n'ay encoire riens veu de luy [...] i'estime qu'aurez veu les deux ou trois volumes que ledit Aldrovandus a mis en lumiere. J'ay encoire plusieurs autres sortes d'animaux, que pense n'avoir esté encoire mis en lumiere si en desires encoire avoir, mande le moy, ie vous donneray de la besogne à quoy passer le temps.

It appears that here Plateau overestimated Clusius's knowledge of Aldrovandi's work on birds. Clusius probably had not even read the work. Had he seen it, he certainly would have noticed that his 'sea-peacock' and Aldrovandi's Crowned Crane were in fact the same species.[25]

What did Plateau expect to receive from Clusius in return? Not only the seeds and dried plants that Clusius had promised to send, or Clusius's recent books ('Je n'ay iamais veu le livre des Observations de Belon traduit en Latin par vous, ne le second livre de Monardes mis nouvellement en lumiere'),[26] but more specifically the honour of being named and quoted in Clusius's works. All this is based on a tacit agreement: Clusius mentions and quotes his correspondent (it is probable that Clusius's description of the *alcatraz* copies Plateau's 'quelque escriture' under the bird's picture), thus giving him eternity in print, in exchange for unknown naturalia. Plateau's conclusion contains a subtle hint not

[25] It is noteworthy that Clusius was not aware that this bird was already illustrated by Adriaen Collaert in his *Avium vivae icones* (Antwerp: 1580).

[26] Both Belon and Monardes were translated by Clusius into Latin.

to break this silent agreement and warns him, between the lines, not to disturb the network on which they both depend:

> Monssr le docteur Monelle est un peu faché contre vous pour cause que n'aves faict encore mention en votre livre de son Colchicum, que Monssr Lobel en son histoire des plantes appelle Colchicum Monelli. Il est fort amateur de sa plante et m'en parle souventesfois. L'ayant une fois perdu, ie luy en ay furny d'une autre plante de laquelle il en fait grand cas, il est bonne personne, n'ayant qu'un fils lequel est aussi docteur. Parquoy si vous fait quelque adionction de plantes en vos livres des Exotiques en porez faire mention, si le trouve bon.

Plateau implicitly demands to be mentioned and quoted in the *Exoticorum* too. It appears that the network had its negative as well as its positive side: whereas Plateau will be mentioned (although not always wholehearted, it seems), Monelle's good character ('il est une bonne personne') and promising son are not sufficient reason to earn a place in the *Exoticorum*.[27]

Testimonies and cabinets de curiosités

More and more species of toucans became known during the decades after Aldrovandi. The bird ceased to be a *rara avis* in the *cabinets de curiosités*. They are found in the famous collections of Rudolph II (who possessed six toucan bills),[28] Federico Cesi (founder of the *Accademia dei Lincei*),[29] Ole Worm and Albert Seba. Jan Brueghel the Elder, who had access to the collections and menageries of Rudolph II and of the Archdukes Albert and Isabella, depicted the same species (Red-billed Toucan, *Rhamphastos monilis*), probably even the same specimen, again and again in his animal paintings.[30] In their highly authoritative natural histories of Iberian America, authors like Juan Eusebio Nieremberg

---

[27] Monelle, however, was named in one of the Appendices of another book: Carolus Clusius, *Rariorum plantarum historia* (Antwerp: 1601), "Altera appendix" (oral communication, Sylvia van Zanen).

[28] Haupt *et al.*, *Le Bestiaire de Rodolphe II* 450.

[29] Freedberg D., *The Eye of the Lynx: Galileo, His Friends, and the Beginnings of Modern Natural History* (Chicago: 2002), index s.v. "Toucan". For another collection, see Mason P., *Infelicities. Representations of the Exotic* (Baltimore: 1998), index s.v. "toucans".

[30] Brueghel's toucan has been imitated by Jan van Kessel the Elder in 1666 (oral communication, Peter Mason).

(1635), Willem Piso and Georg Marcgraf (1648) gave such precise descriptions that it is often possible to determine the species.[31] Thus, the *Aracarí* described and illustrated by Marcgraf in his *Historia naturalis Brasiliae* is a Black-necked Aracari (*Pteroglossus aracari*), as is confirmed in a beautiful hand-coloured illustration in the exemplar of the book in Leiden University Library.[32]

Hornbills are much less frequent in collections. At the end of their chapter on Brazilian birds, Marcgraf and Piso (*Historia* 218) mentioned a horned bill of unknown origin:

> habeo et rostrum avis itidem levissimum et magnae molis, quod superiori parte eminens cornu habet e superiori parte rostri enatum et cavum, coloris incarnati, uti et rostrum: non possum autem affirmare utrum e Brasilia an vero India Orientis sit allatum: et nostri qui in Brasilia vixerunt non meminerunt se ibidem vidisse: Tucana autem ibidem frequentissima est [...].
>
> [I also have a very light and very large bird's bill, which has on the upper part a prominent horn emerging from the upper part of the bill, and which is hollow and red-coloured, especially at the bill. I do not know if this bill comes from Brazil or from the East Indies. Those who have been in Brazil cannot remember having seen it, whereas the toucan is a very common bird there [...].

Although the hornbill's bill became increasingly common, its origin remained mysterious for a long time.

This brings us once again to the bizarre description of the hornbill given in the *Delices de Leide*. We have already noticed that the *Delices* copied whole passages from Aldrovandi, but the anonymous author did so by leaving out a number of other passages. The absurd consequence of these omissions is that the names of Hesychius and Varinus are quoted as direct authorities for the Battle of Lepanto. However, these philological blunders are not wholly due to the author of the *Delices*, because he literally translated from a Latin work, *Collectio monumentorum, rerumque maxime insignium, Belgii Faederati* [...] (1674) by Phileleutherus Timaretes (pseudonym of Theodoor Jansson van Almeloveen (1657–1712)). In this work Timaretes gives, in the form of a catalogue, a description of all

[31] On Nieremberg, Piso and Marcgraf, see Asúa M. de – French R.K., *A New World of Animals: Early Modern Europeans on the Creatures of Iberian America* (Aldershot: 2005) 115–123; 162–170.

[32] Georg Marcgraf – Willem Piso, *Historia naturalis Brasiliae* [...] (Amsterdam: 1648); University Library of Leiden, shelfmark 1407 B 3.

the curiosities of the Leiden Anatomical Theatre. In his translation of Aldrovandi's text, Timaretes not only truncated but also added a minor but interesting detail: the form of the bill is compared to a moon – or as the *Delices* puts it more precisely: a 'croissant', *i.e.* the crescent moon. Retrospectively, Timaretes's words *quasi luna* (which did not appear in Aldrovandi) may explain why Aldrovandi's informants attached such importance to the bird: the bill's form is that of the Turkish moon, and the appearance and killing of the bird can therefore be considered as a *prodigium* foretelling the imminent defeat of the Turkish army. Apparently, Aldrovandi was not very receptive to this latent form of 'emblematic' signification of natural phenomena.

Indirectly, the *Delices* informs us about the quality of scientific research in Leiden at the beginning of the eighteenth century. Internationally, Leiden rapidly dropped behind. The Leiden Anatomical Theatre, once an international scientific highlight, soon degenerated into a tourist attraction,[33] as can be deduced from the travel journals of French and English tourists in the eighteenth century.[34]

Willughby's and Ray's Ornithology

Although anterior to the *Delices*, the scientific value of the *Ornithology* of the seventeenth-century ornithologists John Ray (1627–1705) and Francis Willughby (1635–1672) is vastly superior. This work was originally written by Willughby; after his death it was continued, much enlarged and published by his tutor, friend and companion Ray, first in Latin (1676), then in English (1678).[35] Although they put toucans and hornbills together in the chapter of the 'Birds of the Crow-kind', they clearly distinguish between them: the hornbills are considered as ravens, whereas the toucans are seen as magpies. They carried out extensive bibliographical research on the toucans. Not only do their sources include Thevet, Léry, Aldrovandi, Nieremberg etc., but they

---

[33] See Otterspeer W., *Groepsportret met Dame II. De vesting van de macht. De Leidse universiteit, 1673–1775* (Amsterdam: 2002) 126–128.

[34] See Strien-Chardonneau M. van, "Leiden, metropool der muzen en tuin van Holland. Leiden in Jean Nicolas Parivals *Les Délices de la Hollande*", *De zeventiende eeuw* 22 (2006) (171–184) 181–183.

[35] Francis Willughby – John Ray, *The Ornithology* [...] *in three books. Wherein all the birds hitherto known, being reduced into a Method suitable to their Natures, are accurately described* [...] (London: 1678).

are also the first to quote Oviedo, probably via Oviedo's commentator, the German naturalist Johannes Faber.[36] Their quotations are not uncritical, as can be seen in their remarks on Oviedo's statements on the toucan's bill: '*Oviedo* makes it very heavy, and to weight more than the whole body besides, which is certainly a mistake' (*Ornithology* 129). They also question the hardness of the bird's bill, alleged by Oviedo. They do so by quoting Faber:

> *Faber* doth not undeservedly enquire how, seeing the Bill is so light and thin, the Bird can pierce trees with it? Which difficulty he thus satisfies, that though it be thin and light, yet is it of a bony substance, and therefore it is not to be wondred at that, dextrously used by the living Animal, it should therewith by many repeated strokes pierce a tree, having perchance the instinct to chuse a rotten one, as we see drops of rain wear holes in Flints [...] (*Ornithology* 129).

This description also shows another remarkable aspect: their interest in the bird's behaviour. They quote Faber's observation that toucans wag their tail, 'like wrens' (which is a correct observation), and when Faber observes that the tail of the specimen he saw was very short, Ray explains this phenomenon with common sense and dry humour: 'having been, I suppose, plucked off'. This interest in animal behaviour, and their evident concern for the readability of their text, seem to announce Buffon's ornithological preoccupations, as we shall see.

Their final remark proves that by now the bird and its bill are generally known: 'We have seen in several Cabinets the Bill of this Bird, and ourselves have also one of them' (*Ornithology* 130). As for the Hornbill, things are different. Ray and Willughby can only rely on Aldrovandi and on the description of a hornbill given by the Dutch Jacobus Bontius, which for a long time will be the only known description of the bird as a whole:

> This horned Bird as it casts a strong smell, so it hath a foul look, much exceeding the *European Raven* in bigness. It hath a thick Head and neck, great Eyes; the Bill but moderate in respect to the body: The longer and more acuminate part bending downward argues the Bill to be made and designed for rapine. But the upper part, which is shorter, thicker, and bending upward doth resemble a true Horn, both to fight and touch

[36] Johannes Faber, *Animalia Mexicana* [...] (Rome: 1628) 697.

> [...] It lives upon Carrion and Garbage, *i.e.* the carcasses and Entrails of Animal (*Ornithology* 127).[37]

Besides this rather unspecific description (it remains unclear which species Bontius is describing), Willughby and Ray can only rely on the bills they have seen in collections:

> Of this sort of Bill we have seen three varieties, all which we have caused to be engraven and exhibited to the Readers view.

The authors provide two plates. The first, Tab. XVII [Fig. 7], represents four birds: the first two are bills 'engraven' from autopsy;[38] the third is the bird's head from Aldrovandi; and the fourth is an unidentifiable *corvus indicus* of unknown origin, taken from Jacobus Bontius's *Historiae Naturalis & Medicae Indiae Orientalis* (1658 or another edition).[39] The second plate, Tab. LXXVIII – the very last table of the book – shows the bill of another 'Indian Raven', probably a female Plain-pouched Hornbill (*Rhyticeros subruficollis*), and seems to have been added to the book at the very last moment [Fig. 8].[40]

Willughby and Ray's *Ornithologia* is important not only for its descriptions, its (not always) accurate illustrations, and the attention given to all aspects of the bird, including behaviour, but also for their classification, mainly based on the form of the birds' bills and feet. All birds find their place in a well-conceived and motivated classification, visualized in conveniently arranged and presented tables [Fig. 9]. From now on, toucans and hornbills are classified as distinct families. This can be deduced from the very practical survey of the most important classifications by Michael Walters:[41] Latham 1781–1785; Schaeffer 1789; Lacépède 1799; Illiger 1811; Wagler 1827. It underlines once again

---

[37] This passage corresponds to Jacobus Bontius, *Historiae naturalis et medicae Indiae Orientalis*, in Willem Piso, *De Indiae utriusque re naturali et medica* (Amsterdam: 1658) 63.

[38] The first is a Great Hornbill (*Buceros bicornis*), the second is a Rhinoceros Hornbill (*Buceros rhinoceros*).

[39] This *Corvus indicus* is described on p. 126: 'Bontius his Indian Raven' (see Bontius, *Historiae naturalis et medicae Indiae Orientalis* 62). The *Corvus indicus* should be distinguished from 'The *horned* Indian Raven or Topau, called the Rhinocerot Bird' (my emphasis), described on the following page (*Ornithology* 127). Curiously, Willughby and Ray's description of the *Corvus indicus* comes from Bontius's '*horned* Indian Raven', whereas their illustration is a copy of Bontius's 'Indian Raven'.

[40] I thank Hein van Grouw (Naturalis, National Museum of Natural History, Leiden) for his help with the identification of the birds.

[41] Walters M., *A Concise History of Ornithology. The lives and works of its founding figures* (London: 2003) 176–235.

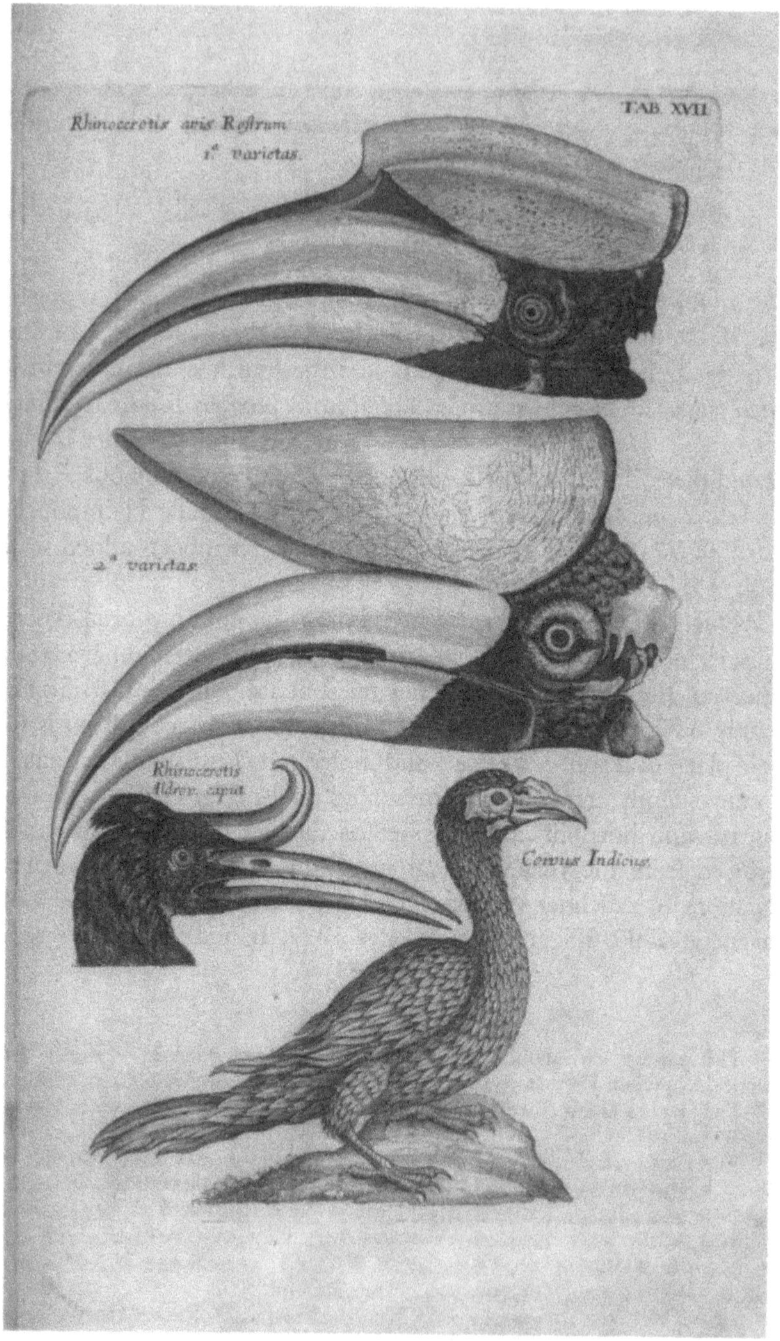

Fig. 7. Four species of hornbills. From Francis Willughby – John Ray, *The Ornithology* [...] *in three books* [...] (London: 1678) Tab. XVII. The Hague, Royal Library.

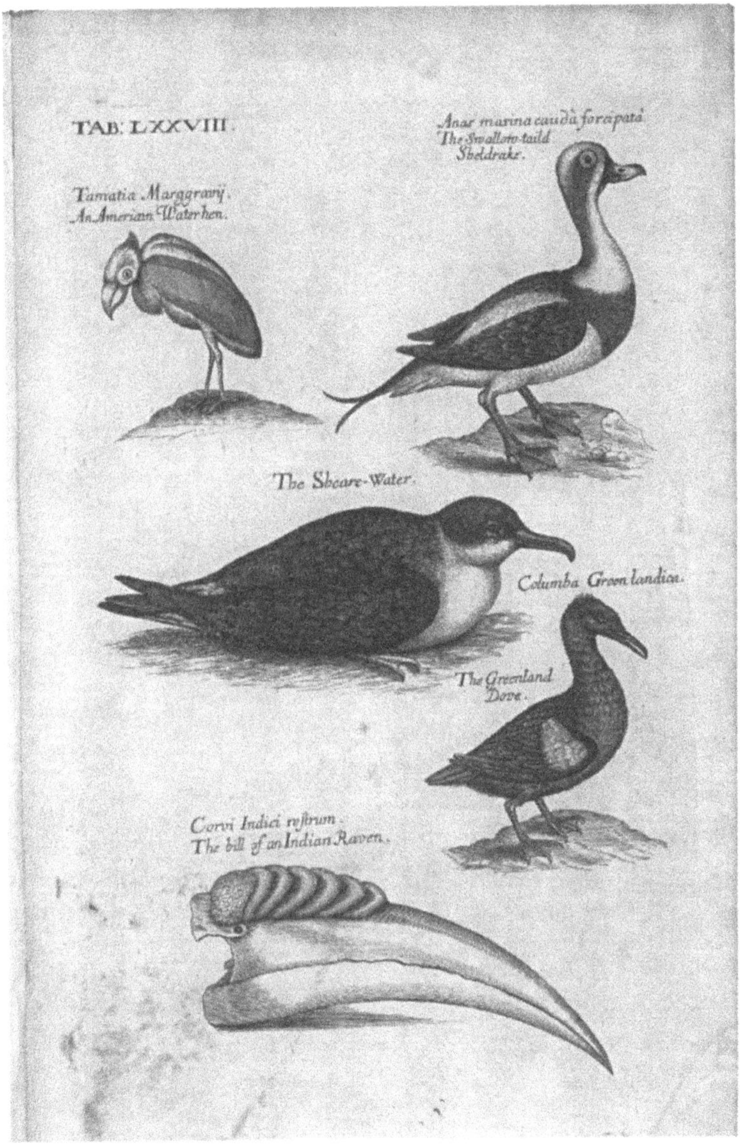

Fig. 8. Bill of a Plain-pouched Hornbill from Thailand and Malaysia. Other birds represented: a male Long-tailed Duck (*Clangula hyemalis*), a shearwater (*Puffinus spec.*), a Black Guillemot (*Cepphus grylle*) and an unidentifiable 'American Waterhen'. From Francis Willughby – John Ray, *The Ornithology* [...] *in three books. Wherein all the birds hitherto known, being reduced into a Method suitable to their Natures, are accurately described* [...] (London: 1678) Tab. LXXVIII. The Hague, Royal Library.

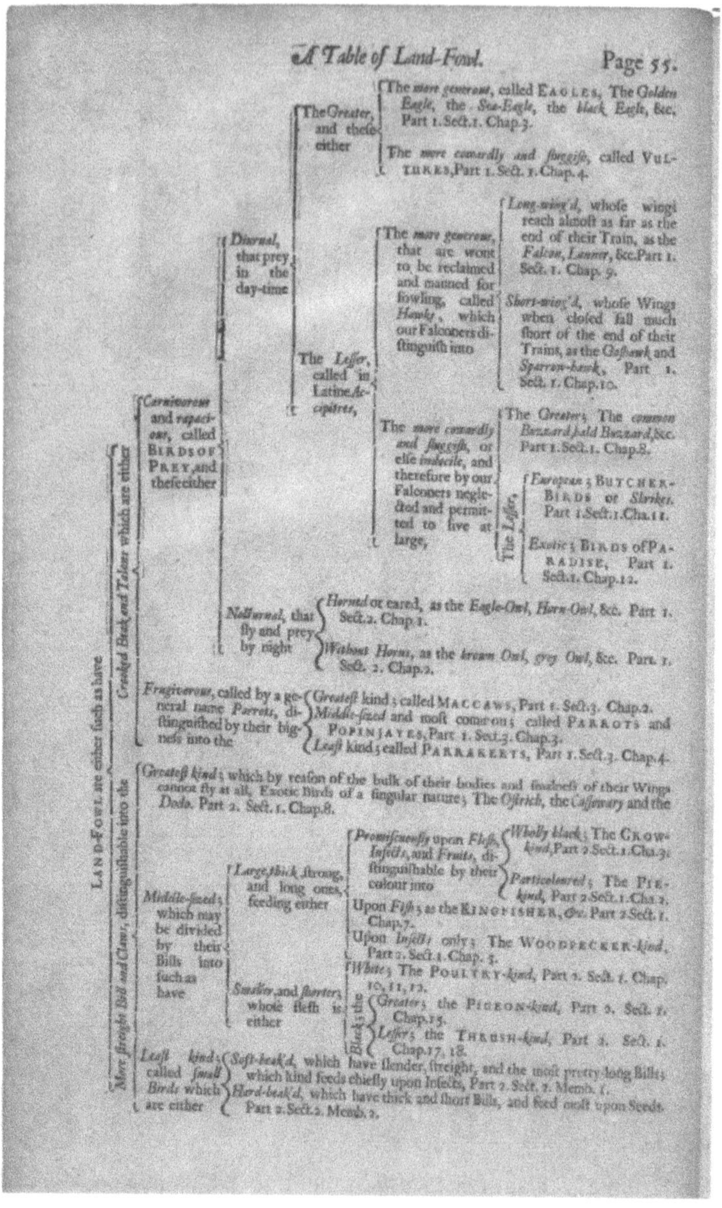

Fig. 9. A Table of Land-Fowl. Hornbills and toucans are put in the lower half of the Table: the hornbills are included in the "Crow-kind"; the toucans belong to the "Pie-kind". From Francis Willughby – John Ray, *The Ornithology* [...] *in three books* [...] (London: 1678) 55. The Hague, Royal Library.

Willughby's and Ray's modernity, announcing not Buffon this time, but Linnaeus and Brisson, as we shall see in the next section.

Brisson's systematizations

For the highlights of eighteenth-century ornithology, we must return to France, to the works of Mathurin Jacques Brisson (1723–1806) and his rival and successor Georges-Louis Leclerc, Count of Buffon (1707–1788). Both ornithologists were heavily dependent on the big aristocratic collections of their time. Brisson's *Ornithologie ou Méthode contenant la division des oiseaux en Ordres, Sections, Genres, Espèces & leurs Variétés* (1760)[42] was based on the collection of René Antoine Ferchauld de Réaumur (1683–1757), a collector who occupies an important place in the history of ornithology. Réaumur explains why ornithology made only little progress:

> The study of birds, has remained as yet very imperfect, nor has it yet made them sufficiently known to us, because no considerable collections have hitherto been made of them; and those who had begun to make any soon became weary of going on, having had the mortification to see them every day destroyed by ravenous insects, in spite of all the care that had been taken to preserve them against their teeth.[43]

Without totally solving the problem, Réaumur found a quick and effective method to conserve birds in the tropics: by drying them in an oven. By this method, promulgated in an anonymous treatise,[44] his correspondents succeeded in sending him good specimens, from which he built up the largest bird collection in Europe. This enormous and permanent collection enabled its curator Brisson not only to develop a new classification in which all bird species would fit, but also to provide meticulous descriptions of the birds and to have them illustrated with coloured lithographs. After Réaumur's death in 1757, the collection was confiscated by the French King, Brisson was fired, and replaced

---

[42] Mathurin Jacques Brisson, *Ornithologia, sive synopsis methodica sistens avium divisionem in ordines, sectiones* [...] *varietates – Ornithologie ou Méthode contenant la division des oiseaux en Ordres, Sections, Genres, Espèces & leurs Variétés* (Paris: 1760), 6 vols.

[43] Quoted from Walters, *A Concise History of Ornithology* 53.

[44] *Différens moyens d'empêcher de se corrompre les oiseaux morts qu'on veut envoyer dans des pays éloignez et de les y faire arriver bien conditionnez* [1745]. I quote the title of this text, now extremely rare, according to Balis J., *Van diverse pluimage. Tien eeuwen vogelboeken* (Antwerp – The Hague – Brussels: 1968) 57.

by Buffon, who, as we shall see, was in many ways the antipode of Brisson.

Brisson's *Ornithologie* begins with a very interesting historical survey of ornithology, which allows Brisson to position himself in respect to his predecessors. Brisson begins by judging his predecessors on their classificatory merits. Thus, according to Brisson, although his information is somewhat old-fashioned, Belon 'est le premier qui ait donné quelque ordre à cette partie de l'Histoire Naturelle' (*Ornithologie* I viii). Brisson criticizes Gesner's alphabetical order, 'qui a proprement parler ne mérite pas le nom [of 'ordre']' (*Ornithologie* I ix). Aldrovandi is useful but too wordy, whereas Jonston only copied others. Brisson is full of praise on the subject of Ray and Willughby: 'Enfin parut celui que je regarde comme le vrai Méthodiste, *Ray*, qui fut l'Editeur de l'ouvrage de Willughby. Ce Savant a rangé les Oiseaux dans un nouvel ordre, & établi des caracteres propres à les faire connoitre.' Other ornithologists we have not encountered, such as Barrère (1745; 'de la pire espèce')[45] and Klein (1750; 'confusion horrible'; 'tout s'y trouve pêle-mêle') are summarily and rather sarcastically executed. And on Linnaeus's ornithological work, he observes dryly that his tenth edition is better than the previous ones, but is still not specific enough. Brisson himself comes up with a classification based, like that of Willughby and Ray, on the form of bill and feet. For our birds this implies that they are more clearly separated from each other: the toucans are put in Ordre XIII, Classe III, Section V, whereas the hornbills find their place in Ordre XIV, Classe III, Section V.

Another innovative aspect of Brisson's method concerns the description of the individual species. This description is invariably done following the same pattern (nomenclature, literature, origin, size, physical description from beak to tail) in a dry, unornamented style. In his introduction he evaluates the pros and contras of this descriptive method: 'J'avoue que cela produit une monotonie, qui deviendrait insupportable dans un ouvrage fait pour être lû de suite' (*Ornithologie* I xv). But the advantage of this descriptive *degré zéro* is that every reader can compare at a single glance the differences between the various

[45] 'Il semble qu'il range ses Oiseaux dans une armoire, & que, voulant ménager le terrein, il fait remplir par les petits le vuide que laissent nécessairement les gros' (*Ornithologie* I xi).

species. This makes the book a forerunner of modern ornithological handbooks and field guides.

One of the other novelties is that every description is bilingual: French and Latin. Moreover, all described specimens receive brief but precise information about their origins (in this Brisson continues in the direction given by Willughby and Ray). A double asterisk ** means that the description was done from autopsy of the whole bird (in these cases an illustration is given); a single asterisk * indicates that Brisson only saw a part of the bird. In both cases the possessor of the object is mentioned (always in an identical phrasing), which not only assures scientific verifiability, but – as in Clusius's time – reinforces the importance of the network, implicitly underscoring his own position in it. In his descriptions without any asterisk, the sources are printed descriptions by other authors, who are briefly mentioned.

The following tables are intended to visualize the progress that Brisson made compared to his predecessors in the extended knowledge of the different species and in the repetitive but practical methods of describing them. Brisson distinguishes twelve species of toucan, of which six by autopsy (double asterisk); he distinguishes six species of hornbill, of which three marked with a double asterisk, and two with a single asterisk.[46]

number plus asterisk	French name	origins/principal sources
1	Toucan	Several references
2**	Toucan à gorge jaune de Cayenne	'du cabinet de Madame de Bandeville'
3	Toucan à gorge blanche du Brésil	Several references
4**	Toucan à gorge blanche de Cayenne	'On le trouve à Cayenne, d'où il a été envoyé à M. de Reaumur par M. des Essars'
5**	Toucan à gorge jaune du Brésil	'On le trouve au Brésil et à Cayenne, d'où il a été envoyé à M. de Reaumur par M. des Essars'
6	Toucan à collier de Mexique	Hern. Hist Nov Hisp.[46]

[46] Francisco Hernández, *Rerum medicarum Novae Hispaniae thesaurus seu nova plantarum, animalium et mineralium mexicanorum historia* (Rome: 1651).

Table (*cont.*)

number plus asterisk	French name	origins/principal sources
7	Toucan verd de Mexique	Hern. Hist Nov Hisp.
8**	Toucan verd de Cayenne	'On le trouve à Cayenne, d'où il a été envoyé à M. de Reaumur par M. des Essars'
9**	Toucan verd du Brésil	'On le trouve au Brésil, à Surinam & à Cayenne, d'où il a été envoyé à M. de Reaumur par M. Artur
10**	Toucan à collier de Cayenne	'On le trouve à Cayenne, d'où il a été envoyé à M. de Reaumur par M. des Essars'
11	Toucan jaune	Nieremberg
12	Toucan bleu	Nieremberg

number plus asterisk	French name	origins/principal sources
1**	Calao	'On le trouve aux Moluques, d'où il a été envoyé à M. l'abbé Auboy, qui le conserve dans son cabinet
2*	Calao des Philip[p ?]ines	Opening sentence: 'Je n'ai jamais vû que la tête & le bec de cet oiseau'; Final sentence: 'M. de Reaumur n'en avoit que la tête & le bec dans son cabinet'
3	Calao d'Afrique	Willughby
4*	Calao des Indes	Opening sentence: 'Je n'ai jamais vû que la tête & le bec de cet Oiseau'; Final sentences: On le trouve dans les Indes Orientales. M. de Reaumur n'en avoit que la tête & le bec dans son cabinet'
5**	Calao à bec noir du Sénégal	'On le trouve au Sénégal, d'où il a été envoyé à M. de Reaumur par M. Adanson'
6**	Calao à bec rouge du Sénégal	'On le trouve au Sénégal, d'où il a été envoyé à M. de Reaumur par M. Adanson'

Brisson's strict method implies some remarkable lacunae. Because he solely relies on dried and stuffed animals, he is not able to give any inner morphology of the bird. Nor is he interested in the bird's behaviour, even if his sources give him information on this item. And the tables show that the order of the described birds within a Section is not explicitly motivated, and may be more or less arbitrary. It is to those shortcomings that Buffon will react.

Buffon: de-classification, behaviour, style

The scope of this article does not allow us to expatiate on the immense ornithological work of Buffon, published under the general title *Histoire naturelle des oiseaux* (10 vols., 1771–1786).[47] We have to limit ourselves to the two bird families under discussion, and to look at how Buffon's treatment of these families informs us about his method, and how sometimes the great French zoologist makes use of it in order to illustrate some of his general ideas on zoology. To start with the latter aspect: one of Buffon's most important issues is his criticism of all kinds of classification pushed too far (as preached by Willughby and Ray, Brisson and Linnaeus). In his view, classification does not express Nature's reality but rather the classifier's mind, who tries to get a grip on Nature and therefore distorts her:

> Plus on augmentera le nombre des divisions naturelles des productions naturelles, plus on approchera du vrai, puisqu'il n'existe réellement dans la nature que des individus, et que les genres, les ordres, les classes n'existent que dans notre imagination.[48]

That is why he seems to take some pleasure in subversively undermining Brisson's classification by introducing numerous changes. Thus he juxtaposes the two families which have been separated since Ray and Willughby. He also corrects Brisson's classification of the toucans by making a subdivision between toucans and aracaris on the basis of their size, and adding two other birds, which are presented as intermediate

[47] Our quotations from Buffon come from the electronic edition of his work, available on Internet (www.buffon.cnrs.fr).

[48] Buffon, *Discours sur la manière de traiter l'histoire naturelle*, quoted and commented by Foucault M., *Les mots et les choses. Une archéologie des sciences humaines* (Paris: 1966) 159–160.

species: the *barbican* and the *cassican*, the former between the barbets (*Capitonidae*) and the toucans, the latter between the icterids (*Icteridae*) and the toucans. The intermediate position of these two species is shown by the names bestowed upon them: 'Comme cet oiseau tient du barbu [barbet] & du toucan, nous avons cru pouvoir le nommer *barbican*' (*Histoire naturelle* 132) and 'Nous avons donné le nom de *cassican* à cet oiseau [...], parce que ce nom indique les deux genres d'oiseaux auxquels il a le plus de rapports: celui des cassiques [casiques] & celui des toucans'.[49] The intercalation of these two species makes it problematic to draw definite lines both within bird families and between families and species. Moreover, the order of the species within a family is arbitrary, according to Buffon. Thus, he does not give any explicit motivation for the order of treatment of the toucans and the aracaris, but it is clear that he begins with the largest species (the Toco Toucan) and finishes with the lesser aracari species. The order of treatment of the ten species of hornbills, on the other hand, is explicitly motivated, and compared with the toucan family the order is inverted, starting with the lesser species:

> En considérant ces dix espèces [...] on reconnoîtra tous les degrés par où la Nature passe pour arriver à cette monstrueuse conformation de bec. Le tock a un large bec en forme de faulx comme les autres, mais ce bec est simple & sans éminence; le calao de Manille a déjà une éminence apparente sur le haut du bec: cette éminence est plus marquée dans le calao de l'île de Panay; elle est très-remarquable dans le calao des Moluques: elle est encore plus considérable dans le calao d'Abyssinie; énorme enfin, dans le calao des Philippines & du Malabar; & tout-à-fait monstrueuse dans le calao rhinocéros (*Histoire naturelle* 140–141).

Buffon problematises not only the notion of family, but also that of species. The first reason for being reluctant to determine the species of a given bird is a practical one: he notes that many ornithologists before him have often considered male and female birds as two different species. The same goes for young and adult birds. He therefore is very satisfied to reduce Brisson's 'quatre prétendues espèces' to one single species, i.e. the *Toucan à gorge jaune*, to which he adds a fifth 'espèce',

[49] Both names still exist in French. Buffon's illustrations indicate that the *barbican* is the Bearded barbet (*Lybius dubius*) of West Africa; and that the *cassican* can be identified as the Pied butcherbird (*Cracticus nigrogularis*) from Australia. The English name 'butcherbird' (also used for shrikes, because they pin their prey on thorns) indicates that Buffon was mistaken in attributing fruit-eating habits to the bird (*Histoire naturelle* 135).

'qui ne diffère que de ceux-ci par la couleur blanche de la poitrine' (*Histoire naturelle* 121). And after his description of the *Calao des Philippines*, Buffon discusses another specimen, saying: 'nous croyons qu'on doit au moins regarder celui-ci comme une variété de l'autre' (*Histoire naturelle* 159). This observation makes clear that Buffon has more fundamental reasons to be suspicious about Brisson's classifications. A species can have variants, and Buffon rightly seems to ask himself what the difference is between species and variant (or 'subspecies' as it is called in modern taxonomy).

Buffon's view of the origin of new species and subspecies is diametrically opposed to Linnaeus's so-called *fixism*, according to which all species remain for ever unchanged. Not that Buffon should be considered as an adherent to Lamarck's *transformationalism* or as a direct forerunner of Darwinism. Buffon's concept is that of *degeneration*: under the influence of its milieu a species can degenerate, and if the degenerate species does not die out, this can only be explained by geographical isolation or a lack of natural enemies. According to this theory, the bills of toucans and hornbills are explained. Like the bills of crossbills and skimmers, they are monstrous excrescences, a nuisance to the bird, without any utility. In order to enlarge his discussion Buffon compares the hornbills to the edentates:

> Nous avons de semblables exemples dans les animaux quadrupèdes, les unaus, les aïs, les fourmilliers, les pangolins, &c. débués ou misérables par la forme du corps & la disproportion de leurs membres, traînent à peine une existence pénible [...] (*Histoire naturelle* 137).

Buffon goes on indefinitely about the monstrosities of the birds. Thus the toucan's bill is disqualified as 'mince & foible'; it serves no purpose whatsoever: it obliges the bird to swallow its food whole; it is useless as a weapon against predators; and it is useless as a counterbalance, because it disturbs the bird in its flight ('lui donnant un air à demi-culbutant, semble le ramener vers la terre lors même qu'il veut se diriger en haut'). The bill is a mistake of Nature: 'les vrais caractères des erreurs de la Nature, sont la disproportion jointe à l'inutilité', it does not belong to 'le grand plan' of mother Nature, but to 'la petite carte de ses caprices'.

The same argumentation is applied to the hornbills, but in an unexpected way: the Old World hornbills are more monstrous than the New World toucans. The opening sentence of the chapter on hornbills gives the following explanation:

> Nous venons de voir que les toucans, si singuliers par leur énorme bec, appartiennent tous au continent de l'Amérique méridionale: voici d'autres oiseaux de l'Afrique & des grandes Indes, dont le bec aussi prodigieux pour les dimensions que celui des toucans, est encore plus extraordinaire par la forme, ou pour mieux dire, plus excessivement monstrueux comme pour nous démontrer que la vieille Nature de l'ancien continent, toujours superieure à la Nature moderne du nouveau monde dans toutes ses productions,[50] se montre aussi plus grande, même dans ses erreurs, & plus puissante jusque dans ses écarts (*Histoire naturelle* 136).

The monstrous uselesness of the bill is enlarged upon in superlatives: 'il n'y a peut-être pas d'exemple dans la Nature d'une arme d'aussi grand appareil & d'aussi peu d'effect (*Histoire naturelle* 137); and most interestingly, the bill's upper and lower parts do not fit: they only touch each other at the bill's point, 'comme si elles n'eussent pas été faites l'une pour l'autre' (*Histoire naturelle* 138).

Another important difference between Buffon and Brisson (and most ornithologists before them, with the exception of Ray and Willughby) is Buffon's interest in the behaviour of the bird. Whenever he has the chance to do so, Buffon describes the animal behaviour from his own observation of animals, mostly in captivity. Thus, he notes extensively the locomotion and behaviour of a *calao de Malabar*, which was offered to him by the Marquise de Pons:[51]

> Cet oiseau sautoit des deux pieds à la fois en avant & de côté, comme le geai & la pie, sans marcher; dans son attitude de repos, il avoit la tête portée en arrière & reculée entre les épaules; dans l'émotion de la surprise ou de l'inquiétude, il se haussoit, se gradissoit & sembloit prendre quelque air de fierté; cependant sa mine en général est basse & stupide, ses mouvemens sont brusques & désagréables: & les traits qu'il tient de la pie & du corbeau, lui donnent un air ignoble, que son naturel ne dément pas (*Histoire naturelle* 153).

Qualifications like 'air de fierté', 'basse & stupide', 'désagréables' and 'air ignoble' show a tendency to humanize the animal. We are dealing here

[50] According to this perspective the animals of the Old World are often superior to those of the New World: the big cats (lion, tiger) are bigger and more dangerous than their counterparts from the Americas (jaguar, puma); the African ostrich is bigger than the South American rhea, and camels are bigger than llamas.

[51] 'qui a eu la bonté de me l'offrir, & à la quelle je me fais un devoir de témoigner ici ma respectueuse sensiblilité' (*Histoire naturelle* 149).This expression of gratitude reminds us of the important of the network, although Buffon, through his function at the Royal collections, is much less dependent on his connections.

not only with a figure of speech – as in Buffon's comparison between the toucan's head and 'ces masques à long nez dont on épouvante les enfans' (*Histoire naturelle* 109) –, but with a much wider notion, as can be read in the opening sentence of Buffon's chapter on toucans: 'Ce qu'on peut appeler physionomie dans tous les êtres vivans, dépend de l'aspect que leur tête présente lorsqu'on les regarde de face' (*Histoire naturelle* 108). Contrary to birds with 'une tête légère avec un bec court & fin' which gives them 'la physionomie fine, agréable & presque spirituelle', toucans 'se présentent avec un air stupide, rarement démenti par leurs habitudes naturelles' (*Histoire naturelle* 108), and further on: 'ce gros bec leur donne une physionomie triste & sérieuse que leurs grands yeux fades & sans feu augmentent encore; en sorte que [...] ils n'en paroissent que plus gauches & moins gais'. Something similar is said of the *calao rhinocéros*: 'ces excès & ces défauts extérieurs semblent influer sur les facultés intérieures de l'animal: ce calao est triste & sauvage; il a l'aspect rude, l'attitude pesante & comme fatiguée' (*Histoire naturelle* 163). These passages show some affinity with current ideas on physiognomy, originating from Aristotle and modernised and promoted by Lavater, among others.

These passages also show Buffon's great attention to the stylistic qualities of the descriptions. In his chapters on toucans and hornbills, Buffon never gainsays his famous, though controversial adage 'le style c'est l'homme même'. He multiplies figures of speech like metaphor, personification, antithesis, and expresses himself in beautifully constructed and well balanced sentences. In assuring the readability of his text, Buffon seems to adopt a position against authors like Brisson, who consciously tries to produce a repetitive, monotonous descriptive mode. This is even visible in the naming of the birds: whereas Brisson's names are descriptive, generally indicating one of a bird's physical characteristics ('Toucan à gorge jaune') and its origin ('de Cayenne'), Buffon delights in inventing exotic neologisms based on linguistic hybridization (like the *barbican* and *cassican*), Indian etymology and onomatopoeia, often all at the same time, as can be seen in the following cases:

- *cochicat*: 'c'est par contraction le nom que cet oiseau porte dans son pays natal au Mexique' (*Histoire naturelle* 124)
- *hochicat*: 'c'est de même le nom, par contraction, que cet oiseau porte au Mexique' (*Histoire naturelle* 125)
- *grigri*: 'à la Guyane où on l'appelle *gri-gri*, parce que ce mot exprime à peu-près son cri qui est aigu & bref' (*Histoire naturelle* 126);

– *koulik*: 'ce petit mot *koulik*, prononcé vite, représente exactement le cri de cet oiseau, & c'est par cette raison que les créoles de Cayenne lui ont donné ce nom' (*Histoire naturelle* 128–129).[52]

And it is with regret that he is forced to accept Brisson's somewhat boring denomination 'toucan à gorge jaune': 'ils [these birds] jettent un cri articulé qui semble prononcer *pinien-coin* ou *pignen-coin*, d'une manière si distincte que les créoles de Cayenne leur ont donné ce nom que nous n'avons pas cru devoir adopter, parce que le toco ou toucan de l'espèce précédente prononce cette même parole, & qu'alors on les eût confondus' (*Histoire naturelle* 122).

One of the illustrative examples of Buffon's stylistic variety can be found in the most daring associative sequences of morphological observation, etymology and metaphor, which are to be found in the passages on the toucan's tongue. Firstly, the strangeness of the bird's tongue is underscored in a hyperbolic way:

> Le bec excessif, inutile du toucan, renferme une langue encore plus inutile, & dont la structure est très-extraordinaire; ce n'est point un organe charnu ou cartilagineux comme la langue de tous les animaux ou des autres oiseaux, c'est une véritable plume bien mal placée, comme l'on voit, & renfermée dans le bec comme dans un étui (*Histoire naturelle* 110).

This comparison with a feather (which Buffon found in Oviedo, probably via Willughby and Ray) unexpectedly brings the author to speak about the bird's name following (but not mentioning) Léry: 'Le nom même de toucan signifie *plume* en langue Brasilienne, & les naturels de ce pays ont appelé *toucan tabouracé*, l'oiseau dont ils prenoient les plumes pour se faire les parures [...]' (*Histoire naturelle* 110). This leads Buffon to speak about the bird's feathers (*Histoire naturelle* 110–111), then about its bill (*Histoire naturelle* 111–113). After this, Buffon returns to the bird's tongue, transforming what was a metaphor ('c'est une véritable plume', *Histoire naturelle* 110) into a zoological reality ('ce sont les seuls oiseaux qui aient une plume au lieu de langue, & c'est une plume dans l'acception plus stricte', *Histoire naturelle* 113). There follows a detailed morphological description of this *plume-langue* (the expression is Buffon's, *Histoire naturelle* 113). Then he informs us about the strange whistle that the birds are believed to make with their tongue ('assez

[52] Both *grigri* and *koulik* have entered French ornithological language.

long temps pour qu'on les ait appelés *oiseaux prédicateurs*') and concludes with the medical use of the tongue: 'Les sauvages attribuent aussi de grandes vertus à cette langue de plume', *Histoire naturelle* 113). It is by this bold, unpredictable, associative way of presenting his material that Buffon hopes to save the reader from boredom.

Conclusions

The numerous paths followed in this article give us some insight into the practice of a variety of describers: zoologists, cosmographers, travellers, collectors. Amateur (i.e. non-professional) zoologists like the cosmographer Thevet and the traveller Léry are more concerned with the bird as a cultural or commercial object. Their descriptions often seem to serve ideological purposes (Léry) or more personal motives of self-promotion (Thevet and Paré). The more professional zoologists like Belon, Gesner, Clusius, Aldrovandi and Marcgraf tried to reconstruct the bird from the scarce material (only a bill) and scarce information available. They were not afraid to venture hypotheses, asking the reader to confirm or falsify them. Some natural historians (Clusius, Marcgraf, Bontius) focused on *exotica*: what counted was the description of the animals, not their classification. For others (Belon, Aldrovandi, Willughby, Ray and Brisson) the insertion of their birds into their systems of classification was a matter of great importance. In the case of Buffon, the birds were used to confirm and illustrate the author's ideas on classification, and more generally on zoology and its discourses. The better the birds became known, the more their description could be formalized – a development culminating in the work of Brisson, who proceeded by external description, moulded in a universally applicable format. Buffon problematised all rigid classification, rejected formalized description in favour of style and readability, and was not interested in the external or internal morphology, but merely in the animals' behaviour. From Belon to Buffon, it is clear that ornithology relied heavily on ephemeral material, conserved, thanks to the developments of taxidermy, in growing collections, with the royal collections in eighteenth-century France as a climax. To keep abreast of the latest information, early modern zoologists needed to maintain a broad network, including not only their fellow zoologists, but also (patrons and) collectors, travellers and other non-professionals. Almost every early modern zoological description contains elements that not only inform us about this network, but are meant to serve as a means of keeping it going.

As a spin-off, there are two final observations to make: firstly, with the sole exception of Timaretes's description of the hornbill at the Battle of Lepanto, I have found no traces at all of the 'emblematized' vision of natural history that William Ashworth took to be typical of Renaissance zoology.[53] As newly discovered birds, our toucans and hornbills definitely belong to a 'disenchanted' world vision (to use Max Weber's terminology).

Neither are there any clear-cut epistemological ruptures, as is described by Michel Foucault in *Les mots et les choses*. One of the main witnesses quoted by Foucault is Buffon and his negative judgment on Aldrovandi: 'qu'on juge après cela quelle proportion d'histoire naturelle on peut trouver dans tout ce fatras d'écriture'.[54] And another often-quoted judgment on this 'prolixité [...] accablante': 'on les réduiroit à la dixième partie si on en ôtoit toutes les inutilités & toutes les choses étrangères à son sujet'. However, one tends to forget that Buffon's general verdict on Aldrovandi is positive, according to the second part of this sentence, which is often omitted: 'ses livres doivent être regardez comme ce qu'il y a de mieux sur la totalité de l'Histoire Naturelle; le plan de son ouvrage est bon, ses distributions sont sensées, ses divisions bien marquées, ses descriptions assez exactes, monotones, à la vérité, mais fidèles'. For Buffon, the differences from Aldrovandi are reduced to a 'défaut', a 'différence de style' and taste: 'nous préférons un petit ouvrage bien raisonné à un gros volume bien sçavant', and he immediatily underlines the dangers of this modern spirit: 'seulement il est à craindre que venant à mépriser l'érudition, nous ne venions aussi à imaginer que l'esprit peut suppléer à tout, & que la Science n'est qu'un vain nom.'

The observations of both Brisson and Buffon continue to draw heavily on classical and Renaissance descriptions, which are of course often discussed, but not rejected as a whole. If there are any clear-cut breaks, it is not with the zoologists of earlier ages, but with their

---

[53] Ashworth W.B. Jr., "Natural History and the Emblematic World View", in Lindberg D.C. – Westman R.S. (eds.), *Reappraisals of the Scientific Revolution* (Cambridge: 1990) 303–332.

[54] Buffon, *Histoire naturelle* 28, quoted by Foucault, *Les mots et les choses* 55.

immediate predecessors and contemporaries: Thevet against Belon, Léry against Thevet, Buffon against Brisson. On the whole, however, at least until Cuvier, ornithology evolved in a continuum, with many side-paths and dead-ends, not by sudden epistemological shifts, great leaps or ruptures.

Selective Bibliography

ALDROVANDI ULISSE, *Ornithologiae, hoc est de avibus historiae libri XII* (Bologna: 1599–1603), 3 vols.

ANONYMUS, *Les Delices de Leide, Une des célèbres villes de l'Europe* [...] (Leiden: 1712).

ASHWORTH W.B. Jr., "Natural History and the Emblematic World View", in Lindberg D.C. – Westmann R.S. (eds.), *Reappraisals of the Scientific Revolution* (Cambridge: 1990) 303–332.

ASÚA M. DE – FRENCH R.K., *A New World of Animals: Early Modern Europeans on the Creatures of Iberian America* (Aldershot: 2005).

AUSTIN O.L. Jr., *Birds of the World. A survey of the twenty-seven orders and one hundred and fifty-five families* (London: 1962).

BALIS J., *Van diverse pluimage. Tien eeuwen vogelboeken* (Antwerp – The Hague – Brussels: 1968).

BELON DU MANS PIERRE, *L'Histoire de la nature des oyseaux. Fac-similé de l'édition de 1555*, ed. Glardon P. (Geneva: 1997).

BIANCACASTELLA A. (ed.), *Les animaux et les créatures monstrueuses d'Ulisse Aldrovandi* (Arles: 2005).

BONTIUS JACOBUS, *Historiae Naturalis et Medicae Indiae Orientalis*, in Piso Willem, *De Indiae utriusque re naturali et medica* (Amsterdam: 1658).

BRISSON MATHURIN JACQUES, *Ornithologia, sive synopsis methodica sistens avium divisionem in ordines, sectiones* [...] *varietates – Ornithologie ou Méthode contenant la division des oiseaux en Ordres, Sections, Genres, Espèces & leurs Variétés* (Paris: 1760) 6 vols.

BUFFON GEORGES-LOUIS, COUNT OF LECLERC, *Histoire naturelle des oiseaux* (Paris: 1771–1786), 10 vols. (www.buffon.cnrs.fr; last consultation April 10, 2007).

CARDANO GEROLAMO, *De subtilitate* (Basle: 1554).

CLUSIUS CAROLUS, *Exoticorum libri decem, quibus animalium, plantarum, aromatum aliorumque peregrinorum fructuum historiae describuntur* (Antwerp: 1605).

——, *Rariorum Plantarum Historia* (Antwerp: 1601).

FABER JOHANNES, *Animalia Mexicana* [...] (Rome: 1628).

FOUCAULT M., *Les mots et les choses. Une archéologie des sciences humaines* (Paris: 1966).

FREEDBERG D., *The Eye of the Lynx: Galileo, His Friends, and the Beginnings of Modern Natural History* (Chicago: 2002).

GEORGE W., "Alive or Dead: Zoological Collections in the Seventeenth Century", in Impey O. – Macgregor A. (eds.), *The Origins of Museums. The cabinet of curiosities in sixteenth- and seventeenth-century Europe* (Oxford: 1985) 179–187.

GESNER CONRAD, *Historia animalium* (Zurich: 1551–1558), 5 vols.

——, *Icones avium omnium, quae in historia avium Conradi Generi describuntur* (Zurich: 1560).

HAUPT H. – VIGNAU-WILBERG Th. – IRBLICH E. – STAUDINGER M., *Le Bestiaire de Rodolphe II. Cod. Min. 129 et 130 de la Bibliothèque nationale d'Autriche* (Paris: 1990).

HERNÁNDEZ FRANCISCO, *Rerum medicarum Novae Hispaniae thesaurus seu nova plantarum, animalium et mineralium mexicanorum historia* (Rome: 1651).

JONSTON JAN, *Historiae naturalis de avibus libri VI* (Amsterdam: 1650).

LÉRY JEAN DE, *Histoire d'un Voyage Faict en la Terre du Brésil* (Geneva: 1585). (http://www.mafua.ufsc.br/BT1730002.html; last consultation April 10, 2007).

MARCGRAF GEORG – PISO WILLEM, *Historia naturalis Brasiliae* [...] (Amsterdam: 1648).

MASON P., *Infelicities. Representations of the Eexotic* (Baltimore: 1998).

NIEREMBERG JUAN EUSEBIO, *Historia naturae, maxime peregrinae, libris XVI. distincta* [...] (Antwerp: 1635).

OTTERSPEER W., *Groepsportret met Dame II. De vestiging van de macht. De Leidse universiteit, 1673–1775* (Amsterdam: 2002).

Oviedo Gonzalo Fernández de Valdés, *Natural History of the West Indies*, transl. and ed. Stoudemire S.A. (Chapel Hill: 1959).

Paré Ambroise, *Des monstres et prodiges*, ed. Céard J. (Geneva: 1971).

Schulze-Hagen K. et al., "Avian taxidermy in Europe from the Middle Ages to the Renaissance", *Journal für Ornithologie* 144(4) (2003) 459–478.

Smith P.J., "Wandelen in de *Délices de Leide* (1712)", *De zeventiende eeuw* 22 (2006) 185–208.

Strien-Chardonneau M. van, "Leiden, metropool der muzen en tuin van Holland. Leiden in Jean Nicolas Parivals *Les Délices de la Hollande*", *De zeventiende eeuw* 22 (2006) 171–184.

Thevet André, *Cosmographie universelle* (Paris: 1575).

——, *Histoire d'André Thevet, de deux voyages par luy faits aux Indes Australes et Occidentales*, eds. Lestringant F. et Laborie J.-C. (Geneva: 2005).

——, *Les Singularitez de la France antarctique* (Paris: 1558).

Willughby Francis – Ray John, *The Ornithology* [...] *in three books. Wherein all the birds hitherto known, being reduced into a Method suitable to their Natures, are accurately described* [...] (London: 1678).

Appendix

Fragment from Anon., *Les Delices de Leide, Une des célébres villes de l'Europe* [...] (Leiden: 1712)

Le Bec de l'Oiseau Rhinocerot ou Toupan; autrement nommé Pie de Bresil. *Présent de* Mons. Adr. de Pauw

Le Bec de cette *Pie de Bresil* est d'une substance très-deliée, comme celle d'une membrane, mais osseuse, claire & reluisante, creuse, & qui donne facilement passage à l'air. C'est pourquoi, contre l'ordinaire de tous les autres Oiseaus, elle n'a pas de Narines; parce qu'étant d'une substance très mince, très-déliée, & diaphane, l'air y peut facilement entrer. Que si la Nature y eût laissé quelque endroit ouvert pour le passage de l'air, cela eût rendu ce bec sujet à être rompu, par quelque accident survenu exterieurement. Or afin que l'air coulat aisément dans ce Bec, & s'y glissat continuellment par le palais, la même Nature en a fait les bords en forme de scie, & comme dentelés, pour l'empêcher. Et afin que ce Bec qui est d'une si grande longueur, & d'une si grosse épaisseur, soit bien apuyée & soutenu, la tête en est à proportion de toute la masse du corps, grande & grosse. [...]

Les Americains naturels appellent cet Oiseau Toucham, les Latins *Pie Barbare*. D'autres le nomment *Ramphast*, & Oiseau *Mange-Poivre*. Les Allemans *Pfeffer Vogel*, car il s'en remplit avec une avidité si grande qu'il le rend tout aussi tôt crud, & sans être digeré. C'est pourquoi les Americains l'estiment davantage, parce, disent-ils, que la chaleur rude & âpre du poivre est corrigée par ce moyen. C'est ce qu'*André Thevet* rapporte dans sa *Cosmographie*. Le *bec de l'Oiseau Rhinocerot ainsi nommé* par Aldrovandus au Liv. XII. du *Traité des Oiseaux*, au Chapitre 20. *Parce que comme l'Animal qui s'appelle du même nom il porte une corne à son front*. C'est un Oiseau qu'on voit très-rarement. Hesichius & Verinus *disent qu'en Ethiopie se trouvent & un Animal & un Oiseau qui s'appellent Rhinocerots*. Ils rapportent *que les Chrétiens le virent en l'air, lors que dans l'Année 1571 ils remporterent cette memorable Victoire sur les Turcs* dans le Golfe de Lepante ou de Corinthe, sous les auspices & la conduite de Jean d'Autriche, *où ayant percé d'un dart en volant, il tomba mort au milieu des combattans*. Sa Tête, disent les Histoires, étoit d'une prodigieuse grandeur, ayant à peu près deux pieds de circonference, les ailes noires & enrichies de touffes de plumes, qui pendoient en bas. Son Bec environ long de quatre pieds, un peu courbé, non toutefois à la maniere des Oiseaux de proye, mais en façon d'un arc, ou d'un croissant. Le dessous du Bec étoit de couleur jaune titant sur le blanc; Le dessus vers la tête étoit d'un beau vermillon, & le reste étoit un peu jaunâtre, comme nous nous avons dit que le dessous étoit un peu blanchâtre. Ce Bec en dedans, tant dans la partie inferieure, que dans la superieure, étoit dentelé comme une scie. Une Corne également grande & pesante, & qui avoit à sa racine environ une paume de largeur, lui sortoit du front, & étoit adhérente à la partie superieure de son Bec, fort peu differente de celle de l'animal Rhinocerot, si ce n'est qu'elle étoit tant soit peu recourbée à son extrémité. La couleur de cette Corne étoit en haut & en bas de vermillon, & sur le milieu de couleur jaune. Cette description convient en tout au prodigieux Bec dont nous parlons.

BIOLOGY AND THEOLOGY IN FRANZIUS'S *HISTORIA ANIMALIUM SACRA* (1612)

Vibeke Roggen

In 1612, the protestant theologian Wolfgang Franzius published a book entitled *Historia animalium sacra*. Franzius (Franz or Frantze, 1564–1628) was Probst [Rural Dean] of Wittenberg and professor of theology at the university [Fig. 1].[1] Wittenberg – the city where Luther had initiated the Reformation in 1517 – was still a stronghold for the Reformation, but the Catholic Counter-Reformation was a threatening reality. The *Historia animalium sacra* was widely read: A long series of editions appeared from 1612 until 1671, and an English translation was published in London in 1670. To later readers, Franzius's work was presented together with extensive commentaries written by Johannes Cyprianus (1642–1723), professor of physics and later of theology in Leipzig. Cyprianus's edition first appeared in 1687 and was reprinted as late as 1712.[2]

The *Historia animalium sacra* was much used and it was referred to with honour and respect. For example, Franzius is mentioned among the ichthyologists in the *Encyclopédie Française*.[3] And Georg Caspar Kirchmayer (1635–1700) refers to Franzius's work many times in *Hexas disputationum Zoologicarum* (Wittenberg: 1661). Among the animals treated by Kirchmayer is the Phoenix: 'To our mind the Phoenix is a pure fig-

---

[1] An earlier version of this paper was presented at a Symposium in Honour of Ann Moss, arranged by The Cambridge Society for Neo-Latin Studies, 20–22 September 2004 at Clare College, Cambridge. An important primary source to Franzius's life is the funeral speech given by the headmaster of the university, Jacob Martini (1570–1649), in Henning Witte, *Memoriae theologorum nostri saeculi clarissimorum renovatae decas prima (-sexta)* (Königsberg – Frankfurt a.M.: 1674–1675) 311–316.

[2] Editions of the *Historia animalium sacra* appeared in Wittenberg 1612, 1613, 1616, 1621, 1624, 1633, 1642, 1659; Amsterdam 1643, 1654, 1665; Frankfurt a. M. 1671. Cyprianus's edition was first published in 1687 in Dresden, and reprinted there the following year. Cyprianus celebrated the centenary for the first edition of *Historia animalium sacra* through an edition with a new preface; Frankfurt and Leipzig 1712. The editions from 1613, 1616, 1654 and 1665 have been examined as preparation for the present paper, in addition to Cyprianus's 1712 edition. The Latin quotes hereunder are taken from the 1613 edition.

[3] In the article "Ichthyologiste", *Encyclopédie, ou dictionnaire raisonné des sciences, des arts et des métiers*, vol. VIII (Neuchâtel: 1765) 485.

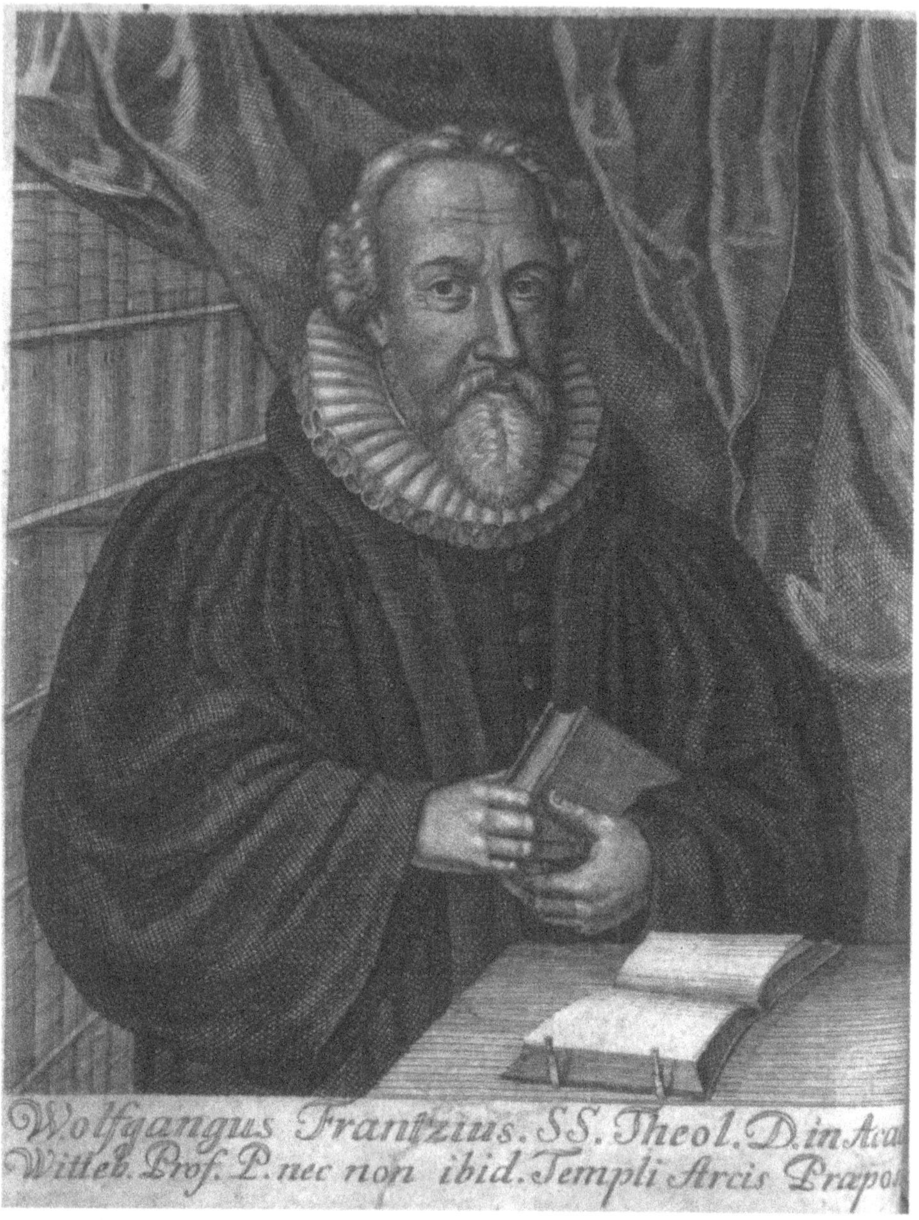

Fig. 1. Wolfgang Franzius, portrait on copper plate. From Franzius, *Historia animalium Sacra* (Frankfurt – Leipzig, Gottofredus Leschius: 1712). Translation of the inscription: 'Wolfgangus Frantzius, doctor of the sacrosanct theology, professor in the academy of Wittenberg, and Rural Dean of the castle church'. University Library of Oslo, Department of Biology.

ment and nonentity. Long ago this was the belief of such great men as Herodotus, Pliny, Gesner, Aldrovandus, Franzius, and Sperlingius'.[4] The company in which Franzius is included demonstrates the high esteem he enjoyed at the time. Another example is found in a treatise by Isaac Schoockius (d. 1681) on a fable regarding the shape of newborn bears. Even if Franzius is not among the authors quoted in the very brief treatise, he is recommended for further reading: 'In addition to the quoted authors, more information can be obtained on the subject of the bear in Aldrovandus, Gesner, Franzius [...]'.[5]

However, Franzius's work *Historia animalium sacra* has not been met with much interest during the last two centuries. Though it has not been completely forgotten, no critical edition of the *Historia animalium sacra* has appeared, no translation after the one from 1670, and no commentary after the one by Cyprianus. It has been included among the targets in Andrew Dickson White's study from the late 19th century on 'the warfare of science with theology'. On the general situation, White says that the religious use of natural science went on after the Reformation, following Luther's example.[6] He devotes some lines to the *Historia animalium sacra* and Franzius, but his treatment is very superficial and confined to only one example – namely the 'natural dragons' (see below). But of course, a work of 1000 pages should not be judged on the basis of one chapter. One century after White, in an article entitled "The Virtues of Animals in Seventeenth-Century Thought", Peter Harrison relates Franzius's way of reasoning to a medieval way of thinking. Harrison's discussion of Franzius seems to be based primarily on examples from one article (on the lion), but the statements he makes are general, namely that the *Historia animalium sacra* 'is a typical early seventeenth-century work of natural history', and that 'there is what seems to us a rather credulous acceptance of the traditional sources'.[7] In the German standard work on the history of biology, *Geschichte der Biologie*, which has appeared in various editions since the first one in 1982, Franzius was not mentioned at all

[4] Kirchmayer is quoted from a collection of 17th-century treatises, translated from Latin by E. Goldsmid and published under the title *Un-Natural History* (Edinburgh: 1886) 49.

[5] Isaac Schoockius in Goldsmid, *Un-Natural History* 106.

[6] White A.D., *A History of the Warfare of Science with Theology in Christendom*, vol. I (Gloucester, Massachusetts: 1978 [1896]) 37–38.

[7] *Journal of the History of Ideas* 59 (1998) 463–489 (469–470). Harrison refers to the English translation from London 1670.

to begin with, when the approach was mainly Marxist.[8] In the latest editions, however, the treatment of biology in the early modern period has been expanded, and there is a new chapter on 'Tierbücher unter dem Einfluß der christlichen Religion', including the *Historia animalium sacra*.[9] This very brief chapter (3/4 page) is based upon the work of Änne Bäumer on Franzius and other representatives for the Biblical zoology (see below).[10]

A full study of Franzius's treatise on animals would exceed the limits of an article; the aim of the present paper is to give a general presentation of the work, and to answer the following questions: What was Franzius's project, as a theologian, for writing a book on animals? What were his sources of knowledge and what were his methods? How did he argue and how did he make up his mind on disputed questions? Priority is given to an examination of the introductory parts, Franzius's classification system, a selection of animal chapters, and his treatment of certain problematic questions, particularly fable animals and spontaneous generation.

The Historia animalium sacra – *structure and presentation*

The *Historia animalium sacra* is a book of about a thousand pages in octavo. It was dedicated to the Duke of Saxony, Johannes Georg I (1585–1656, prince-elector from 1611), who was also the man who accepted the book for printing, according to information on the title page. Through the title, Franzius connected his work to two famous predecessors, Conrad Gesner (1515–1565), whose *Historia animalium* in five large volumes was published in the period 1551–1587, and Aristotle. And by adding *sacra*, Franzius emphasized his ideological basis in the Christian religion as well.

The *Historia animalum sacra* has no illustrations, except for an image on the title page. The layout of the title page differs, particularly when different printing houses are involved. In the edition from Wittenberg 1613, printed by Johann Gormann, there is an image of a bird in flight [Fig. 2]. There is a text written above its head: 'Jahve' in Hebrew letters.

---

[8] Jahn I. – Löther R. – Seuglaub K. (eds.), *Geschichte der Biologie* (Jena: 1982).

[9] Hoppe B., „Botanik und Zoologie in der Zeit der Renaissance und des Humanismus", in Jahn I. (ed.), *Geschichte der Biologie* (Hamburg: 2004) (161–195) 189.

[10] Bäumer Ä., *Geschichte der Biologie*, vol. II (Frankfurt a. M.: 1991) 156–168. An article that appeared two years later is a reprint (very slightly revised): „Biblische Zoologie (Hermann Frey, Wolfgang Franz, Heinrich von Hoevel)", in Kattenstedt H. (ed.), *"Grenz-Überschreitung": Festschrift zum 70. Geburtstag von Manfred Büttner* (Bochum: 1993) 3–17.

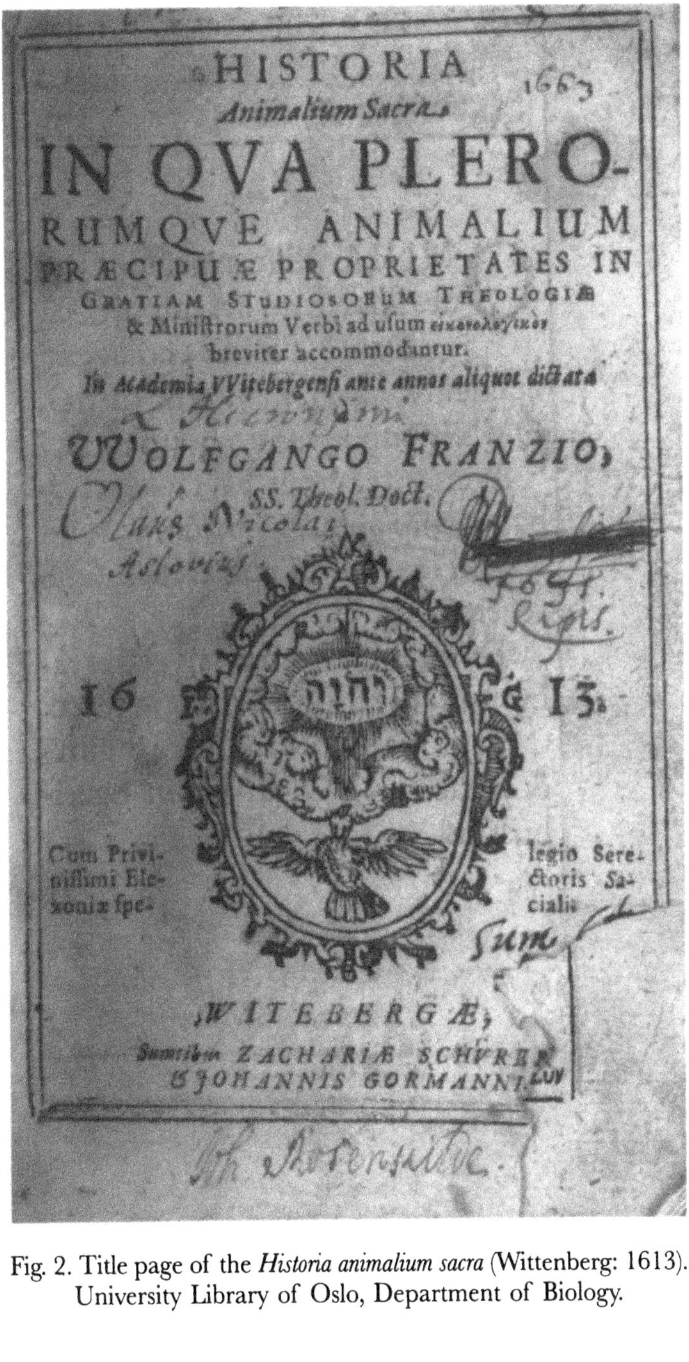

HISTORIA
Animalium Sacra
IN QVA PLERO-
RUMQVE ANIMALIUM
PRÆCIPUÆ PROPRIETATES IN
GRATIAM STUDIOSORUM THEOLOGIÆ
& Ministrorum Verbi ad usum εἰκονολογικὸν
breviter accommodantur.
In Academia VVitebergensi ante annos aliquot dictata
VVOLFGANGO FRANZIO,
SS. Theol. Doct.
יהוה
16 13.
Cum Privi- legio Sere-
nissimi Ele- ctoris Sa-
xoniæ spe- ciali.
WITEBERGÆ,
Sumtibus ZACHARIÆ SCHURERI
& JOHANNIS GORMANNI.

Fig. 2. Title page of the *Historia animalium sacra* (Wittenberg: 1613). University Library of Oslo, Department of Biology.

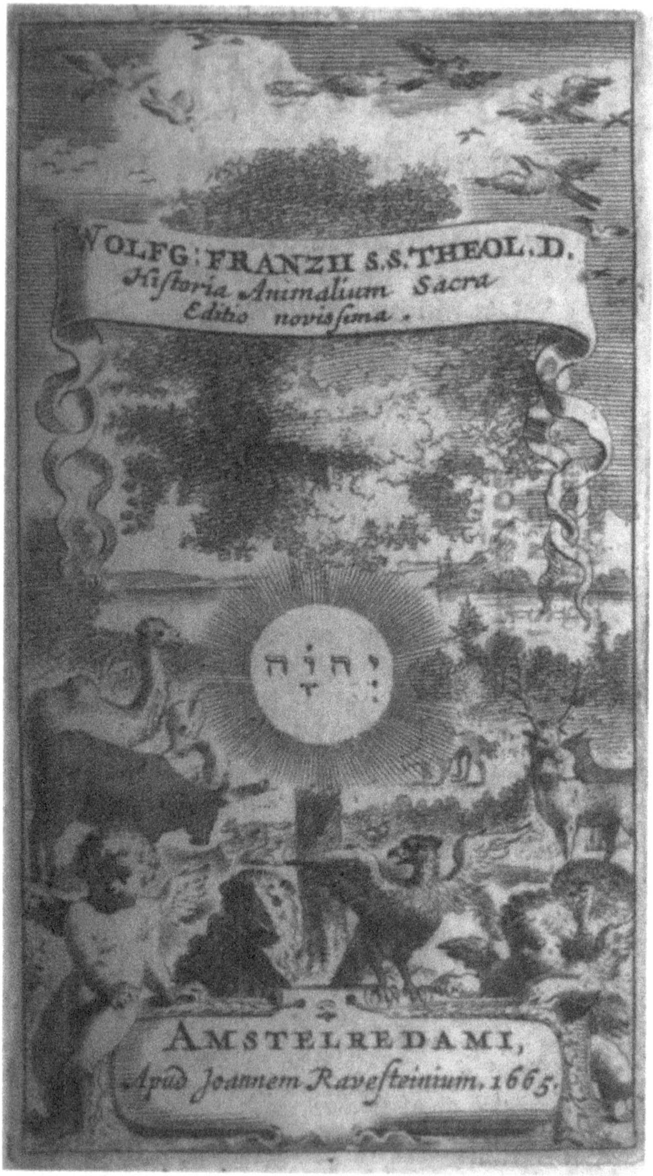

Fig. 3. Title page of the *Historia animalium sacra* (Amsterdam, Joannes Ravenstein: 1665). In this image, God's name is the center around which there is a landscape with various animals. University Library of Oslo, Department of Biology.

One possible interpretation is that the bird is a dove, a symbol of the Holy Spirit. Another possibility is that the bird is a representative of the animals that praise the Lord.[11] The title page of the 1616 edition, which was printed in the same printing house as the preceding edition, has the same layout. The word 'Jahve' is an element also in the title page of the edition from Amsterdam 1665 [Fig. 3]. The title page of the edition from Amsterdam 1654 gives the illusion of three-dimensionality, with Adam and Eve standing on a stele [Fig. 4]. A common element in the title page images is the connection between nature and God.

There are more than 80 pages of paratexts in the *Historia animalium sacra* [see the table below]. On the verso of the title page there is an extract from Psalm 104 with the primary content 'glory to the Lord' and 'God rejoices over his labour'. In Franzius's fairly long *Praefatio* (42 pages) this subject is emphasized very strongly through a phrase that is

---

A table of contents, constructed from the 1613 edition

Title page (see Fig. 2)
Extract from Psalm 104, fol. a v
Dedicatory preface, fols. a2 r–c6 v (42 pp.)
To the pious reader, fols. c7 r–c8 v (4 pp.)
A physical tract on the nature of many brute animals, pp. 1–888
  Chapters 1–3, on classification of animals and other general questions, pp. 1–31
[Part 1, On fourfooted animals], pp. 31–324
Part 2, On birds, pp. 324–584
Part 3, On fish, pp. 584–706
  Chapter 1, On fish in general, pp. 584–609
  [...]
Part 4, On serpents and on insects, pp. 706–888
  Chapter 1, On serpents in general, pp. 706–722
  [...]
  Chapter 7, On insects in general, pp. 755–770
  [...]
  Chapter 16, Conclusion, p. 888
Table of contents, listing all the chapters, 4 pp. (unpaginated)
Index rerum, 28 pp. (unpaginated)

---

11 The layout on the title pages of books printed by Gormann varies; for example, the title page of Franzius's *Augustanae confessionis articuli fidei XXI* (1611) has text only. Thus, it seems that the image is related to the content of the work.

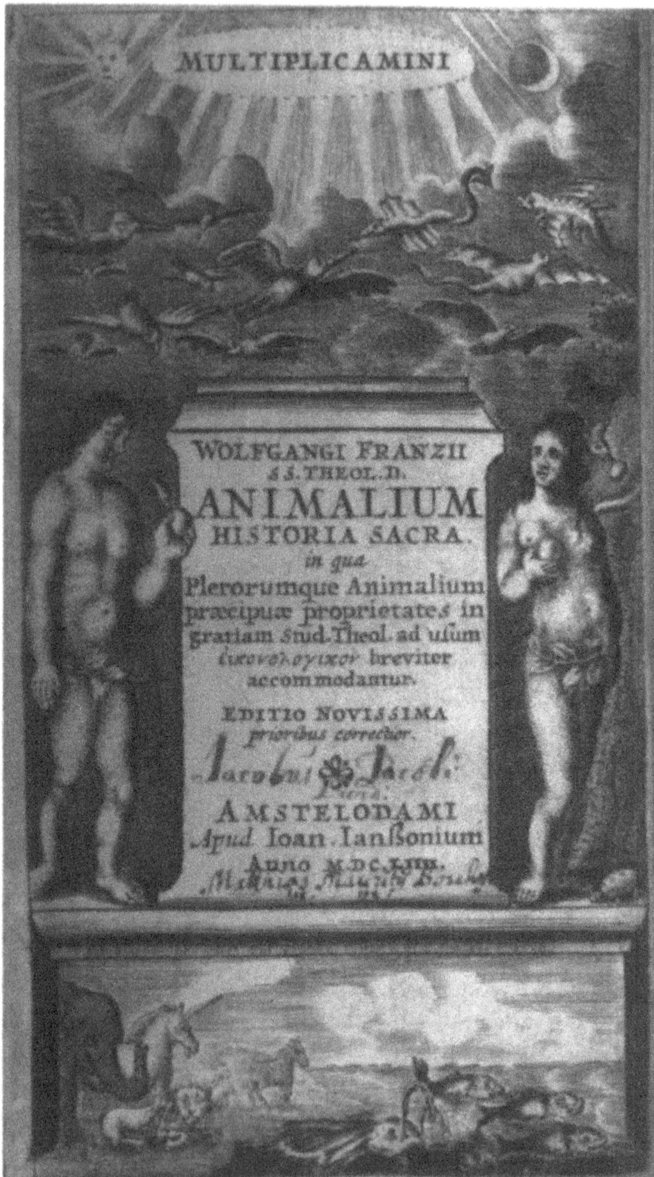

Fig. 4. Title page of *Animalium historia sacra* (Amsterdam, Johannes Janssonius: 1654). Above Adam and Eve is the sky with birds and the Latin word 'multiplicamini' ('multiply') uttered from above (*Genesis*, Chapter 1, 22; 28). The lower section of the page shows a landscape with animals, among them a unicorn. University Library of Oslo, Department of Biology.

repeated again and again: 'Vox confitentium DEUM CREATOREM' ('the voice of those who confess that God is the creator'). The preface demonstrates Franzius's ability as a writer; for example, in the following he describes how each species has its own particular nature,

> to the extent that no cuckoo ever seeks to compete with the cocks' crowing, or lions with the pigs' grunting, or dogs with the horses' whinnying, or hens with the bees' honey-making art, so that never does the cat reveal a thief at night, never do the little birds sing together with the yapping dogs; so that never does the desire for meat sneak into sows, the desire for corn into swallows, the desire for bones into horses, the desire for clover and thyme into dogs; so that you can never accustom the hare or the rhinoceros to the manger, to the yoke and to ploughing the fields like oxen or horses; so that dogs and cats never leave their masters and homes and clothe themselves in the savagery and tendency to range about in the wilderness of wild pigs, wolves or lions [...].[12]

Franzius's explanation of the many differences – which obviously impress him – was that this is God's plan, through the creation of the world. He dryly commented that many learned works arise from these many differences (*Historia animalium sacra*, fol. a4 r).

The plan for the work and the relationship between the animal world and theology is further explained in the preface 'to the pious reader' – the last part of the prefatory material. The *Historia animalium sacra* was meant for theologians, meant for priests, meant for teaching: when you want to teach a certain doctrine to a simple audience, so that they will be able to evocate to their minds the sight of this animal or that, you can find many such parables or similes in this book ['copiam talium seu parabolarum seu similitudinum'] (*Historia animalium sacra* fol. c8 r).

[12] *Historia animalium sacra* fol. a3–a4: 'adeo ut nunquam ullus cuculus gallorum cantilenas, nec leones grunnitum porcorum, nec equorum hinnitum canes, nec apum mellificandi artem gallinae aemulari tentent, ut numquam felis prodat furem nocturnum, nunquam avicula noctu cum canibus latrantibus canant, ut nunquam ovibus carnium, hirundinibus frumentorum, equis ossium, canibus trifoliorum et thymorum obrepant desideria, ut nunquam leporem aut rhinocerotem ad praesepia, ad jugum et ad agros arandos more boum aut equorum assuefacere queas, sicut nunquam canes et feles dominis et domibus derelictis aprorum aut luporum aut leonum feritatem et in desertis circumvagandi naturam induent [...]'. The translations from Latin are mine.

Treatment of categories and species

In the main part of the work, the first three chapters have introductory character. In Chapter 1, 'On the argument of this tract', Franzius argues that even if natural history is part of the 'doctrina physica' (*Historia animalium sacra* 1), it should not be looked upon as primarily the field of the physicians, 'but of each and every scholar, and above all theologians' (*Historia animalium sacra* 2).[13] The subject is useful, firstly, *ad disserendum*, which is to be able to discuss matters regarding animals and nature with other scholars, and present material related to animals in speeches and preaching without making mistakes; secondly, it is useful in order to explain sacred and profane texts;[14] and, thirdly, it is helpful in praising God. As it seems, the first aim is directly related to nature and animal life, whereas aim two is related to interpretation of texts; partly texts on animals *per se*, partly texts where animals and animal behaviour are used in an abstract sense. Chapter 2, "On the definition of the irrational animal", contains some general statements – for example, that small animals are the most intelligent. In the same chapter, Franzius has treated the question of how life arises (see below). Chapter 3, on classification, begins with a summary of the scholarly situation. No less than ten (alternative) sets of criteria for classification are listed, one of which divides animals into those that give birth to living animals and those that lay eggs – *animalia vivipara* and *ovipara*. Franzius's opinion is shown in a logical table [Fig. 5].[15]

[13] Bäumer, on the other hand, says in *Geschichte der Biologie*, vol. II 162, that the animal world 'sei eigentlich die Aufgabe der "Physica"', which seems inaccurate. At this point it has to be said that whereas Bäumer refers to the edition from 1612, I have used the edition that appeared, also in Wittenberg, in 1613. My discussion with Bäumer (who does not quote Franzius in Latin) is based upon the assumption that the second edition was a reprint, which seems likely, since Bäumer's references coincide with the page numbers in the 1613 edition of the *Historia animalium sacra*.

[14] *Historia animalium sacra* 3: '2. ad explicandum scriptores sacros et profanos'; Bäumer's translation is inaccurate: '2. zur Erklärung der Heiligen Schrift'; see *Geschichte der Biologie* vol. II 162.

[15] Bäumer presents a German version of the table; see *Geschichte der Biologie*, vol. II 160.

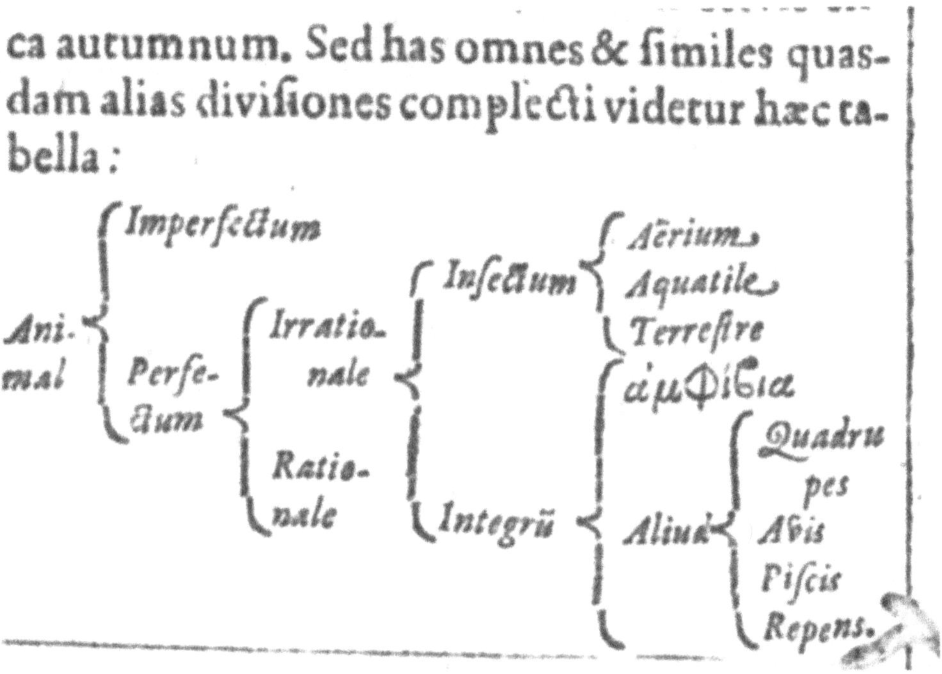

ca autumnum. Sed has omnes & similes quasdam alias divisiones complecti videtur hæc tabella:

Fig. 5. Logical table showing Franzius's classification of the animals (*Historia animalium sacra* 30). The superordinate concept *animal* is divided into 'imperfect' animals (e.g. zoophytes) and 'perfect' animals, the last category in the 'rational animal' (man) and in 'irrational animals' (all other animals). These are divided in insects ('cut in') and *animalia integra* – non-insects. University Library of Oslo, Department of Biology.

Each part of the *Historia animalium sacra* has introductory chapters. Part 3, "On fish", begins with a discussion of twelve general matters, for example the similarities between fish and other groups of animals. There are analogies between the bodily parts of birds and fish: feathers and wings correspond to scales and fins. Moreover, fish move through contractions of the body, like serpents. The big animals in the ocean do not lay eggs, but have genitalian parts like animals on land; likewise, they have lungs. Franzius concluded: 'they are classified by some among the terrestrial animals, but since they always stay in the water, it is better to call them fish'.[16] Which scholars does Franzius refer to? Most famous among the 16th-century ichthyologists are Pierre Belon, who gave a good description of the lungs and brain of the marine animals (1551), and his contemporary Guillaume Rondelet (1554). But according to Hoppe, they both classified the whale among fish.[17] Ilse Jahn claims that Aldrovandi was the first scholar who separated fish and whales, in a posthumous volume, published 1612–1613.[18] But this must be wrong, since this volume was hardly published before the *Historia animalium sacra*; cp. Franzius's words, 'they are classified by some among the terrestrial animals'. How this may be, it seems that all these scholars had a fairly good comprehension of the nature of marine mammals, and the classification could be looked upon as a matter of the *definition* of the concept *fish*. Franzius's knowledge was partly based upon personal observation: 'In big marine animals there is no spine, but solid and very big bones, as people observe with astonishment in this city, in the castle church'.[19] Franzius was *praepositus* in the castle church[20] and surely it was him who had organised the exposition, driven by his interest in zoology and his wish to teach the people of Wittenberg.

---

[16] *Historia animalium sacra* 586: 'ab aliquibus annumerantur terrestribus animalibus, sed tamen quia semper manent in aquis, commodius appellantur pisces'.

[17] Hoppe in Jahn (ed.), *Geschichte der Biologie* 192–193.

[18] Jahn I., „Naturphilosophie und Empirie in der Frühaufklärung (17. Jh.)", in eadem (ed.), *Geschichte der Biologie* (197–230) 226.

[19] *Historia animalium sacra* 585: 'In magnis tamen piscibus marinis non sunt spinae, sed solidissima ossa eaque maxima, quemadmodum cum admiratione talia conspiciuntur in hac urbe, in templo arcis'.

[20] Jacob Martini in Henning Witte, *Memoriae theologorum nostri saeculi* 311.

The main part of the *Historia animalium sacra* is the treatment of the species, one by one or in small groups. Franzius has described the differences between his treatise and the works of his predecessors:

> I have deliberately not taken special consideration of what is usually, and in a very learned manner, taught about the nature of animals, the size of their bodies, their outer and inner parts, their natural actions to maintain life or propagate their kind, the voice of each individual animal, its senses, nourishment, hiding places, movements, health, diseases, growth, conception, birth, age, usefulness [...] but just the things that particularly concern its way of managing its life, and to offer something useful to men of my profession, students of theology and servants of the Word [...].[21]

For example, the chapter on the peacock begins with a description of the bird's appearance, the beauty of its feathers and its slender body, 'since God gave it the palm of victory among the birds'.[22] But the species is also known to have five vices, namely that it is 'cunning, ambitious, envious, malevolent, licentious'.[23] Its ambitions are connected with the fact that the bird is too aware of its beauty, and it is described as envious because it swallows its own excrement – which is useful in medicine, according to Franzius. It is malevolent because it does not love its own children: the male bird destroys the eggs if it finds them. Franzius saw many ways in which the peacock's behaviour and attitudes could be compared with those of certain human beings; he lists about a dozen analogies. For instance, in the same way the peacock spontaneously begins to revolve its tail from time to time, some people may suddenly burst out in praise of themselves.

---

[21] *Historia animalium sacra* fol. c: 'In illo autem non ego ex professo respexi ista praesertim anima, quae de animalium naturis, corporum quantitatibus, partibus externis et internis, naturalibus actionibus pro vitae ipsorum conservatione, vel speciei propagatione, animantis cujusque voce, sensu, alimento, latibulo, motu, sanitate, morbis, educatione, conceptu, partu, aetate, de usibus [...] sed tantum illa, quae morum regendorum rationem cumprimis spectabant, et meae professionis hominibus, Theologiae studiosis et Ministris verbi commoditatem aliquam praebere'.

[22] 'utpote cui Deus palmam dedit inter aves'. Cp. the chapter "On the peacock", in *Historia animalium sacra* 359–365.

[23] 'callida, ambitiosa, invida, malevola, libidinosa'.

Chapter 11 treats two species, unicorn and rhinoceros, and discusses the question whether or not unicorns exist. Franzius's point of departure is the notion that 'some very learned men have denied that the unicorn exists in nature'.[24] But the Bible uses learned images related to the unicorn, and Franzius's conclusion is that the unicorn *does* exist.[25] Bäumer claims that Franzius 'sie [die Existenz des Einhorns] aber schon deshalb als bewiesen ansieht, weil dieses Tier in der Bibel erwähnt werde'.[26] But in addition to the Bible, Franzius based his view upon a report from an alleged eyewitness: 'Add to this that the very serious author Julius Caesar Scaliger, in his *Exercises against Cardanus*, quotes testimony from a certain friend of his who has seen a unicorn'.[27] In the case of the phoenix, Franzius concluded that the bird is fictitious, even though many early Christian authors accepted it.[28] Franzius's arguments were related to the bird's life: whereas other birds live in couples, the phoenix lives alone – which contradicts the story of the Flood; whereas other birds make nests in order to reproduce their young, the phoenix does so to prepare its death; whereas other birds love the sun, the phoenix is burnt and killed by it. When dragons are concerned, such creatures exist in nature, according to Franzius, but they are merely very old and very big serpents; they do not differ from other serpents, except in size (*Historia animalium sacra* 747). However, after this rational explanation, he contradicted it by saying that there are 'two kinds of real dragons' – 'verorum Draconum duae sunt species' – a statement which, as we have seen, was criticized by White.

---

[24] 'doctissimi aliqui viri negarunt esse in rerum natura unicornem'. See the chapter "De Monocerote et Rhinocerote", in *Historia animalium sacra* 109–118.

[25] B. Faidutti's dissertation from University Paris XII 1996 treats the unicorn: *Images et connaissance de la licorne (Fin du Moyen-Age–XIXème siècle)*. In a survey of authors and their points of view regarding the existence of the unicorn, Franzius has been placed among those who say 'Non ou plutôt non', but this is not correct.

[26] Bäumer, *Geschichte der Biologie*, vol. II 163.

[27] *Historia animalium sacra* 109: 'Huc accedit, quod gravissimus author Julius Caesar Scaliger in suis exercitationibus contra Cardanum citat testimonium amici sui cujusdam, qui vidit unicornem'.

[28] In *Historia animalium sacra* 353 he writes: 'satis de hac fabulosa ave' ['enough about this fictitious bird'].

A difficult question: spontaneous generation

The assumption that life can arise through spontaneous generation had a long tradition in ancient Greek thought, even before Aristotle added his authority to it. He saw a combination of heat and humidity as the environment in which certain insects, for example flies and lice, came into being.[29] Pennington sums up the situation regarding spontaneous generation:

> One of the most firmly held of the scientific teachings of Aristotle, expanded by many medieval writers, was spontaneous generation. To question this, Sir Thomas Browne had said, 'is to question reason, sense, and experience'. Johann van Helmont, the discoverer of carbon dioxide, produced a recipe for making mice out of cheese and dirty linen. Everyone knew that putrefied meat generated maggots; but some accounts of the spontaneous appearance of bees and frogs reached almost the same level of mythology as the phoenix. Harvey was uncertain about it. It was not until 1668 that Francesco Redi, after experiments and observations that included the not very difficult one of protecting the meat from flies, claimed that all life arises from reproduction.[30]

Thus, in Franzius's time, all leading scholars accepted spontaneous generation – even Thomas Browne, the author of *Pseudodoxia Epidemica* (1646), often called "Vulgar Errors", an attempt to evaluate vulgar 'truths' in natural history. The common opinion was that there were three ways of reproduction: some species reproduce through sexual intercourse, some take their origin from excrement, and some may be reproduced in either way, for instance mice (*Historia animalium sacra* 19–20). Franzius proceeds:

> But although this is how matters are presented, we shall leave it to more learned men to judge whether everything is bred through the reciprocal custom of the sexes, but nevertheless in such a way that most smaller animals give birth to eggs or something that resembles an egg and lay it in a place where it can be hatched with the help of an external heat.[31]

---

[29] Harig G. – Kollesch J., „Naturforschung und Naturphilosophie in der Antike", in Jahn (ed.), *Geschichte der Biologie* (48–87) 66.

[30] Pennington D.H., *Europe in the Seventeenth Century* (London – New York: 1993 [1970]) 171–172.

[31] *Historia animalium sacra* 21: 'Etsi vero haec ita disputantur, tamen doctioribus judicandum relinquimus, an non omnia producantur ex mutua sexuum consuetudine, ita tamen ut pleraque minora pariant ova vel ovis aliquid simile, illudque hinc inde deponant, ut a calore aliquo externo excludatur'.

Franzius continued his line of thought: One reason why people came to believe that insects originated from, for instance, putrefied material, was the fact that one could not see them: they enter and settle in wood, for instance, through minute pores and, after a short period, breed. Insects settle in water, salted meat, raw meat, in beds, in some trees, in flowers and fruit, and so on. There are no animals without lice, no fish without certain worms. Franzius also argued against the claim that mice arise in ships, and presented an alternative observation to the ancient myth of birds that arise from trees:

> And who can assure us that no mouse has come into the ship together with the goods? So let those who have greater experience consider this more closely. But as for the birds that are said to grow from trees in Scotland,[32] Dutchmen on a sea trip to China saw that they hatched eggs and sat on eggs; so they too are procreated like other birds (*Historia animalium sacra* 22).

True enough, there are examples of spontaneous generation in the Bible. But obviously, such narratives represent *miracles*, very different from regular animal behaviour (*Historia animalium sacra* 767). This must be the case, since God, as Franzius understood him, would not have wanted that one mouse was born from another mouse, whereas others were born from excrement. On this matter, Franzius argued against the view of *all* authorities. However, his point of view has not received much scholarly attention.[33]

Material and methods

It was important to Franzius that his treatment of animal life was correct and corresponding to reality. At his disposal were many early sources: Aristotle and other ancient Greek and Roman authors, *Physiologus*, early Christian authors and medieval bestiaries, and not least, the Bible. But very often, Franzius's starting point for discussing animal life was early modern authorities, whom he referred to as a group with the word *docti*. For example, Julius Caesar Scaliger is characterized as *gravissimus author* (*Historia animalium sacra* 109). In the case of spontaneous generation,

[32] On this myth, see White, *A History of the Warfare of Science* 37.

[33] Bäumer, and Hoppe who follows her, mention Franzius's rejection of spontaneous generation, but relate it to fish only. Bäumer, *Geschichte der Biologie*, vol. II 163; Hoppe in Jahn (ed.), *Geschichte der Biologie* 189.

there was a scholarly agreement. Franzius did not present an open criticism, but wanted the *docti* to think about the arguments that he presented against their view. Through eyewitnesses, he rejected an alleged example of spontaneous generation – birds growing in trees – and when insects in rotten material were concerned, he presented a rational explanation of the phenomenon. In the case of the unicorn, Franzius's discussion was based upon the fact that certain *docti* denied its existence. His primary arguments were the mentioning of unicorns in the Bible, in addition to an eyewitness presented by Scaliger. But even tales from the Bible could be questioned, or explained, as Biblical miracles.

Even more trustworthy than scholarly works were *observations*. When Franzius and the citizens of Wittenberg had seen the skeleton of a big marine animal, they knew what such a skeleton looked like. Also, observations witnessed by other authorities are referred to in the *Historia animalium sacra* and relied upon. An observation by Scaliger is related to the myth that bears are born without shape and licked to shape by their mothers (cp. on Schoockius above): On a bear hunt in the Alps, when a pregnant bear had been killed, the hunters observed that the fetus had the perfect form of a bear (*Historia animalium sacra* 81). In the same chapter, an observation by Pliny the Elder is quoted from circus in Rome. Pliny has also made observations on lions (*Historia animalium sacra* 67). Bäumer, on the other hand, finds that the *Historia animalium sacra* contains interesting zoological information, but claims that this information 'allerdings nicht auf eigenen Beobachtungen beruhen, sondern aus zeitgenössischen Werken entnommen sind'.[34]

Franzius did not use the same kind of critical sense when animals are used in an abstract way. For example, in the chapter on the pelican one finds material from the ancient *Physiologus* that is not trustworthy.[35] After narrating a couple of these fables, Franzius added: 'But even if one thinks that these narratives are not real, but are in fact fictitious, there is nothing that prevents students from using similes from them'.[36] This is a practical and pedagogical attitude; Franzius wanted people to *understand*, and fables could be used in order to achieve this end. As is understood, the criteria for the selection and use of material for pedagogical purposes were quite different from the criteria and use of the

---

[34] Bäumer, *Geschichte der Biologie*, vol. II 164.

[35] *Historia animalium sacra* 354–358, "De Pelicano".

[36] *Historia animalium sacra* 358: 'Etsi autem haec ita sese habere non putantur, imo prorsus fabulosa sunt tamen nihil obstat, quo minus studiosi ex illis similitudine utantur'.

purely biological material. A helpful example is Franzius's treatment of the elk, which he employs for both direct and abstract meaning. This was an animal that he had observed himself: 'Its voice, as I have heard it myself from the animal, is rather sad and miserable; so much that if you did not see it, you would think that it was a child'.[37] The fact that the author is familiar with elks is perhaps the reason why he qualifies Julius Caesar's memorable description of the elk in *De bello Gallico* (Book 6, Chapter 27). Caesar explains that the Germans hunt elks by cutting trees half-way through; since the elk has no knee joints, it sleeps in a standing position, leaning against a favourite tree. And, of course, it is unable to rise again when it falls to the ground and is, therefore, easy to catch. But based upon his observations, Franzius inserts a qualification: 'It has legs, either without joints, or without very mobile joints'.[38] But when the time came to use the elk and its legs in relation to human conduct, Caesar's version was more convenient: 'In the same way the elk has legs without joints, and, consequently, it cannot rise easily when it falls, so Christians are remote from fraud: they do not have splendid positions in this life, they are not armed with external power, and when they meet danger, they are unable to free themselves'.[39]

Among the material in the *Historia animalium sacra* are some proverbs; for example, a German proverb that says that the hare is faithful: 'Er stehet einem bey/wie der Haas beym Trummelschleger: Er bleibet bey seinen Worten/wie der Haas bey seinen Jungen' (*Historia animalium sacra* 203). There was an interest in proverbs and other parts of the popular culture in German areas during Franzius's time, and the process of collecting proverbs and having them published had already begun. Nevertheless, proverbs should be regarded as a primarily oral material.

Animals were also used in the description of famous individuals. In this category, anecdotes and other material related to the fathers of the Reformation should be mentioned. In the chapter on the phoenix, one learns that Melanchthon was called 'Germany's phoenix' and described through a verse, 'Nulla ferent talem secla futura virum' ('No future

---

[37] *Historia animalium sacra* 107: 'Vocem eius ut ipse ex animali audivi, admodum flebilem et miserabilem, adeo, ut si non conspiceretur, existimaretur esse infans'.

[38] *Historia animalium sacra* 107: 'Crura habet aut sine nodis aut sine nimis mobilibus articulis'.

[39] *Historia animalium sacra* 108: 'Quemadmodum enim Alce crura habet sine nodis et sine articulis, et ita cum prociderunt, non possunt facile exurgere: ita Christiani sunt alieni a dolis, et in hac vita non habent splendidas sedes, non sunt armati externa potentia et cum incidunt in periculum seipsos nequeunt liberare'.

epoch will bring forward such a man', *Historia animalium sacra* 348). In the chapter "On the nightingale" (*Historia animalium sacra* 554–565), the myth of the bird's origin is narrated from Ovid's *Metamorphoses*. It is called philomela, from the Greek 'philos' for friend, and 'melos' for song, since 'it cares for nothing but song', and this is related to Melanchthon:

> Since Philip Melanchthon has a name whose initial letters and syllables correspond with the name of this bird, and since he was connected to Luther in a splendid way in the Reformation of the barbarian Papacy, and since he taught and wrote a lot in the fields of theology and philosophy in a very good way, and had innumerable diligent students in this Academy [...], many scholars compare him to the nightingale in their writings and call him Philomela in deed and name.

The link between the bird and the interpretation of the name seems rather tenuous; the connection is made on two levels: that the Reformist 'cares for song' and the literal similarity. The *meaning* of the word Philomela is not related to 'Philipp' or 'Melanchthon' – the latter being a Greek translation of his family name, Schwarzerd. Melanchthon's nickname should be seen in relationship to the interest for word games in the early modern period: acrostics, anagrams, lipograms and so on. Such wordplays were not regarded as merely games. Augarde writes on the anagram: 'This belief in the power of shuffling letters of the alphabet continued from the Middle Ages into the Renaissance. It was thought that character or fate could be discovered by rearranging the letters of a person's name – making an anagram'.[40]

Early modern reception: an example

As a treatise which combines animal life and cultural interpretations, it is not unexpected that the *Historia animalium sacra* was used as a natural history as well as a work in the field of theology and human behaviour. Kirchmayer and Schoockius may serve as examples of the first use; see above. But the primary audience that Franzius had in mind was students of theology and priests, in their work of preaching and teaching. One example of such use is found in a 17th century Norwegian vicar, Nils Thomassøn (1605/06–62), in the commentaries accompanying his

[40] Augarde T., *The Oxford Guide to Word Games* (Oxford, New York: 1986) 71.

Latin poem in *Cestus sapphicus* [A love-girdle in Sapphics], Christiania (Oslo), 1661.[41] Thomassøn's fascinating and original work originated from a wedding poem, and the subject of his rebus poem is *marriage*. Franzius is quoted in the commentary on the first stanza, from his chapter on the cuckoo. This bird's habit of laying eggs in other birds' nests is compared to men who hate marriage and want to avoid the labour of raising children. One may see Thomassøn's high estimation of Franzius's work from his, albeit brief, comment: 'Thus Franzius, by whose authority I congratulate myself that weight has been added to what I said before'.[42] In another context, Thomassøn made use of Franzius's chapter on the vulture; this bird was compared traditionally to people waiting for an inheritance.[43]

What interested Thomassøn in the chapters on the cuckoo and the vulture was Franzius's use of these species in an abstract way; they fitted into his treatment of aspects of marriage and human life. But he also used Franzius as an authority on a philological question – the interpretation of a certain Hebrew word from the Bible, 'Schachak', which had been translated as either cuckoo or seagull; *Vulgata* had *larus*. 'There are many words in the Latin translation of the Bible, says Franzius, which do not correspond to the Hebrew sources'.[44] What Thomassøn wanted was to use the bird as a rebus picture; the letter *c* together with the picture of a *larus* formed the word *clarus* [Fig. 6].

Nils Thomassøn was a vicar on the Norwegian countryside for the last 25 years of his life, and it was in this period that he wrote, or invented, *Cestus sapphicus*. The *Historia animalium sacra* is among the works that Thomassøn quoted so many times that it seems likely that he owned a copy. As a young man, during the years 1625–1630, Thomassøn was a student at the universities of Copenhagen, Rostock, Leiden and – last but not least – Wittenberg. He never met Franzius, who died in 1624, but it was perhaps in Wittenberg that he bought his book on animals.

---

[41] See Roggen V., *Entertainment and learning: Intellectual play – word and picture. A study of Nils Thomassøn's Latin rebus book* Cestus sapphicus. *With edition, translation and a corpus of sources* (Oslo: 2002; orig. Ph.D. thesis, Oslo: 2001).

[42] 'Haec Franzius, cujus authoritate a me supra positis pondus accedere mihi gratulor' (Nils Thomassøn, *Cestus*, in Roggen, *Entertainment* 362–363).

[43] Nils Thomassøn, *Cestus*, in Roggen, *Entertainment* 442–443.

[44] 'Multa sunt, inquit Franzius, vocabula latinae versionis Bibliorum, quae non congruunt cum ipsis fontibus Hebraicis' (Nils Thomassøn, *Cestus*, in Roggen, *Entertainment* 482–483).

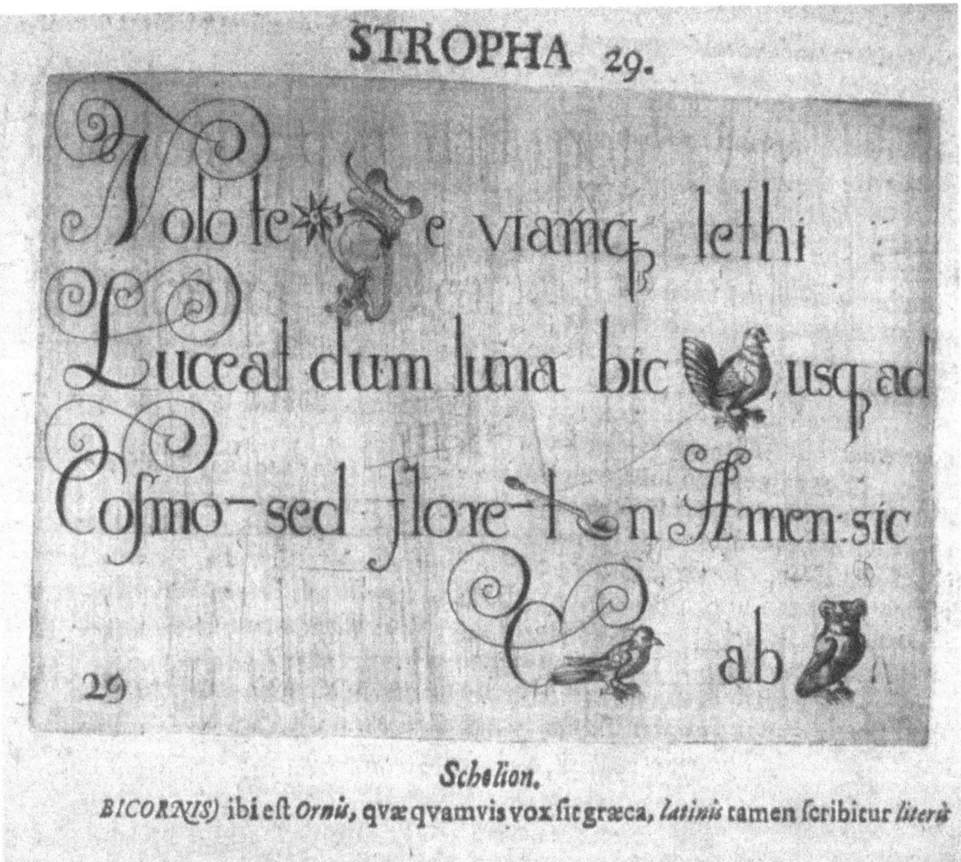

Fig. 6. Stanza 29 in Nils Thomassøn's rebus poem on marriage from 1661, where three birds are used as motifs for rebus pictures. Copper plate engraving by the Norwegian artist Didrik Muus (1633–1706). The rebus figures in the stanza are *calcar* (a spur), *ornis* (hen), *toryne* (spoon), *larus* (cuckoo – see the text), and an *ibis* (long-eared owl). The result of the rebus is a Sapphic stanza: 'Nolo te {calcar}e viamque lethi/ Luceat dum luna bic{ornis}, usque ad/ Cosmo – sed flore – {toryne}n Amen: sic/ C{larus} ab{ibis}' ('I do not want you to tread the road of death under your foot as long as the two-horned moon shines. No, stay healthy until the world is changed into chaos – Amen: then you will pass away gloriously'). National Library of Oslo.

Conclusion

The *Historia animalium sacra* is not the only early modern treatise that combines zoology with theology. But the book demonstrates that Franzius was genuinely interested in the life of animals; as Bäumer observes, he described all groups of animals, whether they were apt for allegorical use or use in sermons or not.[45] And the book found readers – compare the many editions. In his funeral speech on Franzius, the headmaster of the University of Wittenberg, Jacob Martini, emphasized that the *Historia animalium sacra* had the attention of theologians and philosophers alike.[46] It seems likely that 'philosophi' here comprises even natural philosophers – today's biologists.

It would be unhistorical to blame Franzius for his conclusion regarding the existence of unicorns. It is difficult, of course, to prove that a species does *not* exist – particularly in light of the statement of a trustworthy eyewitness. As a result of the European discoveries of new continents during the early modern period, people's image of the natural world had been expanded, and Franzius had no criteria that could help him to decide whether or not the unicorn was fictitious. As he saw it, the life of this species did not contradict experience, reason and logic, as does the life of the phoenix. The label 'credulous', used by Peter Harrison, is quite inappropriate.

Franzius wanted his readers to understand how animals are employed in literature to cultivate abstract meaning. He described his use of animal life in terms of parable, analogy and simile, with reference to their use in the New Testament. 'As a leader in this kind of teaching, you have the Holy Spirit and above all, Christ the Lord'.[47] A passage on the weasel (*mustela*) may serve as an example:

> The weasel with its unbelievable courage is able to investigate the cave of a snake. Consequently, it is said that it smears itself all over with rue [a bitter herb], and that it afterwards, without trembling, attacks and kills even a big serpent and basilisk. This might be an image of Christians,

---

[45] Bäumer, *Geschichte der Biologie*, vol. II 164.

[46] 'Certe *Historiam sacram animalium*, Theologi pariter et Philosophi in oculis habent' (Jacob Martini in Henning Witte, *Memoriae theologorum nostri saeculi* 312).

[47] *Historia animalium sacra* fol. c8 r: 'Habes ducem in hoc ipso genere docendi Spiritum Sanctum et imprimis Christum Dominum [...]'. Concerning the terminology, cp. Lausberg H., *Handbuch der literarischen Rhetorik* (Stuttgart: 1990) 232–233; 869.

> whom the Devil always hates, but by whom he is defeated by the help of rue, that is, the preaching of the Word on Christ's merits and blood (*Historia animalium sacra* 274).

The habits of the animal in question is interpreted in relation to Christian doctrine, or rather Christian life. In this case, the weasel was the Christian and the snake was the Devil, while God's word was represented by the bitter herb. In an allegorical and emblematic interpretation, everything in the universe *possesses* hidden meanings;[48] Franzius had a more pedagogical and practical approach: 'this might be an image of Christians'. In general, his use of animal life may be described by the concept *analogy*.

As stated above, *Historia animalium sacra* has not yet been the object of a thorough study, and some of the available information on the work is erroneous. This is the case with some of Bäumer's conclusions, with the effect that the theological aspect of the *Historia animalium sacra* gets too much weight and the biological too little. But for more than one reason, Franzius's work deserves to be examined properly. His treatment of biological problems is interesting in itself. Moreover, the *Historia animalium sacra* was a work with great influence on the Protestant Northern Europe throughout the seventeenth century and in the beginning of the eighteenth, and as professor of theology at the highest esteemed Protestant university, Franzius had students of many nationalities. And whereas the *Historia animalium sacra* was written in Latin, the content was, surely, handed over to parishes in various vernacular languages.

Franzius was a scholar who had read a lot, and with a fairly open mind. In order to understand why he, a theologian, would want to write a book on animals, one must begin with his own description of his aims, which are related to the public for which the book was written, namely the clergy and the students of theology. Franzius was driven by a wish to find out how the animals *really* live and to teach this to his fellow clergymen and students. In the treatment of his sources, Franzius used reason and logics; he did not accept the conclusions of authorities without testing them. He was not afraid to oppose *all* authorities, as in the case of spontaneous generation. *God* was part of Franzius's method; he asked: Would God have wanted this? Is it compatible with essential

[48] See Ashworth W.B. Jr., "Natural history and the emblematic world view", in Lindberg D.C. – Westman R.S. (eds.), *Reappraisals of the Scientific Revolution* (Cambridge etc.: 1990) 303–332.

parts of the Bible? God, as Franzius saw him, was logical and reasonable, like the world he had constructed. Franzius's argument appears as an interwoven pattern of scientific and theological discussion, and in evaluating him, one should not forget that even for Newton God and nature were aspects of the same matter.[49]

---

[49] Bono J.J., "From Paracelsus to Newton: The Word of God, the Book of Nature, and the Eclipse of the 'Emblematic World View'", in Force R.E. – Popkin R.H. (eds.), *Newton and Religion: Context, Nature, and Influence* (Dordrecht – Boston – London: 1999) 74–76.

Selective Bibliography

ASHWORTH W.B. JR., "Natural history and the emblematic world view", in Lindberg D.C. – Westman R.S. (eds.), *Reappraisals of the Scientific Revolution* (Cambridge: 1990) 303–332.

AUGARDE T., *The Oxford Guide to Word Games* (Oxford – New York: 1986).

BÄUMER Ä., *Geschichte der Biologie* (Frankfurt a.M. – Bern – New York – Paris: 1991).

——, „Biblische Zoologie (Hermann Frey, Wolfgang Franz, Heinrich von Hoevel)", in Kattenstedt H. (ed.), *‚Grenzüberschreitung'. Festschrift zum 70. Geburtstag von Manfred Büttner* (Bochum: 1993) 3–17.

BONO J.J., "From Paracelsus to Newton: The Word of God, the Book of Nature, and the Eclipse of the 'Emblematic World View'", in Force J.E. – Popkin R.H. (eds.), *Newton and Religion: Context, Nature, and Influence* (Dordrecht – Boston – London: 1999) 45–76.

CARDANO GEROLAMO, *De subtilitate*, in Idem, *Opera omnia*, 10 vols. (Leiden: 1663), vol. III 357–672.

"Ichtyologiste", in *Encyc lopedie, ou dictionnaire raisonné des sciences, des arts et des métiers. Par une société de gens de lettres. Mis en ordre et publié par Mr.* ***, (Neuchâtel: 1765) vol. VIII 485–486.

CYPRIANUS JOHANNES, *Wolfgangi Franzii Animalium, cum commentariis et supplementis observationum ex recentiori historia naturali, similitudinum, emblematum, hieroglyphicorum, ad usum oratoriae tum civilis tum ecclesiasticae*, 2 vols. (Frankfurt – Leipzig: 1712).

FAIDUTTI B., *Images et connaissance de la licorne (Fin du Moyen-Age–XIXème siècle)*; diss. Paris XII, Paris: 1996.

FRANZIUS WOLFGANGUS, *Augustanae confessionis articuli fidei XXI et articuli abusuum VII* (Wittenberg: 1611).

——, *Historia animalium sacra* (Wittenberg: 1613 [1612]; 1616; Amsterdam: 1654 and 1665).

HARIG G. – KOLLESCH J., "Naturforschung und Naturphilosophie in der Antike", in Jahn I. (ed.), *Geschichte der Biologie* (Hamburg: 2004) 48–87.

HARRISON P., "The Virtues of Animals in Seventeenth-Century Thought", *Journal of the History of Ideas* 59 (1998) 463–484.

HENRY J., *The Scientific Revolution and the Origins of Modern Science* (Basingstoke: 1997).

HOPPE B., "Botanik und Zoologie in der Zeit der Renaissance und des Humanismus", in Jahn I. (ed.), *Geschichte der Biologie* (Hamburg: 2004) 161–195.

JAHN I. – LÖTHER R. – SEUGLAUB K. (eds.), *Geschichte der Biologie: Theorien, Methoden, Institutionen, Kurzbiographien* (Jena: 1982).

JAHN I. (ed.), *Geschichte der Biologie: Theorien, Methoden, Institutionen, Kurzbiographien* (Hamburg: 2004).

——, "Naturphilosophie und Empirie in der Frühaufklärung (17.Jh.)", in Eadem (ed.), *Geschichte der Biologie* (Hamburg: 2004) 196–230.

KIRCHMAYER GEORG CASPAR, "On the Basilisk"; "On the Unicorn"; "On the Phoenix"; "On the Behemoth"; "On the Dragon"; "On the Spider", in Goldsmid E. (ed.), *Un-Natural History, or Myths of Ancient Science; Being a Collection of Curious Tracts on the Basilisk, Unicorn, Phoenix, Behemoth or Leviathan, Dragon, Giant Spider, Tarantula, Chameleons, Satyrs, Homines Caudati, &c.* (electronic reprint, s.l.: 2000 [Edinburgh: 1886]) 5–87.

MARTINI JACOB, "Programma, in funere perquam Reverendi atque Clarissimi Viri, Domini Wolfgangi Frantzii", in Henning Witte (ed.), *Memoriae theologorum nostri saeculi clarissimorum renovatae decas prima (-sexta)*, (Königsberg – Frankfurt a.M.: 1674–1675).

PENNINGTON D.H., *Europe in the Seventeenth Century* (London – New York: 1993 [1970]).

ROGGEN V., *Entertainment and learning: Intellectual play – word and picture. A study of Nils Thomassøn's Latin rebus book* Cestus sapphicus. *With edition, translation and a corpus of sources* (Oslo: 2002; orig. Ph.D. thesis, Oslo: 2001).

SCALIGER JULIUS CAESAR, *Exotericarum exercitationum liber XV de subtilitate ad Cardanum* (Frankfurt a.M.: 1607 [Paris: 1557]).

SCHOOCKIUS ISAAC, "On Bears licking their Offspring into perfect Shape", in Goldsmid E. (ed.), *Un-Natural History, or Myths of Ancient Science; Being a Collection of Curious Tracts on the Basilisk, Unicorn, Phoenix, Behemoth or Leviathan, Dragon, Giant Spider, Tarantula, Chameleons, Satyrs, Homines Caudati, &c.*, (Electronic reprint, s.l.: 2000 [Edinburgh: 1886]) 103–106.

SLIFKIN N., "The Sweat-Louse, the Dirt-Mouse, and the Dong-Chong: A Study of Spontaneous Generation in Jewish Law" <http://www.aishdas.org/toratemet/dongchong.html>.

THOMASSØN NILS, *Cestus sapphicus* (Oslo: 1661).

WIENER P.P. (ed.), *Dictionary of the History of Ideas: Studies of Selected Pivotal Ideas*, 5 vols. (New York: 1973–1974).

WHITE A.D., *A History of the Warfare of Science with Theology in Christendom*, 2 vols. (Gloucester, Massachussetts: 1978 [1896]).

ZEDLER J.H., *Grosses vollständiges Universal-Lexicon*, 64 vols. (reprint, Graz: 1961–1964 [orig. Halle: 1732–1750]).

BETWEEN EMBLEMATICS AND THE 'ARGUMENT FROM DESIGN'. THE REPRESENTATION OF INSECTS IN THE DUTCH REPUBLIC

Eric Jorink

Introduction

'Everywhere and in the humblest of creatures the traces of divine wisdom and supreme skill are made known', Jan Swammerdam wrote to the Royal Society, after he had dissected a stag beetle in 1673. Amongst other things, Swammerdam had observed with astonishment the creature's genitalia, those of the male 'consisting of no more than a single, long, hollow, innumerably inflected fine thread'.[1] Swammerdam (1637–1680) was the first in Europe to investigate the delicate anatomy of insects in a systematic way. He undertook this extremely difficult enterprise with the aid of a recently invented tool, the microscope. Insects had been a neglected subject of observation and contemplation since Antiquity. With a few exceptions, insects were considered vermin, not worthy of any scientific or artistic attention, but for various reasons Jan Swammerdam considered it his vocation to put these 'less esteemed of God's creatures' on the intellectual agenda.

Swammerdam, who was both a talented observer and draftsman, was very much aware of the pioneering nature of his research. By ignoring insects, or not studying them carefully, men refused to honor God in the most wonderful of his creatures, Swammerdam maintained. In his view, inaccuracy in scientific research was equal to blasphemy, and he never missed an opportunity to pour scorn on his predecessors, ranging from Aristotle to Johannes Goedaert. In fact, there were only a few earlier students of insects who could count on his respect. Referring to one of the earlier observers of the stag beetle, Swammerdam wrote: 'Nothing is more striking in the stag beetle, than the creature's snout or tongue

[1] Henri Oldenburg, *The Correspondence of Henry Oldenburg*. ed. A.R. Hall and M.B. Hall, 10 vols. (London: 1965–1973) IX 413.

[...] This tongue is nicely represented in the pictures of Hoefnagel, which are the best and most widely known until now'.[2]

Here, Swammerdam referred to a work published in 1592, *Archetypa studiaque patris Georgii Hoefnagelii*, a collection of 48 engravings by Joris Hoefnagel, showing *ad vivum* pictures of plants, insects and small animals. These impressive depictions of the 'little world' were intended not merely as representations *per se*, but also as a means to contemplate God's plan of creation.[3] Hoefnagel (1542–1600) was by no means a humble artisan, but a classically educated scholar, who self-consciously called himself 'inventor hieroglyphicus et allegoricus'.[4] In a manner reminiscent of contemporary emblem books, each of the pictures in the *Archetypa* carried a motto. 'Narrabo omnia mirabilia tua, Domine', 'I will tell of all your wonderful deeds, Lord' (Psalm 9.2) was, for example, the capture of a beautifully composed still-life with a may fly, longicorn beetle, spider and grass-leaved iris.

Marginal creatures

The religious significance Hoefnagel attributed to the world of insects was not exceptional to sixteenth-century standards. However, the way in which he represented these creatures certainly was. Both the wide range of species he observed, as the love for detail which characterizes his engravings are remarkable. Until far into the early modern era, conceptions of the world of insects were based on classical and biblical sources, and highly ambivalent.[5] Aristotle spoke of *entoma*, 'those creatures which have insections on their bodies, whether on their underside only, or both on their underside and their backs'.[6] Although the Greek

---

[2] Jan Swammerdam, *Bybel der natuure of historie der insecten. Biblia naturae, sive historia insectorum*, 2 vols. (Leiden: 1737–1738) 282.

[3] Vignau-Wilberg Th., *Archetypa studiaque patris Georgii Hoefnagelii 1592. Natur, Dichtung und Wissenschaft in der Kunst um 1600* (Munich: 1994).

[4] Wilberg Vignau-Schuurman Th., *Die emblematischen Elemente im Werke Joris Hoefnagels*, 2 vols. (Leiden: 1969); Kaufmann T.D., *The Mastery of Nature. Aspects of Art, Science and Humanism in the Renaissance* (Princeton: 1993).

[5] Bodenheimer F.S., *Materialien zur Geschichte der Entomologie bis Linné*, 2 vols. (Berlin: 1928).

[6] Aristotle, *Historia animalium* (here quoted after the translation by A.L. Peck, Cambridge Mass.: 1965) 487 a 32–33; 523 b 13.

appears to have studied some species quite attentively, he did not make a sharp distinction between the respective stages of development (egg, larva, chrysalis). Important in this respect is the fact that Aristotle assumed that most insects originate from spontaneous generation, that is to say, that they grew out of putrefying plants or flesh. This idea was to become extremely influential. In the centuries to come, insects were ranked lowest in the 'great chain of being'.[7] Notwithstanding his conviction that most insects lack the complicated anatomy of the higher creatures, Aristotle wrote that every animal, irrespective of its size, was worth studying. This was an idea Pliny the Elder further elaborated upon. The whole of book XI of his monumental *Historia naturalis* was devoted to *insecta*, 'Creatures of immeasurably minute structure – in fact some authorities have stated that they do not breathe and also that they are actually devoid of blood'.[8] Pliny too assumed that most species originated from spontaneous generation, but was lyrical about their visible external anatomy. 'But we marvel at the elephant's shoulders carrying castles, and bull's necks and the fierce tossings of their heads, at the rapacity of tigers and the manes of lions, whereas really Nature is to be found in her entirety nowhere more than in her smallest creations'.[9] 'Natura nusquam magis quam in minimis tota sit' was to become a *topos* in the study of insects. Their minute structure seemed to imply the existence of a higher being. Moreover, some species taught moral lessons to mankind: the industry of ants, the destructive gluttony of grasshoppers and, most of all, the class-based society of the bees which, it was assumed, stood under the rule of a king.

With the advance of Christianity, this moral and symbolic significance of insects was to become dominant. The book of *Exodus* told how Jahwe punished the Egyptians with gnats and locusts. Other biblical passages described how God not only used insects as instruments of divine retribution, but as means to contemplation as well. 'Go to the ant, thou sluggard' the wise king Solomon said,

---

[7] Lovejoy A., *The Great Chain of Being. A Study of the History of an Idea* (Cambridge Mass.: 1936) 236–240.

[8] Pliny, *Naturalis historia* (here quoted after the edition by H. Rackham, Cambridge Mass.: 1940) XI, 1.

[9] Ibid.

> Consider her ways, and be wise:
> Which having no guide, overseer, or ruler,
> Provideth her meat in the summer, [and] gathered her food in the harvest.
> How long wilt thou sleep, O sluggard?
> When wilt thou arise out of thy sleep?[10]

The symbolic, allegoric and typological potential of lines like these was enormous, and was further elaborated upon by, for example, the *Physiologus*, the collection of religious descriptions of animals dating from the third century AD.

Sixteenth-century ideas on insects (and on nature in general) were firmly based on the legacy of classical and biblical antiquity. Although both the Bible and the classics seemed to imply that God manifested himself in the lowest of his creatures, only very few species were observed, depicted and contemplated upon: the bee, the ant, the grasshopper and, of course, the butterfly, whose mysterious metamorphosis was seen as a symbol of the Resurrection. On paintings, engravings, in emblem books and in the massive natural histories of the sixteenth century, insects were wrapped up in an amazing world of textual references and hidden meanings. Both in, for example, Joachim Camerarius II's *Symbola et emblemata* and Ulysse Aldrovandi's monumental *Historia Naturalis*, insects were represented as a means to learn moral and religious virtues. The way in which the insects were depicted was not considered of major importance, nor were the descriptions of their external anatomy, behavior and life cycle. Instead, sections on each creature consist, in typical humanist style, of long enumerations of the etymology of each creature's name, the biblical and classical sources in which it is mentioned, fables, tales and proverbs on the creature, its symbolic value and moral significance, representations of it on paintings, medals and so on.[11]

[10] *Proverbs* 6:6–9 (King James Version).

[11] Harms W., "Bedeutung als Teil der Sache in zoologischen Standardwerken der Frühen Neuzeit (Konrad Gesner, Ulysse Aldrovandi)" in: Boockman H. (ed.), *Lebenslehre und Weltentwürfe im Übergang vom Mittelalter zur Neuzeit* (Göttingen: 1989) 531–553; Ashworth W., "Natural History and the Emblematic Worldview", in Lindberg D.C. – Westman R. (eds.), *Reappraisals of the Scientific Revolution* (Cambridge: 1993) 303–323.

Early modern Dutch conceptions of insects

The manner in which Dutch authors and artists represented insects at the beginning of the seventeenth century was much in the spirit of the learned Joris Hoefnagel. This was surely no coincidence. Hoefnagel was the uncle of the highly influential diplomat and poet Constantijn Huygens (1596–1687), who during his long life stimulated both artists and scientists. When Hoefnagel died in 1600, a part of his artistic heritage was bequeathed to the Huygens family, were it was seen by another artist, Jacques de Gheyn II. Bergström has convincingly argued that Hoefnagel stood at the basis of the typical Dutch genre of still-life paintings with flowers, shells and insects.[12] The *Archetypa* series were widely known in the young Republic, and offered excellent models for *imitatio* and *aemulatio*.

However, Dutch artists were only interested in the small corpus of symbolically laden species of insects. Not only on still-life paintings, but also in other works of art, only a few sorts were represented. For example, in a wordy poem, the minister and naturalist Petrus Hondius (1578–1621) described the flora and fauna of his country estate 'Moufeschans' (1621). It is interesting to note that, although the place must have been crawling with vermin, the only insect Hondius thought worth lauding was the intelligent and ingenious bee.[13] In a similar poem, Jacob Cats (1577–1660), by far the most popular Dutch poet of the seventeenth century, observed to his annoyance the insects which 'as a kind of evil worms, besiege the trees in summertime'.[14] Only one species deserved his praise: the bee, of course. 'What do I learn here, What do I learn from this diligent creature: We are sluggards, in comparison to this industrious animal!'[15] In his emblem book *Silenus Alcibiadis* (1618), Cats also presented some other insects. In Emblem XL, 'Non intrandum, aut penetrandum', a spider is depicted, in whose web all kinds of insects are caught. In the accompanying text, Cats

[12] Bergström I., *Dutch Still-Life Painting in the Seventeenth Century* (London: 1956); Idem, "On Georg Hoefnagel's Manner of Working with Notes on the Influence of the Archetypa series of 1592" in Cavalli-Björkman G. (ed.), *Netherlandish Mannerism* (Stockholm: 1985) 177–188.

[13] Petrus Hondius, *Dapes inemptae, of de Moufe-schans, dat is de soeticheydt des buytenlevens, vergeselschapt met de boucken* (Leiden: 1621).

[14] Jacob Cats, *Alle de Wercken* 2 vols. (Amsterdam and The Hague: 1726) II, 367.

[15] Ibid.

refers to the Bible, the classical writers and the fathers of the Church, explaining the virtues of moderation, soberness, prudence and chastity. In the same manner, a butterfly emerging from its cocoon is depicted in Emblem LII, 'Amor elegantiae pater'. Here, Cats meditates on purity, love and the hope of resurrection.[16]

To be sure, it was not only painters and poets who were caught in the web of humanist preferences for certain species. In 1597, Dirck Outgersz Cluyt (1546–1598), an apothecary who was appointed keeper of the newly founded *hortus botanicus* of Leiden university, published a solid monograph on the bee.[17] It was actually one of the first books ever written in Europe devoted to one single insect. In the classical form of a colloquy, Cluyt lectured his interlocutor on the origin, nature and use of the bee. Cluyt repeatedly stated that he not only wrote down 'what I read in the Classics or heard from other people, but, most of all, what I learned from my own experience'. Indeed, Cluyt made some original remarks on the bee. Amongst other things, he refuted the idea that bees had their origin in spontaneous generation. This conviction, however, was based more on *a priori* reasoning than critical observation. Cluyt could simply not believe that such a noble creature had its origin in putrefying meat, like dung flies or other kinds of vermin had. Many other chapters of his book testify to the power of tradition as well.

In conformity with humanist traditions, Cluyt started his colloquy with a lesson in etymology. The bee, *apis*, derived its name from the fact that it was born without feet, *a-pes*. The social order within the beehive was a moral example for mankind, and the rule of the wise king-bee was so remarkable 'that it appears to be supernatural'. Of course, Cluyt mentioned the biblical passages which spoke of bees, and he took great pains to identify the 'wild honey' which St John the Baptist ate in the wilderness (Matthew 3:4; Mark 1:6). He concluded his book with a reference to Psalm 104: 'O LORD my God, thou art very great; though art clothed with honour and majesty [...] The LORD shall rejoice in his works'.

---

[16] Jacob Cats, *Silenus Alcibiadis dat is, Sinne- en minnebeelden*, ed. H. Luijten, 2 vols. (The Hague: 1996) II, 342–249; 757–772.

[17] Dirck Outgersz. Cluyt, *Van de byen, hare wonderlicke oorspronc, natuer, eygeschap, crachtige, ongehoorde ende seldsame wercken: Waerin gemerct werden hare wonderlicke Politien en ordentelicke Regeringe, die sy met haren Coning ende onder melcandern onderhouden. Vervat in drie Boecken: Seer genuechlich om te lesen* (Leiden: 1597).

Another interesting monograph on a symbolic insect was published by Cluyt's son, Augerius Clutius (1577–1636), in 1634.[18] Trained as a physician, it was Clutius' intention to write a small booklet on the alleged therapeutic effects of the ephemeron of may fly. The complicated life cycle of this insect had fascinated authors since Antiquity. As we know now, may flies are aquatic in their preparatory stages. The eggs are dropped into the water by the female in large masses and sink to the bottom of the river or lake. There after a considerable time – up to three years – the young larvae hatch. When the insect finally has reached its full growth, it emerges from the water. This usually occurs during a few days in June. Reaching the surface, the thorax splits and the winged form appears, and the creature flies, usually in huge swarms. The insects live from a few hours to a couple of days, they mate and die, after which the life cycle starts anew. Both Cicero and Pliny had already meditated on the short life of the may fly, and the creature would become one of the favorites of humanist scholars and artists such as Hoefnagel.[19] Clutius elaborated further on this theme. Like his father, he claimed to have based his book on his own observations. However, this treatise was firmly rooted in the humanist tradition as well. Clutius started with the creature's name, identifying Aristotle's *ephemeron* with the animal described by Pliny, and in Holland known as 'haft' or 'oever-aas'. Then, he continued with the relevant proverbs. After an untimely death, the Romans used to say: 'ephemeri vita'. When a place is densely crowded, the Dutch say: 'Het isser soo dicht als haft' – it is like a swarm of may-flies. Clutius does not seem to have studied the creature *in vivo*. Instead, he stressed the moral lesson taught since Antiquity: life is miserable and short, and all creatures are condemned to the same fate (*fatum*): Death.

The persistence of this highly symbolic view on insects could perhaps best be illustrated by the way a revolutionary new instrument was employed in the first decades after its invention.[20] Around 1620,

---

[18] Augerius Clutius, *Opuscula duo singularia. I. De nuce medica. II. De hemerobio sive ephemero insecto et maiali verme* (Amsterdam: 1634). See also Francissen F.P.M. – Mol A.W.M., *Augerius Clutius and his 'De hemerobio'. An Early Work on Ephemeroptera* (Marburg: 1984).

[19] Cicero, *Tusculanae disputationes* I 39: 'Confer nostram longissimam aetatem cum aeternitate: in eadem propemodum brevitate, qua illae bestiolae reperiemur'; Pliny, *Naturalis historia* XI 42. Cf. Hoefnagel, *Archetypa* Pars II, 10.

[20] Jorink E., "'These wonderfull glasses'. Dutch humanists and the microscope", in Grob B. and Hooijmaijers H. (eds.), *Who needs scientific instruments? Conference on Scientific Instruments and their Users, 20–22 October 2005* (Leiden: 2006) 115–124.

the Dutch inventor Cornelis Drebbel (1572–1633) had constructed the first compound microscope, consisting of two lenses in a tube. He demonstrated his device to the young Constantijn Huygens, who was lyrical about this instrument: 'It is really as if one stands before a new theatre of nature, as if one is on a different earth'.[21] It is in this 'New World' of the smallest possible things that we encounter the dedication of the divine Architect, Huygens cheered:

> Nothing can compel us to honor more fully the infinite wisdom and power of God the creator unless, satiated with the wonders of nature that up till the present day have been obvious to everyone (for usually our astonishment cools as we become familiar with Nature through frequent contact) we are led into this second treasure-house of nature, and in the tiniest and most disdained of creatures meet with the same careful labor of the Great Architect, and everywhere we will come upon the same ineffable Majesty.[22]

Huygens quickly realized the instrument's potential and tried to persuade his friend Jacques de Gheyn II (1565–1629) to produce a set of microscopic drawings to give a new impetus to the somewhat faded interest in the wonders of God's Creation. But, alas, De Gheyn passed away before these plans could be realized. One wonders which kinds of insects the artist would have depicted if he had lived longer. Presumably it would have been the traditional bees, ants and butterflies. In any case, these were the creatures his contemporaries first came to mind when they employed the expensive and hard to handle microscope.

Recently, historians of science have spoken in somewhat derogatory terms of the way in which humanists like Huygens used this new instrument, considering it more like *Spielerei* than a truly scientific enterprise.[23] However, seen from contemporary concepts of *scientia* it was obvious to study nature with the Bible and the classics in one hand, and a microscope in the other. Nature was, after all, not an unknown territory waiting to be explored, penetrated and mapped, but God's second revelation.[24] As the articles of faith of the Dutch Reformed

---

[21] Constantijn Huygens, *Mijn jeugd*, ed. C.L. Heesakkers (Amsterdam: 1987) 132.

[22] Ibidem.

[23] Fournier M., *The Fabric of Life. Microscopy in the Seventeenth Century* (Baltimore: 1996) 30; Ruestow E., *The Microscope in the Dutch Republic. The Shaping of Discovery* (Cambridge: 1996) 5; 39.

[24] Jorink E., *Het Boeck der Natuere. Nederlandse geleerden en de wonderen van Gods Schepping, 1575–1715* (Leiden: 2006); an English translation, *Reading the Book of Nature in the Dutch Golden Age* will appear in the series *Brill's Studies in Intellectual History*.

Church explicitly stated: we know God in two ways. Firstly, through Creation, 'which is before our eyes as a most elegant book, wherein all creatures, great and small, are as characters'. And secondly, God can be known through Scripture. Seen from this perspective, it was quite natural to observe only the insects mentioned in (or associated with) the Bible. The aim of observation was to contemplate God, not to penetrate into Nature's secrets. And in studying the smallest letters of the book of nature, the microscope proved an excellent tool. In the first decades after the invention of the microscope, the instrument was extensively used by humanist scholars, as a private means to meditate on the wonders of the Creator. No 'scientific' treatises were published on insects, nor (as far as we know) were drawings made.

Instead, Huygens recommended the study of the limbs of ants, in order to marvel at Gods creation.[25] His friend Jacob Westerbaen (1599–1670) observed lice, butterflies and bees, and exclaimed that God reveals Himself more clearly in a gnat than an elephant – explicitly referring to Pliny.[26] Another classically trained *virtuoso*, the Dordrecht minister Andreas Colvius (1594–1672), was fascinated by the microscope as well. It was Colvius who brought the young Christiaan Huygens in contact with a local lense-grinder called Calthoff, and was rewarded for his good offices with a microscope manufactured by Huygens himself – an instrument of excellent quality, no doubt. 'Is it not true what Pliny says', Colvius rhetorically wrote to Christiaan Huygens, 'Natura nusquam magis quam in minimis tota sit?'.[27] Like many of his contemporaries, Colvius seems to have been fascinated mostly by symbol-laden insects. He studied and collected the may-fly, which was endemic in the rivers around Dordrecht. In 1669, Jan Swammerdam would describe a species he had received 'from the honorable D. Colvius'.[28]

---

[25] Constantijn Huygens, *De gedichten van Constantijn Huygens*, ed. J.A. Worp, 9 vols. (Arnhem: s.d.) III 80–81.

[26] Jacob Westerbaen, *Arctoa Tempe. Ockenburgh. Woonstede van den Heere van Brandwyck, in de Clingen buyten Loosduinen* (The Hague: 1654) 140–145.

[27] Christiaan Huygens, *Oeuvres complètes de Christiaan Huygens* 22 vols. (The Hague: 1888–1950) I 322.

[28] Jan Swammerdam, *Bybel der natuure* 219.

Changing attitudes: Goedaert

In the course of the seventeenth century, observers of insects in the Dutch Republic broadened their scope. From the 1640's on, other insects than the well known corpus were studied and even depicted, including wasps, dung flies and all kinds of beetles. It is important to note that this process was not the outcome of the increasing use of the microscope or any other form of 'scientific progress'. On the contrary, it seems that this was the result of an increasing stress on what since early Christianity had been the major motivation for the study of the bee, ant and grasshopper, namely the conviction that the Maker revealed himself in *all* of His works. Moreover, seventeenth-century conceptions were strongly influenced by the classical lines by Pliny: 'Natura nusquam magis quam in minimis tota sit'.

The first to embark on a systematic study of the generation of insects, not only in the Netherlands but in Europe as a whole, was Johannes Goedaert (1617–1668). This now largely forgotten figure earned his living as a painter of still-lifes in the spirit and style of Hoefnagel, depicting bouquets of flowers adorned with butterflies and caterpillars. Contrary to Hoefnagel, Goedaert was an artisan in the first place. Although he lacked a classical education, he was deeply religious and well-versed in the Bible. According to his own testimony, he started to study insects around 1635, driven by the conviction

> that there is no creature so small, or by the attentive observation thereof one can find immediate cause and ample reasons to glorify God, and wonder at His marvelous wisdom and providence. The truth of this I have discovered by the observation of worms, caterpillars, maggots and other crawling animals [...] Which is confirmed by the divine psalmist who has said that the benevolence of God is manifest in all of His works, and that He has created them with wisdom.[29]

The pious Goedaert brought the lines of *Genesis* 1–7 and *Psalm* 104 to its logical conclusion: if God was the creator of everything, why would He not care for the insects? And thus Goedaert started to collect,

---

[29] Johannes Goedaert, *Metamorphosis naturalis, ofte historische beschrijvinge van den oirspronk, aerdt, eygenschappen ende vreemde veranderingen der wormen, rupsen, maeden, vliegen, witjens, byen, motten en diergelijke dierkens meer; niet uit eenige boeken, maar aleenlyck door eygen ervaretntheyt uytgevonden, beschreven ende na de konst afgeteykent* 3 vols. (Middelburg: s.d. [1660–1669]) I, "Dedicatie".

observe and draw all the insects he could find around his native town Middelburg. He scrutinized his own garden, walked through woods, marshes and dunes, and even visited graveyards at night, armed with a torch, in order to catch insects wherever he could. Then, he put them in glass jars, fed them with whatever they seemed to like, and closely observed their behavior and transformation from larvae or pupae to insect. According to their nature, he gave them nicknames like 'Runny nose' or 'Ugly sluggard'. This somewhat anthropocentric approach clearly becomes manifest in the pictures Goedaert drew of the respective stages of each insect. Some of the maggots have unmistakable human features. It is somewhat surprising that Goedaert seems to have employed the microscope only occasionally.

Nevertheless, while Goedaert broadened the scope of insect observation on an empirical level, he was not able to approach these creatures from a new theoretical framework. Like all of his contemporaries, Goedaert was convinced that insects emerge from spontaneous generation or metamorphosis. To him, the emergence of each one of them was a kind of divine miracle. For example, he described with astonishment how he once observed how two identical caterpillars died. From the one corpse a swarm of dirty 'flies' emerged, while the other gave birth to a beautiful butterfly! It must have been a wondrous sight indeed, although we tend to suspect the activities of a hymenopterous parasite instead of the direct intervention of God. To Goedaert, the world of insects was the world of deeply religious meanings. Following Christian tradition, he considered the metamorphosis from caterpillar into butterfly a symbol, even a proof, of the Resurrection:

> As it is miraculous to our eyes, but at the same time clearly evident, that from dead caterpillars emerge living animals; so it is equally true and miraculous, that our dead and rotten corpses will rise from the grave.[30]

The fact that insects were attracted to burning candles at night was considered a divine warning against *curiositas*. Goedaert quoted at length 1 *Timothy* 6: 15–16:

> Which in his times shall shew, [who is] the blessed and only Potentate, the King of kings, and Lord of lords; Who only hath immortality, dwelling in the light which no man can approach unto; whom no man hath seen, nor can see: to whom [be] honour and power everlasting.

[30] Jan Goedaert, *Metamorphosis* II 140.

And thus Goedaert concluded that it was good to rejoice at the marvels of Gods creation, but wrong to try to penetrate them – which might explain why he refrained from the use of the microscope.

It took some decades before Goedaerts works were published. Between 1660 and 1669, three parts of his *Metamorphosis naturalis* appeared, both in Dutch and Latin. Although by no means revolutionary in itself, the works caused some excitement amongst European scientists and *virtuosi*, since they were the first in which native insects were described and depicted in a systematic way. The series went through many reprints, and was translated into French and English. A new Latin edition was published by Martin Lister, Fellow of the Royal Society, who grudgingly remarked that Goedaert's illustrations of insects were quite satisfying, but that the artist 'seemes rather to have diverted with them, than to have given himselfe the trouble of well understanding them'.[31]

Indeed, even before the last part of the book appeared posthumously in 1669, the intent of the work was attacked from several angles. First of all, there was the publication of Robert Hooke's magnificent *Micrographia* in 1664. Although the book did not contain many descriptions of insects, it was considered a landmark in the history of the visual representation of these creatures. After perfecting the compound microscope, Hooke was able to give stunning engravings of a gnat, a louse and even the facet-eyes of a bee. 'The book that popularized the microscope', as Wilson recently labelled it.[32] In Italy, Franscesco Redi used this optical instrument it in order to study the procreation of insects, in 1668 publishing his *Esperienze intorno alla generatione degl'insetti* (1668) in which he questioned the idea of spontaneous generation. A year later, his fellow-countryman Marcello Malpighi was able to dissect a silkworm. Using the microscope, Malpighi observed, described and depicted the internal parts of this creature. Applauded (and sponsored) by the Royal Society, the results were published in the autumn of 1669.[33] Notwithstanding Malpighi's reputation as being one of the pioneers

---

[31] Jan Goedaert, *Of insects. Done into English, and Methodized, with the Addition of Notes* (York: 1682).

[32] Wilson C., *The Invisible World: Early Modern Philosophy and the Invention of the Microscope* (Princeton: 1995) 75; 86. See also Harwood J.T., "Rhetoric and graphic in *Micrographia*" in Hunter M. – Schaffer S. (eds.), *Robert Hooke: New Studies* (Woodbridge: 1989) 119–147.

[33] Cobb M., "Malpighi, Swammerdam and the Colourful Silkworm: Replication and Visual Representation in Early Modern Science", *Annals of Science* 59 (2002) 111–147.

of microscopy, his amazing study on the intestines of the silkworm would remain his only published work on insects. While *De Bombyce* was in print, an ambitious Dutchman wrote the last lines of what would become the classical study on the generation of insects.

Swammerdam and new arguments for the existence of the 'Great Designer'

As had his predecessor Johannes Goedaert, Jan Swammerdam had an open eye for every kind of insect. But unlike Goedaert, he worked from a new theoretical framework. Moreover, he would bring the microscopic study of insect anatomy on a level which was not surpassed until the nineteenth century. Swammerdam would become the key figure in the whole process of transformation of the moral significance of insects and, more generally, of the increasing popularity of "the argument from design" in early modern intellectual culture. For Swammerdam, insects had completely lost their allegoric and symbolic value. Insects were, by their minute nature and delicate anatomy, the most convincing evidence of the existence of a godly masterplan, of an intelligent designer. 'Wonder', Daston and Park observed recently,

> Was now displaced almost entirely to commonplace objects praised as marvels of divine handiwork. Late-seventeenth- and early-eighteenth-century entomology was particularly rich in such natural theological expressions of wonder at the ordinary. The Dutch naturalist Jan Swammerdam, for example, thought the humble ant deserved as much admiration as God's largest and gaudiest creations.[34]

In the wake of Swammerdam's work, the study of insects became extremely popular. It is certainly no coincidence that in the last quarter of the seventeenth century Dutch collectors, scientists and artists were struck by a real insect-craze. Echoing both Pliny and Swammerdam, the Dutch author Bernard Nieuwentijt (1654–1718) rhetorically wrote in his exceptionally popular work on natural theology, *Het regt gebruik der wereldbeschouwingen* (1715) that 'the Maker shows himself no less adorable in the creation of a gnat, a fly, a flee or a cheese-mite than in the biggest elephant'.[35]

---

[34] Daston L. – Park K., *Wonders and the Order of Nature, 1150–1750* (New York: 1998) 323.

[35] Bernard Nieuwentijt, *Het regt gebruik der wereltbeschouwingen, ter overtuiginge van ongodisten en ongelovigen* (Amsterdam: 1715) 560; 563.

Much the victim of later, somewhat romantic historians, Swammerdam is usually portrayed as a rather isolated, somewhat erratic mystic, who after a short scientific career felt prey to an attack of religious frenzy from which he never recovered.[36] This, however, is contrary to the evidence now available.[37] Suffice it to say that, with only a short interruption in 1675, Swammerdam continued his research until his death in 1680, and that he was much more communicative than often assumed – as for example his letters to the Royal Society testify.

After his matriculation at the famous faculty of medicine of Leiden university in 1661, Swammerdam excelled in anatomical dissections of (mostly) dogs and frogs. Although a talented observer, he was by no means a mere empiricist. Swammerdam was strongly guided by philosophical and religious ideas, and much of his work must be viewed as an effort to prove certain axioms. One idea Swammerdam vehemently opposed was the concept of spontaneous generation. This would imply, Swammerdam wrote, that creation was contingent, and that certain parts of the universe were excluded from God's will. It would be ridiculous, not to say blasphemous, to think that man and the higher creatures were created by God, while insects were the result of the rotting of flesh or the decay of plants. Believing in spontaneous generation was equal to believing in blind chance or fate, and was thus a rejection of the power and glory of the almighty Architect. Gods creation, Swammerdam repeated time after time, was uniform and stable.

Swammerdam's ideas on the order of nature were much influenced by the natural philosophy of René Descartes, which was then in its heyday among Dutch intellectuals.[38] In his *Discours de la méthode* (1637) and subsequent works, Descartes had reduced all processes in nature to three underlying laws. Everything in the universe obeyed to these laws of nature, and nature itself was orderly, obeying to fixed laws and

---

[36] Two recent examples are Ruestow, *Microscope* 105–145; Wilson, *Invisible world* 186–190.

[37] Jorink E., "'Outside God, there is Nothing'. Swammerdam, Spinoza, and the Janus-Face of the Early Dutch Enlightenment", in Bunge W. van (ed.), *The Early Enlightenment in the Dutch Republic, 1650–1750. Selected papers of a conference held at the Herzog August Bibliothek, Wolfenbüttel 22–23 March 2001* (Leiden: 2003) 81–103.

[38] Much literature is available on this topic. Good introductions are: Bunge W. van, *From Stevin to Spinoza. An Essay on Philosophy in the Seventeenth-Century Dutch Republic* (Leiden: 2001); Verbeek Th., *Descartes and the Dutch: Early Reactions to Cartesian Philosophy, 1637–1650* (Carbondale: 1992).

open to rational explanation. Of equal importance with his physics was Descartes's epistemology. Only rational concepts would lead to true understanding. Descartes rejected all bookish knowledge, regardless of its status or standing, implicitly creating a sharp contrast between natural philosophy and the biblical account of creation. For Swammerdam this epistemology implied that not only the massive natural histories of the sixteenth century could be left unread, but Aristotle and Pliny as well. Ironically, it was Swammerdam who gave Pliny's *dictum* 'Natura nusquam magis quam in minimis tota sit' new relevance.

Swammerdam thought according to the lines sketched by Descartes. He was totally convinced that the generation of all creatures obeyed to the same laws. After studying the reproductive organs of men and women, he took a logical next step, and started to study the generation of other creatures, mostly insects. Soon he made a stunning discovery. In 1668, Swammerdam observed that the "king" bee was actually a queen, since he discovered egg masses in the creature's intestines! This discovery greatly stimulated his pioneering research on the anatomy and generation of insects in general, on which he would focus his research for the rest of his life. When the Florentine prince Cosimo de' Medici visited Swammerdam in June 1669, he was shown that in the entrails of a caterpillar, rudiments of limbs and wings of the future butterfly could be discerned. This was a revolutionary discovery, since it was commonly assumed that first the caterpillar would die, after which by a mystical process of metamorphosis, the butterfly would resurrect. Swammerdam's general theory on the generation of insects was revealed only some months later. In November 1669 he published his pioneering *Historia insectorum generalis, ofte, algemeene verhandeling van de bloedeloose dierkens* (translated into French in 1682, and into Latin in 1685).[39] Including some last-minute references to Malpighi's *De Bombyce*, Swammerdam self-confidently stated he had solved a mystery that had puzzled mankind since the days of Aristotle. He refuted three traditional assumptions concerning insects. Firstly, that they lack internal anatomy; secondly, that they originate from spontaneous generation; and thirdly, that they develop by metamorphosis. Swammerdam demonstrated

[39] Jan Swammerdam, *Historia generalis insectorum ofte Algemeene verhandeling van de bloedeloose dierkens* (Utrecht: 1669).

that all insects come from eggs, and that what was usually considered a sudden transformation in fact was a slow growth of limbs. In other words, there was no fundamental distinction between insects and the so-called higher animals: all belonged to the order of nature. *All* creatures had their origin *ex ovo*. Chance and contingency were excluded from Swammerdam's natural philosophy, as being incompatible with the power of God, the almighty Architect. Swammerdam even went so far as to call the theory of spontaneous generation 'the surest way to atheism'.

Much in train with the more radical Cartesians of the day, Swammerdam declared war on vulgar errors, symbolic thinking and superstitious beliefs. To Descartes himself, nature was a huge piece of mechanics, a giant item of clockwork. Physical phenomena were only fully explained when they were explained by the reciprocal physical contact bodies had upon each other. The implication was, of course, that nature was devoid of symbolic meaning. When Descartes explained, for example, the rainbow, he did so with the laws of mechanics, according to which particles of light collided with each other and were refracted. He wasted no words on any divine meaning, nor did he quote Genesis 9, the *locus classicus* in this context in which God promises Noah not to punish mankind with another Flood: 'I do set my bow in the cloud, and it shall be for a token of a covenant between me and the earth'.

Swammerdam vehemently attacked the symbolic interpretation of insects, and spent no words on biblical references or analogies concerning these creatures. In his militantly Cartesian *Historia*, Swammerdam used the work of the recently deceased Goedaert as a sitting target. Swammerdam called his books, which were full of biblical quotations and allegorical digressions, 'foolish', 'laughable' and 'ridiculous'. He overtly ridiculed the traditional metaphor of the bee-hive, since its leader turned out to be a queen and not a king. Describing his revolutionary discovery of the organs of the future butterfly in the intestines of a caterpillar, he remarked: 'So here we witness the digression of those who have tried to prove the Resurrection of the Dead from these obviously natural and comprehensible changes within the creature itself'.[40]

Instead of using dubious analogies as a means to contemplation, Swammerdam pointed to the visible constitution of insects as a reference

[40] Jan Swammerdam, *Historia* 28.

to God, as an incontestable proof of His existence. Describing for example the ant, Swammerdam did not meditate on the creature's industry nor on the words of Proverbs 6, but on the animal's delicate anatomy. The glory of God was revealed by all of his creatures, be it elephants, lions, bees and ants, or gnats, cockroaches, lice or any other insect. There was no longer a fundamental difference between big and small, beautiful and ugly, or common and exotic, since all creatures belonged to the same order of Nature. But the ingenuity of the Creator showed itself most strikingly in the minute scale of the 'most disdained of creatures'. Wonder was now evoked by the marvelous design of insects, and not by their hidden qualities, deeper meaning and symbolic value. Swammerdam, himself a gifted draftsman, offered splendid illustrations of the various stages of the developing insects. Like in Hookes *Micrographia*, the creatures are depicted in incredible detail, without any material or symbolic context.

Swammerdams *Historia* was essentially a programmatic book. Within an impressive theoretical framework, Swammerdam offered many remarkable observations, mostly of the external parts of insects and done with the naked eye. 'Part I', was the promising subtitle of the work. And, indeed, from 1670 until his death in 1680, Swammerdam worked on Part II – posthumously published as *Biblia Naturae – Bybel der nature* in 1737–1738. This massive work is essentially an expansion of the *Historia*, both in scope and in depth. Convinced that all insects were worth studying, Swammerdam collected and observed as many species as he could. Secondly, and most important, he now started to study the intestines of these creatures in a systematic way, using the microscope and highly refined dissecting techniques. Inspired (and perhaps challenged) by the splendor of Malpighi's *De Bombyce*, Swammerdan started to anatomize silkworms, may-flies, ants, stag-beetles, cheese-mites, bees and other insects. In every creature Swammerdam saw God's immense wisdom and omnipotence. Swammerdam's lyrical outburst on the anatomy of the louse – itself a work of incredible ingenuity – would become a *locus classicus* in apologetic literature: 'Herewith I offer you the Omnipotent Finger of God in the anatomy of a louse; wherein you will find miracle heaped on miracle and see the wisdom of God clearly manifested in a minute point'.[41] Taking the

[41] Jan Swammerdam, *Bybel der natuure* 67.

Dutch conception of the Book of Nature as his point of departure, Swammerdam gave its interpretation a new twist. It was no longer biblical lines or preferences for certain creatures which pointed to God; it was each creature's delicate structure. Through his microscope, he observed things no man had seen before. Working on his treatise on the bee, Swammerdam wrote:

> Look, so all-wonderful is GOD, in respect of these small animals, that I dare say, that in the insects GOD'S countless wonders are sealed up, which seals are revealed as one diligently turns over the leaves of the Book of Nature, the Bible of Natural Theology, in which GOD'S invisibility becomes visible; because treasuries of ineffable wonders then manifest themselves; and the hidden Creator becomes so manifest in these small Animals, that the experiences of the same, serve me as the biggest proofs to uphold without yielding his eternal Divinity and Providence against all his detractors.[42]

Swammerdam's researches brought him so close to God that he sometimes remarked that he not only could see Him, but touch Him as well in the entrails of insects.

In the course of 1673, Swammerdam fell prey to a religious crisis. He had initially believed that his research was a tribute to the Creator; but now he feared he was worshipping an idol called *curiositas*. Swammerdam sought the spiritual comfort of the mystic prophetess Antoinette Bourignon (1616–1680) and joined her sect in Schleswig Holstein. Elsewhere, I have described this confusing part of Swammerdam's life.[43] Suffice it to say here that Swammerdam was back in Amsterdam in the spring of 1676. Contrary to what is often maintained, he resumed his microscopic research. 'I have more need of the bridle than the spur', he wrote to Henry Oldenburg, the secretary of the Royal Society. 'I was never at any time busier than in these days, and the chief of all architects has blessed my endeavors'.[44] He completed the manuscript of what later would be known as the *Biblia Naturae* only a few weeks before he died in February 1680. The work was nearly lost, but thanks to the good offices of the Leiden professor Herman Boerhaave at last published in 1737.

---

[42] Ibid. 394.

[43] Jorink, "Outside God, there is nothing". See also Baar M. de, *'Ik moet spreken'. Het spiritueel leiderschap van Antoinette Bourignon (1616–1680)* (Zutphen: 2005) passim.

[44] Henri Oldenburg, *Correspondence* XIII 343–344.

Contrary to what is often maintained, Swammerdam was not an isolated and erratic mystic. Swammerdam made sure that his stunning discoveries, descriptions and illustrations became known in the scientific world, as for example his correspondence with Henry Oldenburg testifies. Colleagues and friends such as Gottfried Wilhelm Leibniz (1646–1716) and Nicolas Malebranche (1638–1715) knew his microscopic *work in progress* and used its results to substantiate their own natural or moral philosophy. Referring to Swammerdam's as yet unpublished study on the bees, father Malebranche wrote in his *Entretiens sur la métaphysique et religion* (1688); 'ces petits animaux sont l'ouvrage d'une sagesse infinie'.[45] The *Historia* was already widely known and applauded, running through several reprints, and being translated into French in 1682 and into Latin in 1685. John Ray (1627–1705), the influential author of amongst other things the apologetic *Wisdom of God manifested in the Works of creation* (1691) and a *Historia insectorum* (1705), wrote that Swammerdam's method 'seems to me the best of all'.[46] In other words, Swammerdam's influence was much greater than is usually assumed. His work was a herald of the natural theology of the eighteenth century.[47] God's grand design became visible in the whole of nature – in the celestial mechanics of the solar system, the variation of the seasons, the geometry of snowflakes and the anatomy of the human eye – but nowhere more strikingly than in the world of insects.

Epilogue

Around 1680, the study of insects had become incredibly popular in the Dutch Republic. A host of books were published on the subject. Antoni van Leeuwenhoek (1632–1723), who since 1673 published his microscopic observations in the *Transactions of the Royal Society*, carefully

[45] Nicolas Malebranche, *Oeuvres complètes*, 12 vols. (second edition, Paris: 1972–1984) XII 230.

[46] British Library, Ms Sloane 4062 f 210.

[47] Bots J., *Tussen Darwin en Descartes. Geloof en natuurwetenschap in de achttiende eeuw in Nederland* (Assen: 1972); Vermij R., *Secularisering en natuurwetenschap in de zeventiende en achttiende eeuw: Bernard Nieuwentijt* (Amsterdam: 1991); Stebbins S., *Maxima in minimis. Zum Empirie- und Autoritätsverständnis in der physikotheologischen Literatur der Frühaufklärung* (Frankfurt am Main: 1980).

described insects and, like Swammerdam, dismissed the theory of spontaneous generation. In 1688 the Cartesian Stephan Blankaart (1650–1704) followed this example in his *Schouw-burg*.[48] Referring to Swammerdam, Blankaart attacked insect-symbolism, gave matter-of-fact descriptions of various species, and gave advice on how to catch, collect, conserve and study them.

Within a very short time, insects found their way to the rich Dutch collections of natural curiosities.[49] Around 1660, these creatures had been virtually absent, with the notable exception of, for example, 'a locust of the sort that St John the Baptist ate in the wilderness', which the young John Ray observed in a Delft collection.[50] Within two decades, great quantities of all kinds of species were exhibited in the famous collections of, for example Nicolaas Witsen, Levinus Vincent and Albertus Seba. These cabinets included beautiful butterflies and all other kinds of rare and exotic sorts from the East- and West-Indies, arranged according to sort, and showing their successive stages of development. 'Swammerdam states', Vincent wrote in a descriptive catalogue of his collection, 'that the smallest of creatures yield in nothing to the biggest'.[51] Elsewhere in this volume, a laudatory poem stressed that even the most evil of atheists would be silenced in view of this magnificent collection of animals.

Collectors like Vincent often hired skilled artists to make drawings of their treasures. The creatures were no longer captured in emblematic mottos or *vanitas* still life paintings, but depicted *ad vivum*. If they were put in any context at all, it is that of the species to which they belong, or that of the plants they live on. The intention of these pictures and collections was clearly to show the structure and richness of God's creation. The wealth of the Dutch collections of overseas insects even provoked the jealousy of English collectors, who were involved in the same program of natural theology. John Ray and Robert Hooke wrestled with the Dutch language in order to understand Swammerdam's writings.

---

[48] Stephan Blankaart, *Schouw-Burg der Rupsen, Wormen, Maden en Vliegende dierkens daar uit voorkomende* (Amsterdam: 1688).

[49] Cf. Bergvelt E. – Kistemaker R. (eds.), *De wereld binnen handbereik. Nederlandse kunst- en rariteitenverzamelingen, 1585–1735* (Zwolle – Amsterdam: 1992).

[50] John Ray, *Observations, topographical, moral & physiological made in a journey through parts of the Low-Countries, Germany, Italy and France* (London: 1673) 25.

[51] Levinus Vincent, *Het tweede deel of vervolg van het Wondertooneel der Natuur* (Amsterdam: 1715) 11.

Sir Hans Sloane, president of the Royal Society, was deeply honored by the species his Dutch fellow-*virtuosi* sent him. Seba even presented him with two impressive folios, containing beautiful watercolors of insects from all over the world, writing 'As in my View in this undertaking has been to render my best Hommage to the Great Author of Nature, by displaying His curious and wonderfull Works'.[52]

However, the changing attitude towards insects is perhaps best expressed by the fact that Dutch virtuosi now started to collect endemic species as well. The Delft burgomaster and physician Hendrik d'Acquet (1632–1709), for example, possessed not only insects from the Indies, but also many specimens from Holland. Mindful of Swammerdam's conviction that God manifested Himself even in the most ordinary Dutch louse, d'Acquet wrote in the description of his collection that he was the proud owner of a beetle he had found 'here in Delft', a butterfly 'from my garden', an unidentified insect 'I caught on the honeysuckle' and a waterlouse 'I found in my rain barrel'.[53]

---

[52] British Library, Ms. Sloane 4047 fol. 13r. See also Ms. Sloane 5273–5274 ('A book containing severall sorts of insects [...] from the collection of Mr. Albert Seba').

[53] Amsterdam, Koninklijk Instituut voor de Tropen, Ms. RG 84 (Henrik d'Acquet, 'Opus magnificentissimum et unicum [...] ad exemplaria naturalia summo studio ultra quinquaginta annos e universo terrarum orbe quaesita').

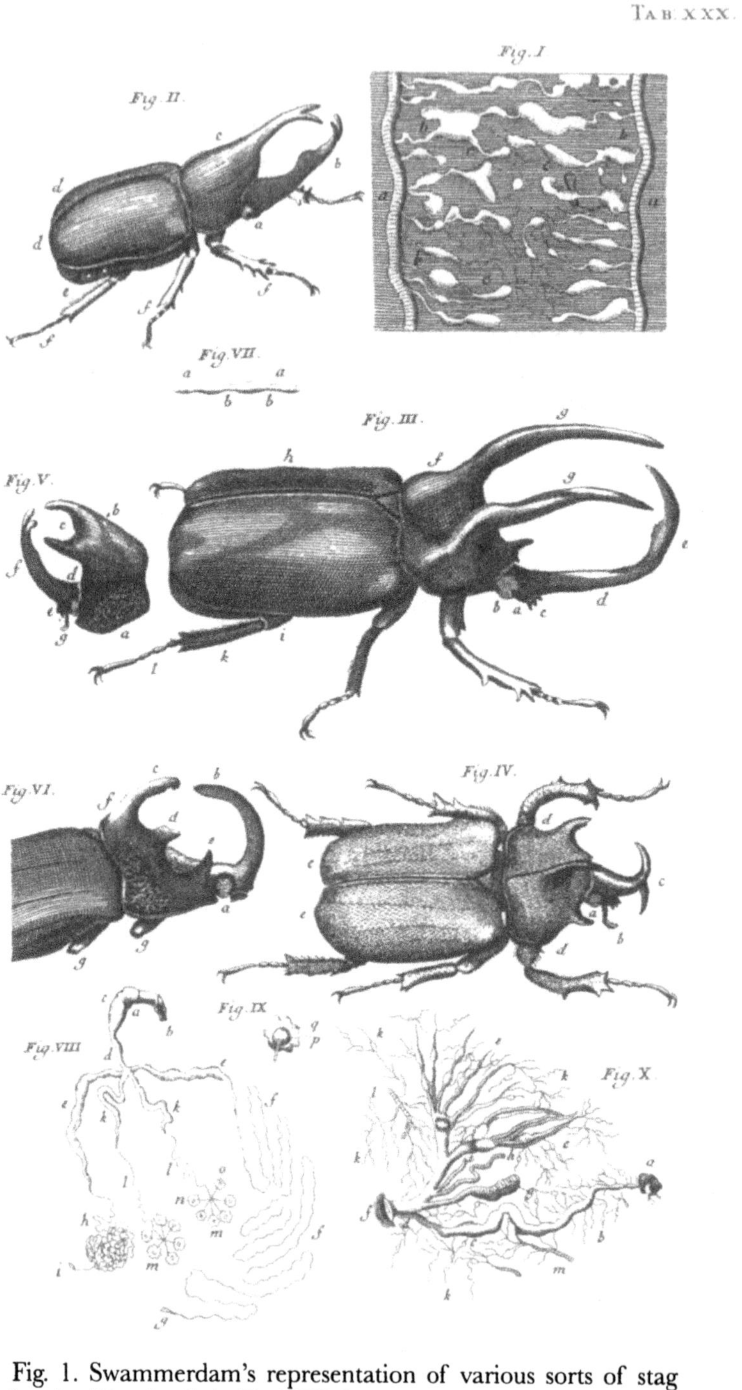

Fig. 1. Swammerdam's representation of various sorts of stag beetles. The details in Fig. VIII show the male genitals, Fig. X the female uterus. Engraving in Jan Swammerdam, *Bybel der natuure of historie der insecten. Biblia naturae, sive historia insectorum* (Leiden: 1737–1738) Tab. XXX.

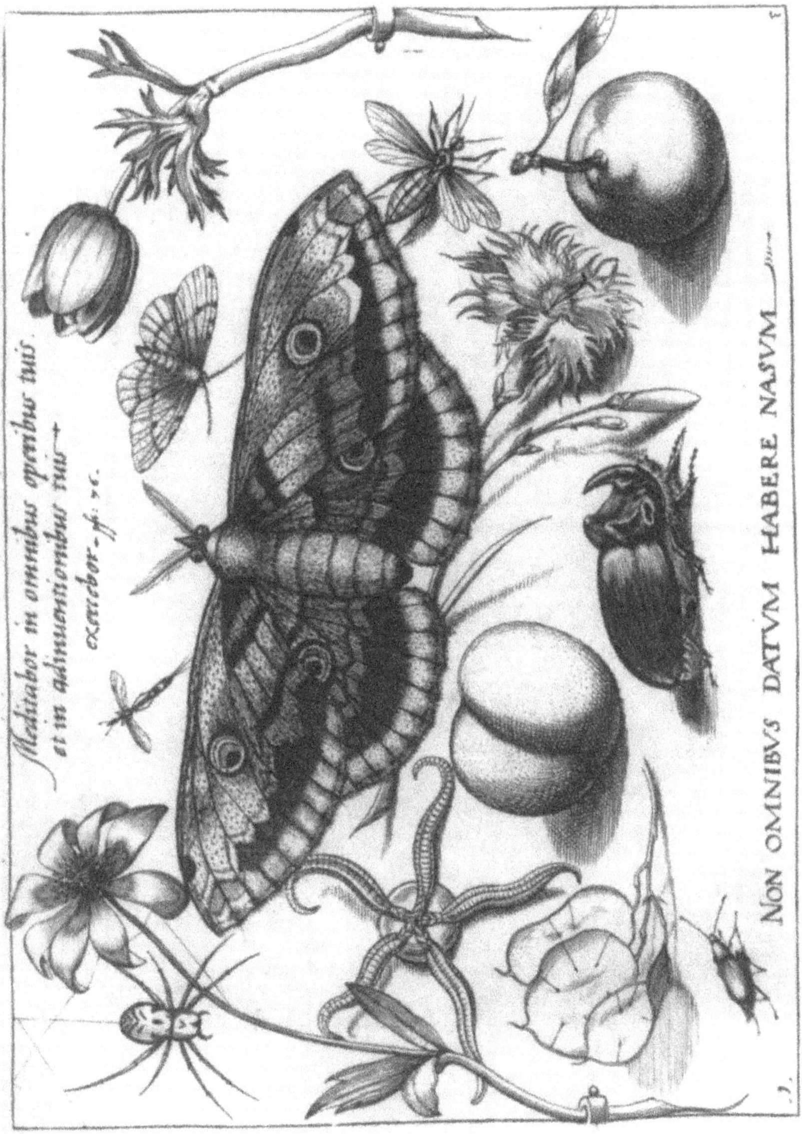

Fig. 2. The stag beetle amongst other creatures, as represented in Hoefnagel's *Archetypae*. The motto is derived from *Psalm* 77:14: 'I will meditate also of all thy work, and talk of thy doings'. From the *Archetypa studiaque patris Georgii Hoefnagelii* (Frankfurt a.M.: 1592) pars I, 8.

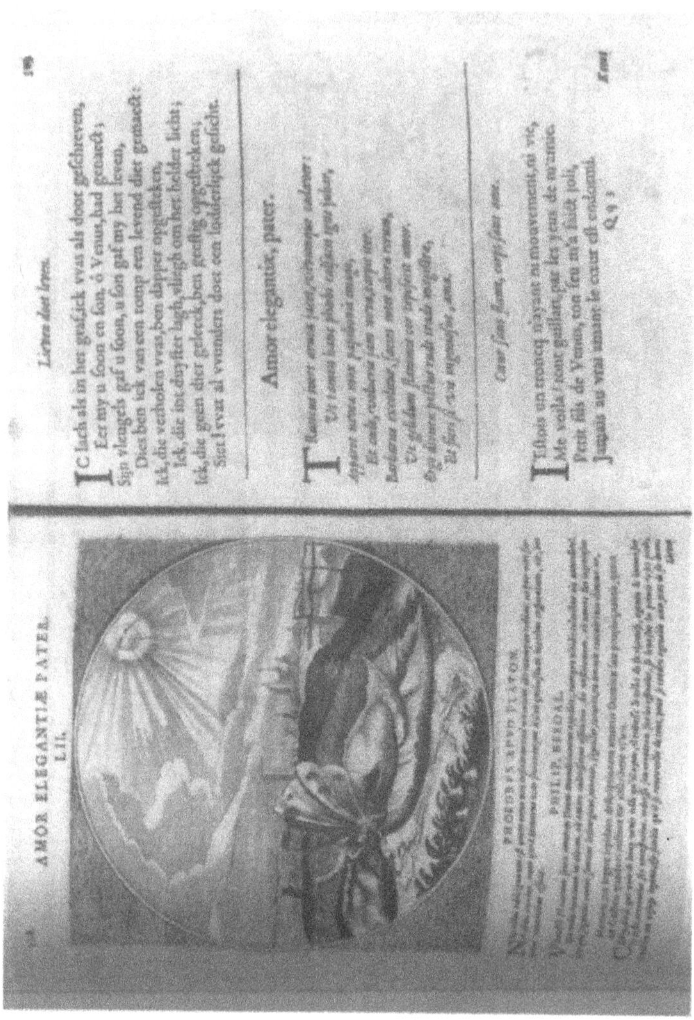

Fig. 3. A butterfly emerging from its cocoon. Emblem LII from Jacob Cats's *Silenus Alcibiadis sive Proteus* [...] (Middelburg: 1618). In the accompanying texts, Cats meditates on, amongst others, the Resurrection of the Dead.

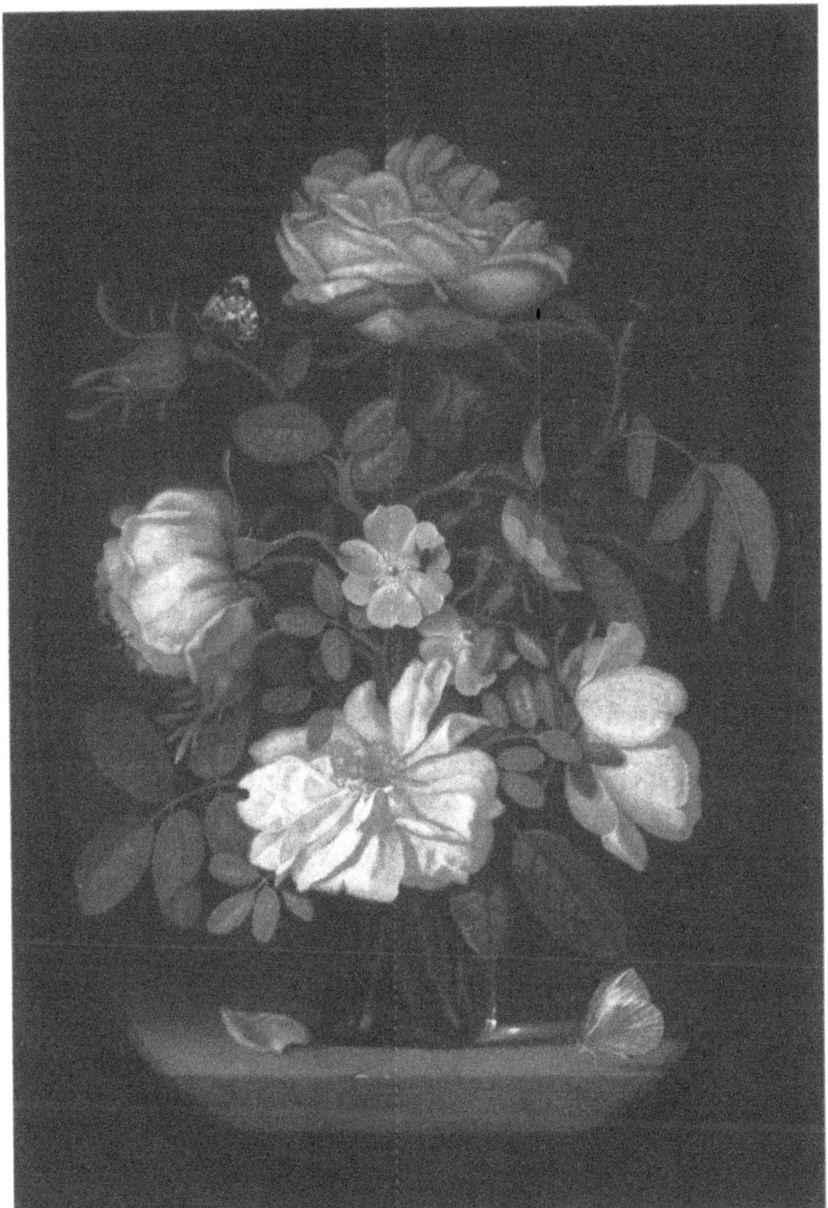

Fig. 4 [Col. pl. I]. Johannes Goedaert, *A bouquet of roses in a glass vase.* Also depicted are various insects, amongst others a bee, a butterfly and a dragonfly. Oil on canvas, date unknown. Collection Zeeuws Museum Middelburg, inv. nr. M96-031.

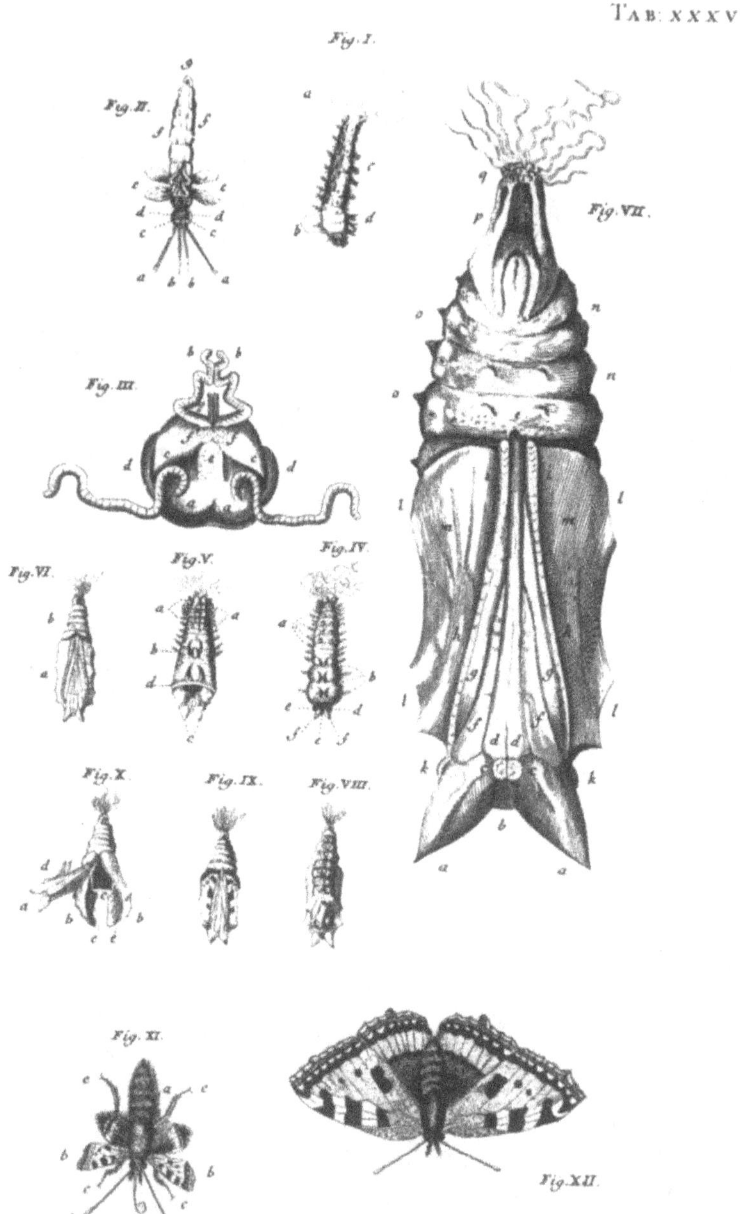

Fig. 5. One of Jan Swammerdams most important discoveries was the fact that parts of the future butterfly were already discernable in the caterpillar. Swammerdam published his discovery in the *Historia insectorum generalis* (1669); this engraving is taken from Jan Swammerdam, *Bybel der natuure of historie der insecten. Biblia naturae, sive historia insectorum* (Leiden: 1737–1738) Tab. XXXV.

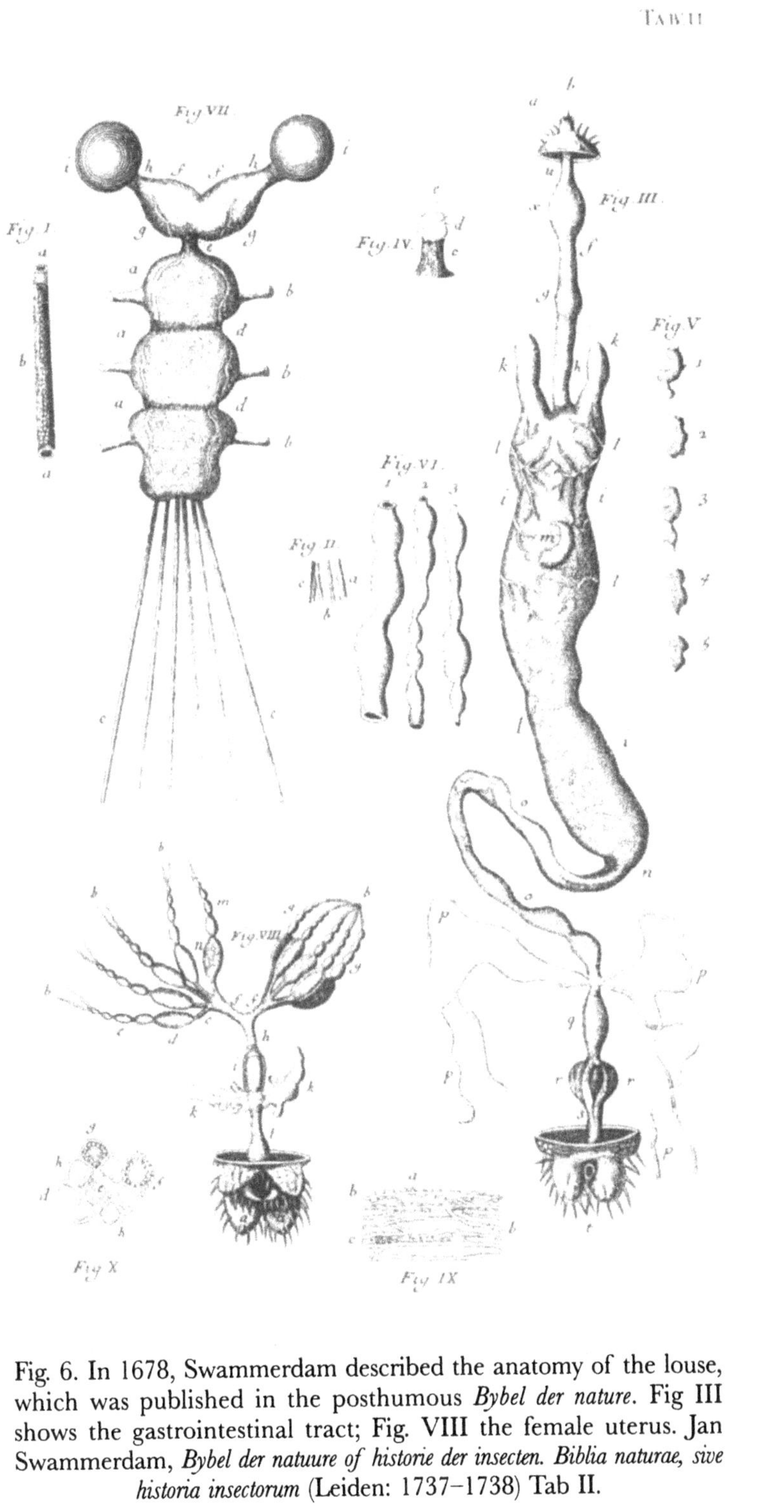

Fig. 6. In 1678, Swammerdam described the anatomy of the louse, which was published in the posthumous *Bybel der nature*. Fig III shows the gastrointestinal tract; Fig. VIII the female uterus. Jan Swammerdam, *Bybel der natuure of historie der insecten. Biblia naturae, sive historia insectorum* (Leiden: 1737–1738) Tab II.

Selective Bibliography

Aristotle, *Historia animalium*, ed. Peck A.L. (Cambridge Mass.: 1965).
Ashworth W., "Natural History and the Emblematic Worldview", in Lindberg D.C. – Westman R. (eds.), *Reappraisals of the Scientific Revolution* (Cambridge: 1993) 303–323.
Baar M. de, *'Ik moet spreken'. Het spiritueel leiderschap van Antoinette Bourignon (1616–1680)* (Zutphen: 2005).
Bergström I., *Dutch Still-Life Painting in the Seventeenth Century* (London: 1956).
——, "On Georg Hoefnagel's Manner of Working with Notes on the Influence of the Archetypa series of 1592", in Cavalli-Björkman G. (ed.), *Netherlandish Mannerism* (Stockholm: 1985) 177–188.
Bergvelt E. – Kistemaker R. (eds.), *De wereld binnen handbereik. Nederlandse kunst- en rariteitenverzamelingen, 1585–1735* (Zwolle – Amsterdam: 1992).
Blankaart Stephan, *Schouw-Burg der Rupsen, Wormen, Maden en Vliegende dierkens daar uit voorkomende* (Amsterdam: 1688).
Bodenheimer F.S., *Materialien zur Geschichte der Entomologie bis Linné*, 2 vols. (Berlin: 1928).
Bots J., *Tussen Darwin en Descartes. Geloof en natuurwetenschap in de achttiende eeuw in Nederland* (Assen: 1972).
Bunge W. van, *From Stevin to Spinoza. An Essay on Philosophy in the Seventeenth-Century Dutch Republic* (Leiden: 2001).
Cats Jacob, *Alle de Wercken*, 2 vols. (Amsterdam – The Hague: 1726).
——, *Silenus Alcibiadis dat is, Sinne- en minnebeelden*, ed. H. Luijten, 2 vols. (The Hague: 1996).
Clutius Augerius, *Opuscula duo singularia. I. De nuce medica. II. De hemerobio sive ephemero insecto, et majali verme* (Amsterdam: 1634).
Cluyt Dirck Outgersz, *Van de byen, hare wonderlicke oorspronc, natuer, eygeschap, crachtige, ongehoorde ende seldsame wercken: Waerin gemerct werden hare wonderlicke Politien en ordentelicke Regeringe, die sy met haren Coning ende onder melcandern onderhouden. Vervat in drie Boecken: Seer genuechlich om te lesen* (Leiden: 1597).
Cobb M., "Malpighi, Swammerdam and the colourful silkworm: Replication and visual representation in early modern science", *Annals of Science* 59 (2002) 111–147.
Daston L. – Park K., *Wonders and the Order of Nature, 1150–1750* (New York: 1998).
Fournier M., *The Fabric of Life. Microscopy in the Seventeenth Century* (Baltimore: 1996).
Francissen F.P.M. – Mol A.W.M., *Augerius Clutius and his 'De hemerobio'. An Early Work on Ephemeroptera* (Marburg: 1984).
Goedaert Johannes, *Metamorphosis naturalis, ofte historische beschrijvinge van den oirspronk, aerdt, eygenschappen ende vreemde veranderingen der wormen, rupsen, maeden, vliegen, witjens, byen, motten en diergelijke dierkens meer; niet uit eenige boeken, maar aleenlyck door eygen ervaretntheyt uytgevonden, beschreven ende na de konst afgeteykent* 3 vols. (Middelburg: s.d. [1660–1669]).
——, *Of insects. Done into English, and Methodized, with the Addition of Notes* (York: 1682).
Harms W., „Bedeutung als Teil der Sache in zoologischen Standardwerken der Frühen Neuzeit (Konrad Gesner, Ulysse Aldrovandi)", in Boockmann H. (ed.), *Lebenslehre und Weltentwürfe im Übergang vom Mittelalter zur Neuzeit* (Göttingen: 1989) 531–553.
Harwood J.T., "Rhetoric and graphic in *Micrographia*", in Hunter M. – Schaffer S. (eds.), *Robert Hooke: New Studies* (Woodbridge: 1989).
Hondius Petrus, *Dapes inemptae, of de Moufe-schans, dat is de soeticheydt des buytenlevens, vergeselschapt met de boucken* (Leiden: 1621).
Jorink E., "'Outside God, there is nothing'. Swammerdam, Spinoza, and the Janus-Face of the Early Dutch Enlightenment" in: Bunge W. van (ed.), *The Early Enlightenment in the Dutch Republic, 1650–1750. Selected papers of a conference held at the Herzog August Bibliothek, Wolfenbüttel 22–23 March 2001* (Leiden: 2003) 81–103.

——, "'These wonderfull glasses'. Dutch humanists and the microscope", in Grob B. – Hooijmakers H. (eds.), *Who needs scientific instruments? Conference on Scientific Instruments and their Users, 20–22 October 2005* (Leiden: 2006) 115–124.

——, *Het Boeck der Natuere. Nederlandse geleerden en de wonderen van Gods Schepping, 1575–1715* (Leiden: 2006).

Huygens Christiaan, *Oeuvres complètes de Christiaan Huygens*, 22 vols. (The Hague: 1888–1950).

Huygens Constantijn, *Mijn jeugd*, ed. C.L. Heesakkers (Amsterdam: 1987).

——, *De gedichten van Constantijn Huygens*, ed. J.A. Worp 9 vols. (Arnhem: s.d.).

Kaufmann T.D., *The Mastery of Nature. Aspects of Art, Science and Humanism in the Renaissance* (Princeton: 1993).

Lovejoy A., *The Great Chain of Being. A Study of the History of an Idea* (Cambridge Mass.: 1936).

Malebranche Nicolas, *Oeuvres complètes*, 12 vols. (Second edition, Paris: 1972–1984).

Nieuwentijt Bernard, *Het regt gebruik der wereltbeschouwingen, ter overtuiginge van ongodisten en ongelovigen* (Amsterdam: 1715).

Oldenburg Henri, *The Correspondence of Henry Oldenburg*. ed. A.R. Hall and M.B. Hall, 10 vols. (London: 1965–1973).

Pliny the Elder, *Naturalis historia*, ed. H. Rackham (Cambridge Mass.: 1940).

Ray John, *Observations, topographical, moral et physiological made in a journey through parts of the Low-Countries, Germany, Italy and France* (London: 1673).

Ruestow E., *The Microscope in the Dutch Republic. The Shaping of Discovery* (Cambridge: 1996).

Stebbins Stephan, *Maxima in minimis. Zum Empirie- und Autoritätsverständnis in der physikotheologischen Literatur der Frühaufklärung* (Frankfurt a.M.: 1980).

Swammerdam Jan, *Historia generalis insectorum ofte Algemeene verhandeling van de bloedeloose dierkens* (Utrecht: 1669).

——, *Bybel der natuure of historie der insecten. Biblia naturae, sive historia insectorum*, 2 vols. (Leiden: 1737–1738).

Verbeek Th., *Descartes and the Dutch: Early Reactions to Cartesian Philosophy, 1637–1650* (Carbondale: 1992).

Vermij R., *Secularisering en natuurwetenschap in de zeventiende en achttiende eeuw: Bernard Nieuwentijt* (Amsterdam: 1991).

Vignau-Wilberg Th., *Archetypa studiaque patris Georgii Hoefnagelii 1592. Natur, Dichtung und Wissenschaft in der Kunst um 1600* (Munich: 1994).

Vincent Levinus, *Het tweede deel of vervolg van het Wondertooneel der Natuur* (Amsterdam: 1715).

Westerbaen Jacob, *Arctoa Tempe. Ockenburgh. Woonstede van den Heere van Brandwyck, in de Clingen buyten Loosduinen* (The Hague: 1654).

Wilberg Vignau-Schuurman Th., *Die emblematischen Elemente im Werke Joris Hoefnagels*, 2 vols. (Leiden: 1969).

Wilson C., *The Invisible World: Early Modern Philosophy and the Invention of the Microscope* (Princeton: 1995).

SHELL COLLECTING. ON 17TH-CENTURY CONCHOLOGY, CURIOSITY CABINETS AND STILL LIFE PAINTING

Karin Leonhard

Shell collecting was one of the favourite pastimes of the 17th century. These treasures from the sea united all of the opposites between art and nature which had become important in the Baroque period: their maritime provenance evoked the fact that all life had its origins in the sea – one reason why Aristotle used shells as an example of his theory of spontaneous generation. According to him, shells could be generated by the sun warming up the mud on the seabed – a *creatio ex nihilo* which brought dead material to life. We know that in the 17th century this theory was very much discussed but in the end proven to the contrary. Nevertheless, molluscs were still regarded as primitive creatures while their patterned and shaped shells were highly sought after by collectors and could fetch incredibly high prices. The baroque curiosity about molluscs was provoked by a series of contrasts: the soft, living creature versus its hard, robust shell; rough and weathered surfaces versus surfaces that were polished as smooth as glass; the perfect geometry of cones, spirals and curves versus their natural provenance – aesthetic qualities versus manifestly primitive ones.[1] However, for the coastal regions of England, France, Italy, Spain and the Dutch provinces, the sea was more than just the source of a powerful morphogenetic force of nature. The sea was also ruled by Poseidon, who patronized commerce, expeditions, sea wars, and conquests, and enabled the import of exotic herbs, spices, plants or animals as well as the colonization of foreign countries and archipelagos. Shells therefore became symbols of maritime and colonial power, and allegorical attributes of a national self-image [Fig. 1].

During the 16th and 17th centuries, these precious and exotic objects mainly served as diversions for members of the aristocracy, who were fascinated by their great diversity and began collecting shells in *curiosity*

[1] See Mauries P., *Shell Shock: Conchological curiosities* (London: 1994) 15.

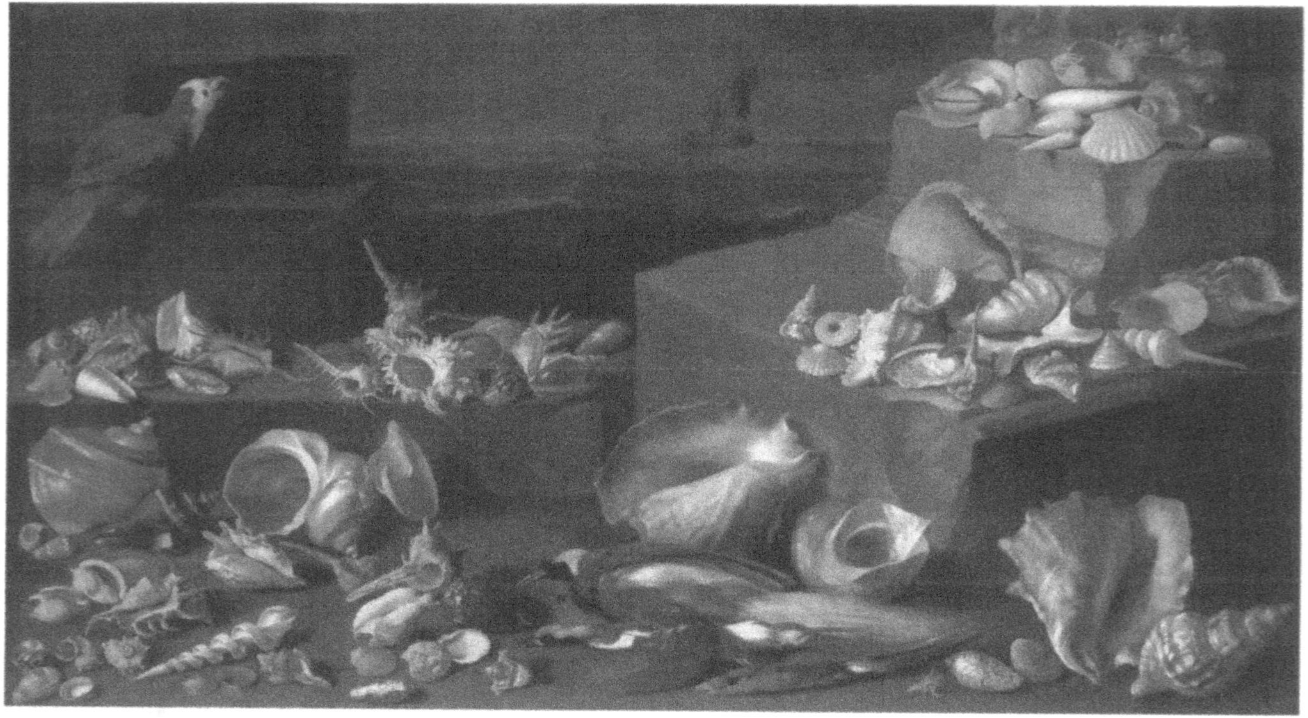

Fig. 1 [Col. pl. II]. Still life painting with shells and birds (ca. 1640/50). Flemish, attributed to Jan Davidsz. De Heem, ca. 1640/50. Oil on canvas, 88.5 × 136 cm. Vienna, Kunsthistorisches Museum, Gemäldegalerie. These species of *Halotidae*, *Strombidae*, *Muricidae*, *Conidae* and *Pectinidae* mainly come from the Indo-Pacific and fascinate by their various sculpturing and shape. The ship in the background evokes the sea trade routes of the Dutch East India Company.

cabinets. Early on, an encyclopedic impulse to create a systematic collection of natural objects rivaled the desire to impress viewers with rare specimen or *mirabilia* – wonders of creation that showed off Nature's artistic capabilities, her inexhaustible *vis plastica*, and the colourful design of her production. As time went on, these two collector's impulses went their separate ways, becoming mutually exclusive. Shell collections in baroque *curiosity cabinets* came to form the basis for the science of conchology. At the same time they presented a showcase for the creations of *Natura*, which were as artful as painted and designed objects.

This essay seeks to sketch some interrelations between the passion for shell collecting in the Baroque era, the beginnings of conchology as a discipline, and the birth of shell still life painting. The initial connections can be found in the 17th-century's intense preoccupation with shells, which in turn was closely related to:

- a formalistic approach to biology and zoology
- theories of spontaneous generation versus theories of biological procreation
- theories of the history of the earth[2]
- the emergence of a dynamic world view
- the adoption and domestication of the unknown
- a political iconography and national self-image.

I will only discuss certain aspects of these interrelationships here. In so doing, one of my main focuses will be early depictions of shells in art and the natural sciences, which raises questions regarding the place of these creatures in the baroque biological hierarchy. In the 17th century, the specific structures of shells and whelks were investigated and classified. At the same time scientists started to study the living animal, its functions and natural behavior.

[2] For additional information regarding the relationships beween conchology, geology and paleontology, see the bibliography.

The formalist approach to nature

Aristotle was a gifted zoologist and the descriptions in his *Historia animalium* were unmatched for centuries – one reason why a great deal of zoological terminology is based on his writings. Aristotle named snails and shells 'shell creatures' (*Testacea, Ostracoderma*) after their most salient characteristic. Initially, these 'shell creatures' were contrasted with a group of noncrustaceans that were termed 'soft creatures' (*Mollusca, Malacia*). However, as a result of comparative anatomical studies realized around 1800, they were combined into a single phylum known as molluscs (*Mollusca*). The name of the discipline devoted to the study of molluscs (*Mollusca*) – malacozoology or malacology – stems from this period, whereas conchology was defined as the study of the shell structures of snails and molluscs.[3]

The increased prevalence of shell display cases in early Renaissance *curiosity cabinets* can be only attributed to a conchological interest, which went hand in hand with the formal study of natural objects. The focus on morphological issues then led to a desire for interchanges between the arts and sciences. The coloured speckles, lines and structures on seashells that often resembled the markings on tigers and leopards, and the iridescence of mother of pearl whose mutable colours were akin to those of peacock feathers and butterfly wings, had already struck Pliny as being painterly and had inspired artists as well. This gave rise to the common saying that their surfaces were 'painted by nature' which had its origins in the literature of antiquity and existed well into the 18th century. The well-known topos of "nature sporting" also allegorized the objects in baroque shell collections. As Barbara Stafford has put it: 'Design is not a separable imprint or impresa stamped on the surface: Design is a real symbol of the development of that medium, radical immanence not transcendence'.[4] And in fact the colouration on mollusc shells is not painted on the surface, but is the result of pigments embedding themselves in the shell as it develops. These pigments are secreted continuously, producing spiral or radial lines and bands, or discontinuously, giving rise to patterns of speckles and patches. The markings were directly associated with their carriers and interpreted as

---

[3] Lindner G., *Muscheln und Schnecken der Weltmeere* (Munich et al.: 1999) 14.

[4] See Stafford B., "Characters in Stones, Marks on Paper: Enlightenment Discourse on Natural and Artificial Taches", *Art Journal* 44 (1984) 233–240.

a diagram of the creature's morphogenesis which was directly inscribed into the body. Thus for example, molluscs became a paradigm for the passage of time and historical growth – space-time made visible. The sculptural forms of gastropod shells fascinated baroque observers owing to the periodic progression of their surfaces, which to some extent resembled processes in sculpture, goldsmithing and pottery.

We now know that mollusc shells grow discontinuously, that their forms mutate in fits and starts, and that the shell develops in two directions: parallel to the periphery and perpendicular to the surface. Thus the periphery of the shell becomes larger and broader while its surface becomes progressively stronger. Periodically, the developmental process either stops or shifts to a different element of the shell's form. The presence of smooth interstitial surfaces indicates that the lip of the mantle abutted the shell without any folds while structural development was in progress. However, the mantle generates folds in most cases, leading to the development of ribs in shells and three dimensional spiral bands in gastropods, while during the fortification period, knobs, humps and spikes form. If axial and spiral three dimensional elements overlap, grid patterns form.[5] The laws governing this morphological process were unknown in the 17th century, leading to a fascination with the mysterious biological "architectural" plan that rhythmically drove this generative process and in so doing sculpted patterns and forms into the surface of the shell.

This process also appealed to several preferences of the Baroque period. The seemingly endless variety of regular patterns and forms etched into the shells recalled machine-made products such as regularly patterned textiles or carefully turned pieces of ivory, of rotating musical clocks, mills, and gears.[6] It was accompanied by an interest in mechanically produced whorls, curves, ellipses, and grids – infinite forms that could be endlessly combined with each other.[7] The 17th

---

[5] Lindner, *Muscheln* 31.

[6] On the sophisticated mechanical dynamics of lathes and the close proximity of nautilus goblets and ivory objects to each other in *curiosity cabinets* see: Maurice K., *Der drechselnde Souverän* (Zurich: 1985). For a picture of a lathe from circa 1600 that produced logarithmic spirals, see Frieß P., *Kunst und Maschine* (Munich: 1993).

[7] On the mathematical iconography of spiral shells (above all nautilus shells) see Mette H.-U., *Der Nautiluspokal: Wie Kunst und Natur miteinander spielen* (Munich et al.: 1995); Pickover C.A., "Mathematics and Beauty. A Sampling of Spirals and 'Strange' Spirals in Science, Nature and Art", *Leonardo* 21,2 (1988) 173–181; Fonseca R., "Shape

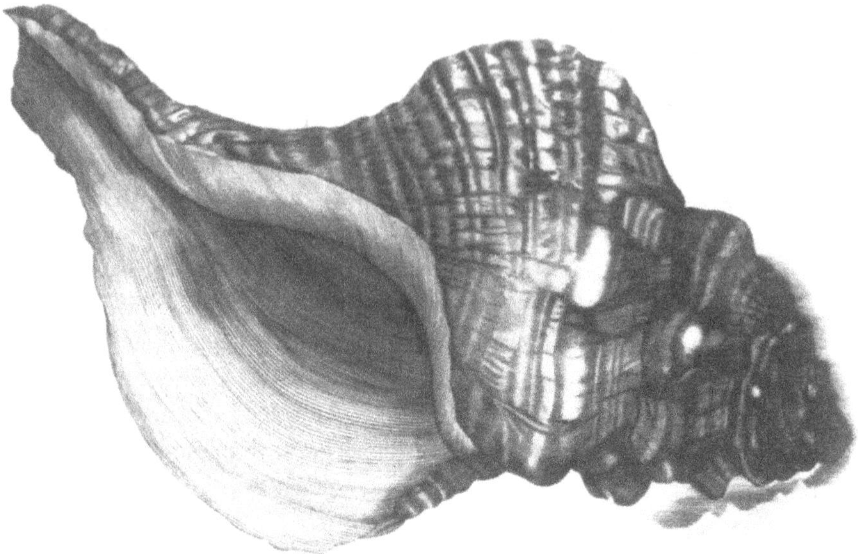

Fig. 2. Univalve gastropod shell. Large-sized engraving from Martin Lister, *Historia Conchyliorum* (London: 1685–1692). All gastropods have spiral forms and a structural affinity to baroque theories on biological and cosmological evolution.

century also adhered to a vertebral theory of the cosmos according to which material could be thickened by means of rhythmically iterative movements in such a way that mass centres, bodies and ultimately whole planets could come into existence. However, the occurrence of this phenomenon necessitated perpetual movement around a central axis. According to the Baroque theory, whorled gastropod shells were based on just such a cosmic principle, making them an allegory of the rotating universe – one that could be accommodated in a cabinet drawer [Fig. 2].

Dutch 17th-century shell collections and still life painting

In 1603 Hendrick Goltzius painted a portrait of the Rederijker and patron of the arts Jan Govertsz van der Aar standing before some shells

and order in Organic Nature. The Nautilus Pompilius", *Leonardo* 26,3 (1993) 201–204; Holländer H., "Nautilus mirabilis", *Aachener Kunstblätter* (1998) 461–464.

of his collection [Fig. 3].[8] Van der Aar is depicted balancing on his fingertips what is perhaps the most precious object in the collection – a *Turbo marmoratus* whose mother of pearl surface gleams in the soft light. On the front edge of the table is a *Charonia* from the Bahamas.[9] A portrait of van der Aar by Cornelisz Cornelis painted four years later also shows the shell collector standing before his treasures, this time encircled by personifications of the arts and sciences.[10] This passionate desire to collect shells amongst the affluent bourgeoisie had been in full swing at least since the middle of the century and ultimately evolved into a veritable collecting mania. The trend rubbed some contemporaries up the wrong way, however: 'It's annoying what people spend their money on',[11] said a caption in Roemer Visscher's *Sinne-Poppen* [Fig. 4], which was critical of the high prices fetched by shells: a spiral *Wentletrap* could cost up to 500 guilders in the 17th century.[12]

The shell collections of baroque curiosity cabinets formed a natural springboard for all future research into molluscs. Upon returning from a trip from Holland, the famous art dealer Gersaint called that country the market leader, writing that 'everyone there was curious'.[13] In 1614, the Dutch poet Philibert van Borsselen composed a lyric entitled "Beach" about the shells in the cabinet of Cornelis van Blijenburgh in Amsterdam, singing the praises of the multifaceted beauty of molluscs in over exuberant verses. The 17th century *curiosity cabinets* of Brayne and Reynst, of Frederik Ruysch, Johan Swammerdam, Albertus Seba and Uyttenbogaert in Amsterdam, the renowned museum of Delft's burgomaster Henri D'Acquet, the early 18th-century *Wunderkammern*

[8] See Bergvelt E. – Kistemaker R. (eds.), *De wereld binnen handbereik. Nederlandse kunst- en rariteitenverzamelingen, 1585–1735* (Zwolle – Amsterdam: 1992) 46, cat. no. 62. I am grateful to Claudia Fritzsche for her insights into the depiction of shells in still lives.

[9] Fritzsche C., *Genuss und Disziplinierung. Zum Umgang mit Besitz und Wohlstand im holländischen Stillen im 2. Drittel des 17. Jahrhunderts* (Leipzig: 2000; unpublished manuscript) 61.

[10] He is holding a specimen of *Hippopus hippopus* in his hand. In front of him are specimens of *Strombus pugilis*, *Trochus niloticus*, *Aulicina vespertilio*, *Harpa doris Röding*, and a *Conus* specimen. See Dance S.P., *Shell Collecting. An Illustrated History* (London: 1966) 34.

[11] Roemer Visscher, *Sinne-Poppen* (Amsterdam: 1614) vol. I, no. 4.

[12] Georg Eberhard Rumphius, *The Ambonese Curiosity Cabinet* (*D'Amboinsche Rariteitkamer*) (Amsterdam 1705), translated, edited, annotated, and with an introduction by E.M. Beekmann (Ann Arbor: 1999); see Dance, *Shell Collecting* 229. Additional information about prices can be found in Jutting G.W.S.S. van Bentheim, "Brief history of the conchological collections at the zoological museum of Amsterdam", *Bijdragen tot de Dierkunde* (1938) 199–217.

[13] Gersaint E.F., *Catalogue raisonné de coquilles, et autres curiosités naturelles* (Paris: 1736).

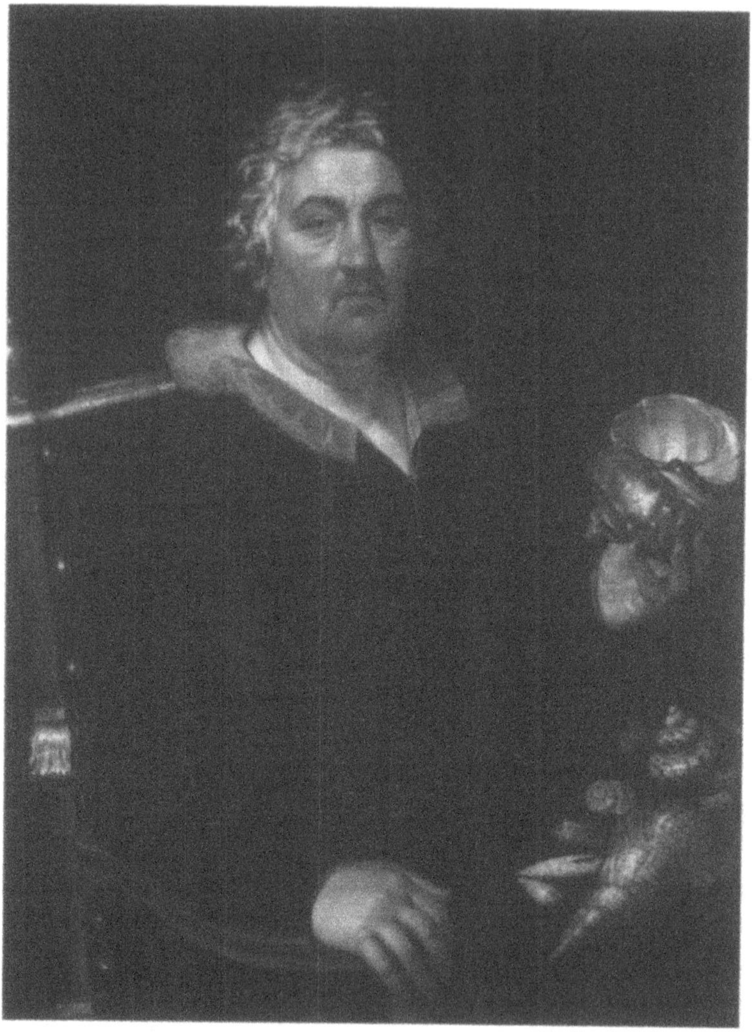

Fig. 3 [Col. pl. III]. A polished *Turbo marmoratus* presented by the Dutch shell collector Jan Govertsz. van der Aar. Hendrick Goltzius, Portrait of Jan Govertsz. van der Aar, 1603, 107.5 × 82.7 cm. Museum Boymans-van Beuningen, Rotterdam.

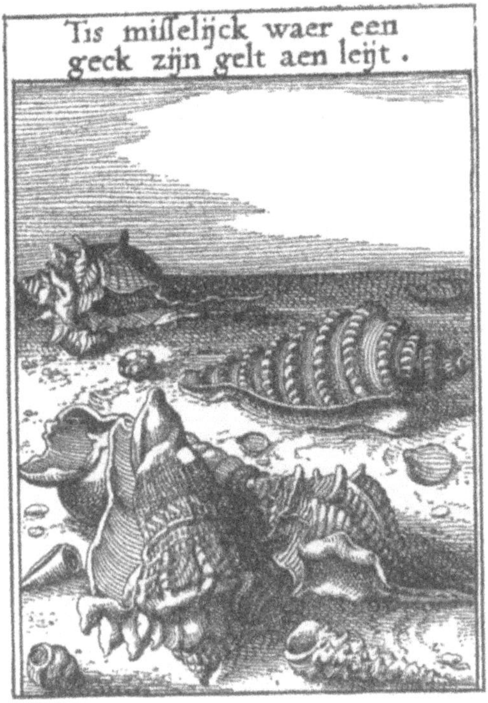

Fig. 4. "It is annoying what people spend their money on". Illustration from Roemer Visscher's *Sinne-Poppen* (Amsterdam: 1614) vol. I, no. 4.

of De La Faille in The Hague or J.C. Brandt and Simon Schijnvoet in Amsterdam contained substantial shell collections, among them such sought-after specimens as *Cedo Nulli*,[14] *Scala pretiosa* (*Wentletrap*) and *Conus gloria maris* (*Glory of the Sea*). In England, William Courten (or Charlton) made his massive shell collection available to natural history enthusiasts, and Martin Lister used the collection as the chief source for his *Historia Conchyliorum*. The collection was later purchased by Henry Sloane, who had brought back to England exotic snails and shells that he had acquired in Jamaica. These specimens also found their way into Lister's *Historia*.

[14] See Jutting, *Brief History* 200: 'Mr. De La Faille, a famous 18th century collector at The Hague, who possessed two specimens, refused to sell one for 6000 French pounds'.

Beginning in the 1630s, Dutch painters started developing a distinctive style of still life painting, the so-called shell still life.[15] The first genuinely Dutch shell still lifes were painted by Balthasar van der Ast, probably in the 1630s. This subject was explored until the end of the century by artists such as Abraham Susenir and Willem Kalf, as well as by Adriaen Coorte in his alluring depictions. Ambrosius Bosschaert, Abraham van Beyeren, Dirck van Delen, Clara Peters and Maria van Oosterwijck all combined the colorfully patterned surfaces of fruit, flowers and shells in their still lifes, elevating this genre to the status of self-reflective trompe l'oeil painting, while Jan Davidsz de Heem integrated shells into sumptuous still lifes, thereby underscoring the national and political iconography of shells as well as the interrelations between commerce and riches in the new Dutch republic.

'The outer form of shell still lifes – a table with shells from various countries and parts of the earth carefully arrayed on a table next to each other – remained basically the same during this period, but the significance of their content gradually changed. Whereas the *curiosity cabinet* is still a palpable presence in van der Ast's depictions of shells, and although a natural science and encyclopedic interest in shells as well as a sense of the evanescence of the aesthetics of the picture prevail, they are treated primarily as a pretext for painting and composition by Willem Kalf, even though the shell motif lends itself to a broader interpretation'.[16] Balthasar van der Ast [Fig. 5] devoted himself to the shell motif for aesthetic reasons as well, and in so doing indirectly cited Pliny by including in his still life a *Conus marmoreus*, two examples of the genus *Harpidae*, several serrated *Murex* shells, a magnificent polished Turban shell and a speckled Cowrie shell (*Cypraea tigris*) whose markings genuinely resemble leopard's spots. The beauty of their colours and patterns and the smoothness of their porcelain-like surfaces made this type of shell one of the most prized collector's items in the 17th century. The value of such precious curiosities was then translated to the canvas, thus heightening the artificiality and mediality of the painting. In this case, the supreme artist, *Natura*, first had a hand in creating the coloured patterns and forms on the shell surface, and the painter merely replicated this creative act. Nature as painter and nature imitating painting are mirror images of each other and are joyfully conjoined

---

[15] For literature on shells in art see the bibliography.

[16] Gemar-Koeltzsch E., *Stillebenmaler*, vol. I 29 <translation K.L.>.

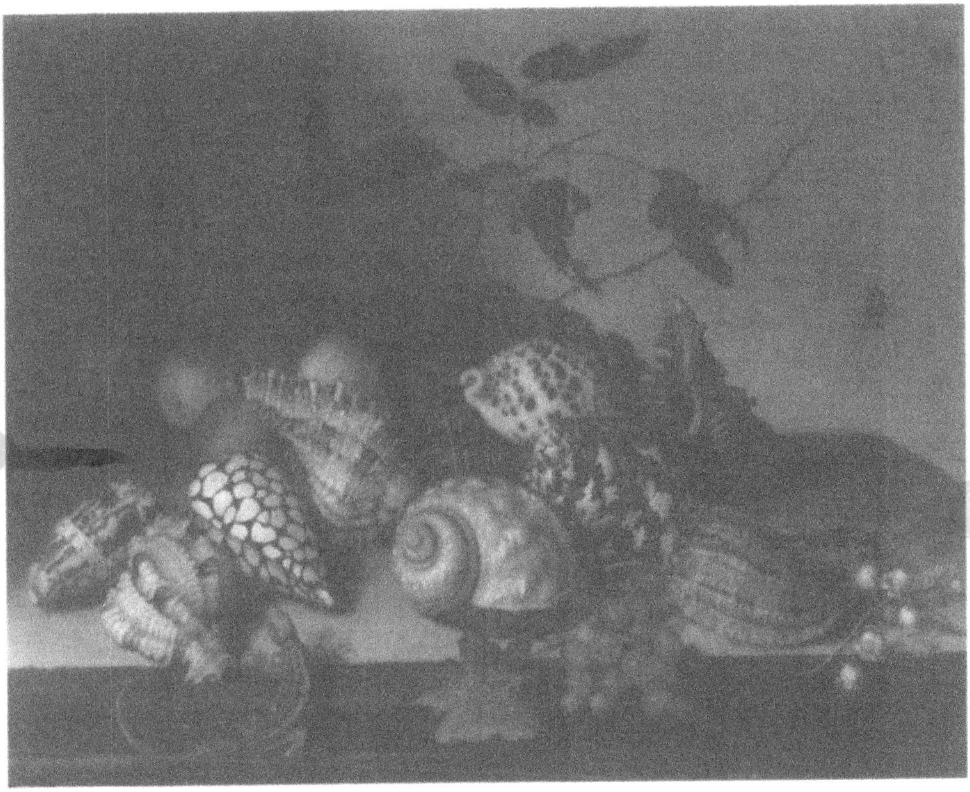

Fig. 5 [Col. pl. IV]. Colours and patterns of *Conus marmoreus, Cypraea tigris, Turbo marmoratus* and *Conus ermineus*. Still life painting by Balthasar van der Ast, Shells and fruits, ca. 1620. Oil on wood, 29 × 37 cm. Dresden, Gemäldegalerie Alte Meister.

at the aesthetic margin of the canvas. In Willem Kalf's paintings, the light only grazes the smooth surface of the shells, which emerge radiantly from the dark background as colouristic bravura pieces. On the other hand, Adriaen Coorte's still lifes focus on the sculptured shape of a few singled out shells. They cast strong shadows and have a more tactile quality than those of any other artist, as if they could be picked up by the beholder and turned in his hand [Fig. 6].

Shell Books

Zoological taxonomy and nomenclature were originally based on purely visual analogies and were an outgrowth of baroque *curiosity cabinets*. Early illustrations of sea animals and shells can be found in Adam Lonitzer's *Naturalis historiae opus novum*[17] (1551–1555) as well as Conrad Gesner's *Icones animalium quadrupedum et oviparorum* and *Historia animalium IV qui est de piscium et aquatilium animantium natura*[18] (1553 and 1558). Pierre Belon's *De aquatilibus libri duo*[19] (1553) contains some well-executed illustrations, while more than 100 molluscs were illustrated in Guillaume Rondelet's *Universae aquatilium historiae pars altera*[20] (1554–1555). Ulisse Aldrovandi's *De reliquis animalibus exanguibus*[21] (1606) is an encyclopedic treatise on molluscs that set the standard for future generations, despite the work's rather crude illustrations. Two excellently illustrated works by Fabio Colonna[22] (1616) set high standards and were the first works not to depict gastropods in reversed mirror images.[23] Jan Jonston's *Historiae naturalis de exanguibus aquaticis*, published in in 1650, was also an important work for Baroque collectors.[24]

---

[17] Adam Lonitzer, *Naturalis historiae opus novum* [...], 2 vols. (Frankfurt a.M., Christoph Egenolph: 1551–1555).

[18] Conrad Gesner, *Icones animalium quadrupedum viviparorum et oviparorum* [...] (Zurich, Christoffel Froschauer: 1553); *Historia animalium liber IV qui est de piscium et aquatilium animantium natura* [...] (Zurich, Christoffel Froschauer: 1558).

[19] Pierre Belon, *De aquatilibus libri duo* [...] (Paris, Carolus Stephanus: 1553).

[20] Guillaume Rondelet, *Universae aquatilium historiae pars altera* [...] (Lyon, Mathias Bonhomme: 1555).

[21] Ulisse Aldrovandi, *De reliquis animalibus exanguibus libri quatuor* [...] *nempe de mollibus, crustaceis, testaceis et zoophytis* (Bologna, J.B. Bellagamba: 1606).

[22] Fabio Colonna, *Fabii Columnae Lyncei Purpura* [...]; *Aquatilium et terrestrium aliquot animalium* [...]. (two treatises bound together) (Rome: 1616).

[23] Dance, *A History of Shell Collecting* 24 and 19.

[24] Jan Jonston, *Historiae naturalis de exanguibus aquaticis libri IV* (Frankfurt a.M., Haeredes Merianei: 1650).

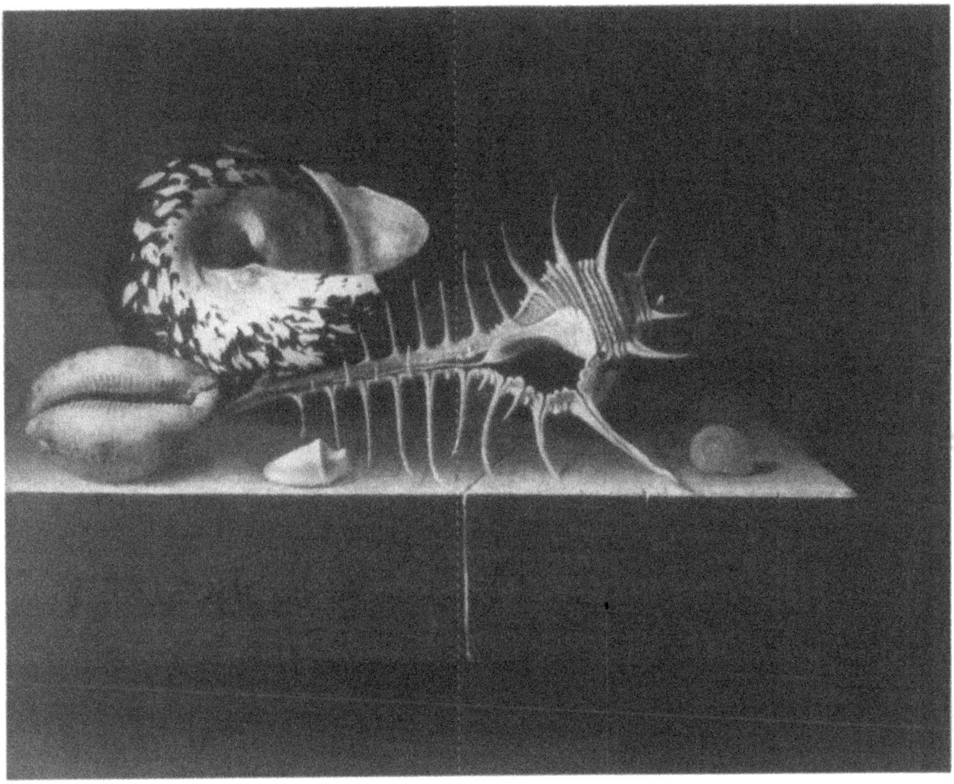

Fig. 6 [Col. pl. V]. *Cypraea arabica*, *Murex troscheli*, *Cittarium pica* and *Syphoma gibbosum* in a still life painting. Adriaen Coorte, *Shells*, 1698. Oil on paper on wood, 17 × 22.5 cm. Boston, private collection. The shells were carefully selected and depicted in their natural size.

The description of Francesco Calceolari's collection published in 1584 counts as the first inventory of a *curiosity cabinet* to mention shells;[25] an additional description of the same collection published in 1622 contains several impressively accurate shell illustrations.[26] In addition, the two accounts of the famous natural collections of Basil und Michael Rupert Besler[27] (1616 and 1642) were frequently cited during the 17th century [Fig. 7]. Shells are mentioned but not illustrated in the account of Ole Worm's collection of natural objects (1655),[28] Ludovico Moscardo's description of his collection (1656) is illustrated with large engravings,[29] while accurately drawn figures, though reversed, are found in Adam Olearius's inventory (1666) of the Duke of Schleswig Holstein Gottorp's collection.[30] In England, John Tradescant, gardener to Charles I and a keen shell collector, established a museum of natural objects whose catalogue was published by Tradescant's son (1656) with a dedication to Elias Ashmole.[31] The numbers of publications on shell collections and curiosity cabinets increased greatly, beginning in the mid 17th century and reaching a peak during its final two decades. It was then, when Filippo Buonanni published the earliest practical illustrated guide for collectors and students of shells, his *Ricreatione dell' occhio* (1681),[32] while Martin Lister's *Historia Conchyliorum* (1685–92) and Georg Eberhard Rumphius's *D'Amboinsche Rariteitkamer* (1705) also appeared during this period.

---

[25] Olivi J.B. *De reconditis et praecipuis collectaneis in Museo Calceolaris asservatis* (Venice: 1584).

[26] Ceruti B. – Chiocco A., *Musaeum F. Calceolarii* (Verona: 1622).

[27] Basil Besler, *Fasciculus rariorum et aspectu dignorum varii generis quae collegit* (Nuremberg: 1616); Michael Rupert Besler, *Gazophylacium rerum naturalium e regno vegetabili, animali et minerali* (Leipzig: 1642).

[28] Ole Worm, *Musaeum Wormianum. Seu historia rerum rariorum* [...] *quae* [...] *in aedibus authoris servantur* (Leiden: 1655).

[29] Ludovico Moscardo, *Note overo Memorie del Museo di L. Moscardo* (Padua: 1656).

[30] Adam Olearius (Ölschlager), *Die Gottorfische Kunstkammer, worinnen allerhand ungemeine Sachen* (Schleswig: 1666).

[31] John Tradescant, *Musaeum Tradescantianum: or a collection of rarities, preserved at South Lambeth, near London* (London: 1656).

[32] Dance, *Shell collecting* 43.

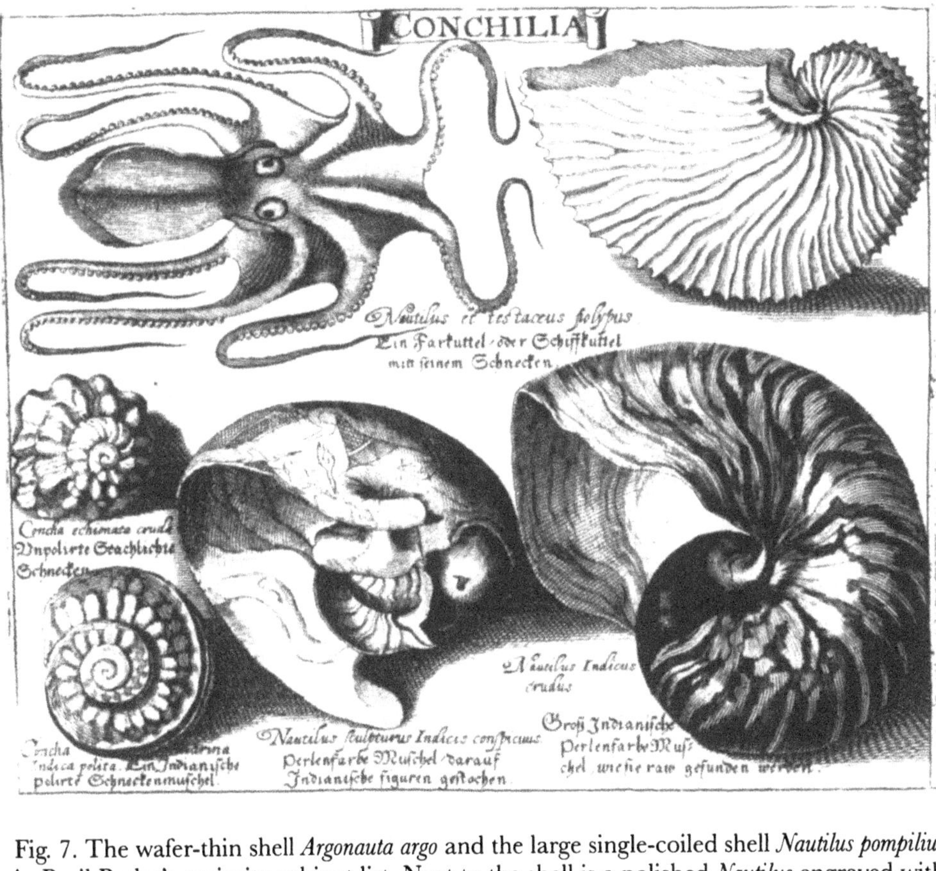

Fig. 7. The wafer-thin shell *Argonauta argo* and the large single-coiled shell *Nautilus pompilius* in Basil Besler's curiosity cabinet list. Next to the shell is a polished *Nautilus* engraved with "Indian figures". From Basil Besler, *Fasciculus rariorum et aspectu dignorum varii generis quae collegit et suis impensis aeri ad vivum incidi curavit atque evulgavit Basilius Besler* [. . .] (Nuremberg: 1616).

Filippo Buonanni's Ricreatione dell'occhio *(1681) versus Antoni van Leeuwenhoek's* Observations on Oysters *(1696)*

By the end of the 17th century the interest in molluscs was overwhelming. Shells and gastropods were studied against the background of theories concerning the origins of life and cosmological evolution; their colours and features gave rise to a debate regarding visuality that was rooted in theoretical reflections on painting and sculpture. In 1681 the two-volume shell atlas *Ricreatione dell'occhio e della mente nell'observation' delle chiocciole* was published in Rome by the Jesuit priest Filippo Buonanni (1638–1725). This was the first treatise entirely devoted to molluscs and containing numerous copper-plate engravings in an extra volume. Like Rembrandt's etching of the *Conus marmoreus* [Fig. 8], the direction of the spirals was reversed due to the printing process. Such mirror images were avoided in later shell compendiums since they were regarded as falsifications of taxonomic properties. The spirals on virtually all snail shells are dextral. 'If the shell is held in such a way that the apex points upward and the entrance is facing the observer, the opening is located on the right side. Otherwise the shell is sinistral. Sinistral spirals are only found in a few species and as rare abnormalities in normally dextral shells'.[33] Kant later noticed this same law, which formed the basis for his ideas about naturally curved phenomena such as bean tendrils, snail shells, hair whorls and so on – Buonanni had previously wondered about the spirals on some shells. However, his illustrations are reversed, although his efforts to describe shell morphology with the greatest precision constitute the most notable achievement of his book. It led Peter Dance to conclude that Buonanni's 'illustrations would have been indispensable to collectors had they not been eclipsed by those in another book published in England a few years later. [...] Buonanni knew less about molluscan anatomy than did Aristotle and his ideas of classification were no more advanced. By and large he added very little to what was known on the subject in his day and upheld many of the superstitious and mythical beliefs of the earlier commentators. On the other hand he stoutly defended the practice of shell collecting, considered a useless pastime by many of his contemporaries [...]. Among the problems he investigated were the spiral form and the colouration of gastropod shells; and he also discussed the economic

---

[33] Lindner, *Muscheln* 45 <translation K.L.>.

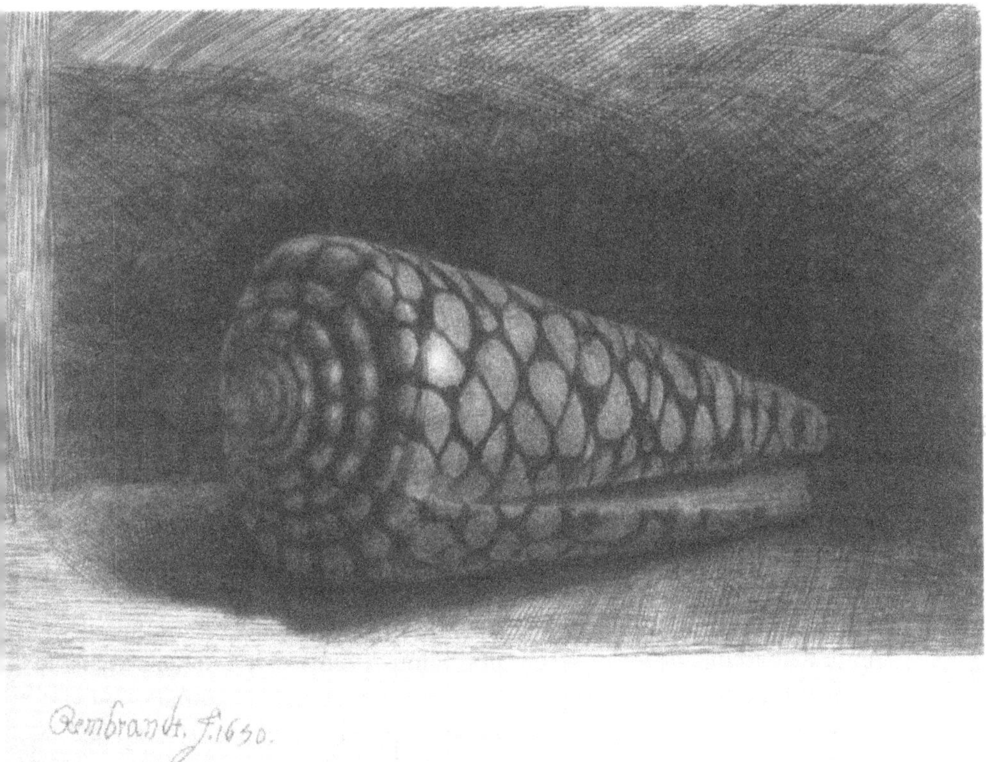

Fig. 8. A *Conus marmoreus* by Rembrandt van Rijn (1650). Engraving, Ashmolean Museum, University of Oxford. Due to the printing process the shell shows a reversed coil.

importance of molluscs'.[34] Buonanni's work is more notable for his exclusive interest in the unusual forms and colours of sea creatures than for classification on the basis of scientific criteria. In his study, he used objects from Athanasius Kircher's collection at the Jesuit College in Rome.[35] The book was intended as an introduction to conchology for shell collectors that regarded shells, as the title stated, mainly as objects of visual interest and 'recreation of the eyes', i.e. as precious ornaments for collectors' museums. However, as the earliest conchology compendium on record, this work made an important contribution in that it laid the foundation for a new discipline. And it is important to note that some of the illustrations depicted the insignificant creature inside the shell, thus giving rise to a new malacological interest [Fig. 9]. Nevertheless, Buonanni still adhered to Aristotle's theory of spontaneous generation.

In Protestant Holland, the microscopist Antoni van Leeuwenhoek broke with scholastic tradition and began searching for biological evidence that could prove Aristotle wrong. Aided by powerful microscope lenses, van Leeuwenhoek found that dissected oysters contained countless minute offspring, thus clearly indicating that procreation occurs naturally and contradicting the theory of spontaneous generation: 'Indeed, if others had seen the dissection of these three big Oysters as I saw them, they would no doubt say that in all Europe there are not as many people as there were living creatures in those three Oysters'.[36] In continuing the description of his findings, van Leeuwenhoek even made fun of himself: 'These Oysters were also as well-fed and white of flesh as ever one could behold Oysters with one's own eyes, and at ten o'clock in the evening I sent them, along with most of the living Animals therein, to my stomach, not being at all worried having such an inconceivably large number of living animals in my stomach, but not with as much pleasure as I should have done had I not seen the Animals and the further dissection of the Oysters. This proved my weakness, about which I laughed at myself'.[37]

---

[34] Dance, *A History of Shell collecting* 43.

[35] See Dance, *A History of Shell Collecting* 29: 'Kircher's natural curiosities were described in another work by Buonanni in which the numerous figures of shells are of better quality than those in his *Ricreatione*'; Filippo Buonanni, *Musaeum Krichenanum sive Musaeum a P.A. Kirchero in Collegio Romano Societatis Jesu* [...] *descriptum* [...] (Rome: 1709).

[36] Anthoni van Leeuwenhoek, *Collected Letters* (1696–1699) (Lisse: 1993) vol. XII 6.

[37] Ibid.

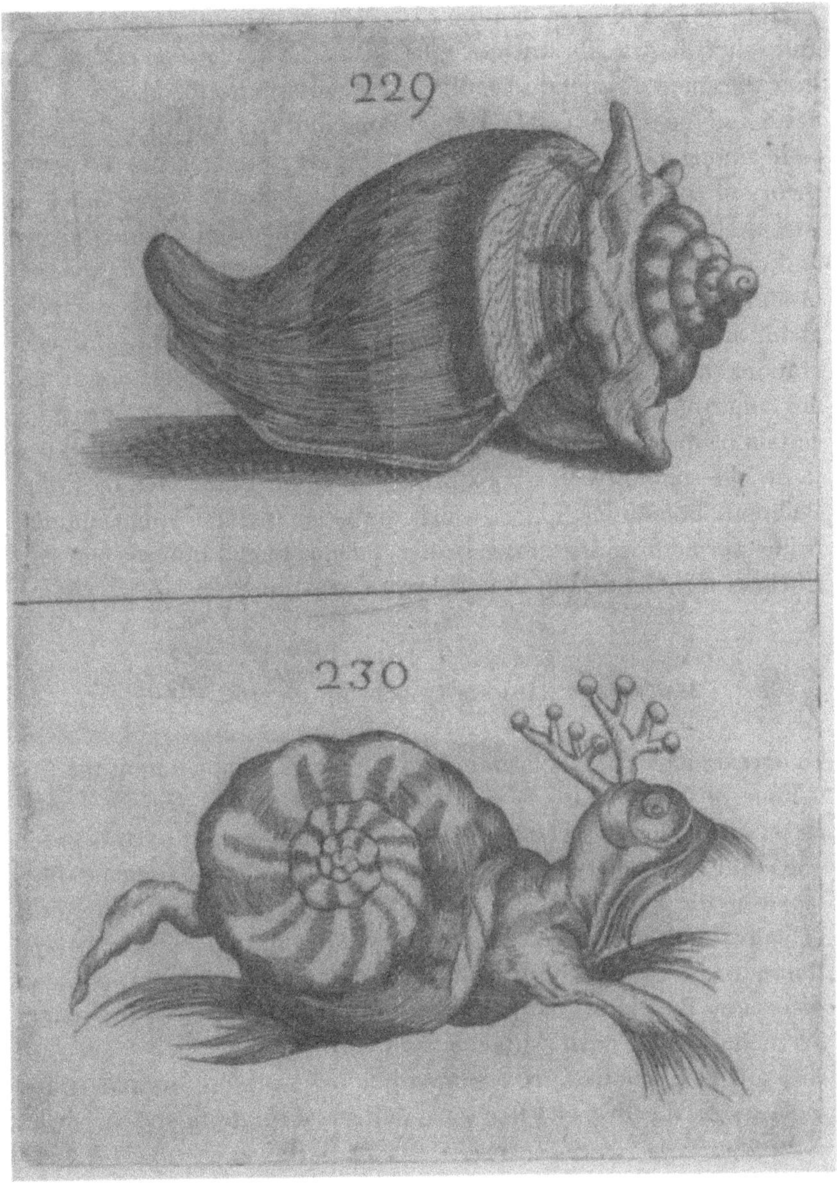

Fig. 9. Shell and living animal in Filippo Buonanni's *Ricreatione dell'occhio e della mente nell'observation' delle chiocciole* (Rome: 1681), Fig. 229. This was the first illustrated book devoted solely to shells.

Here we can observe that the 17th-century curiosity about molluscs and other marine life was not satisfied merely by *curiosity cabinets*, but instead centered around scientific investigations into the history of the earth and the origins of life. Filippo Buonanni and Athanasius Kircher were among the most prominent 17th-century proponents of Aristotle's theory of spontaneous generation, while Antoni van Leeuwenhoek (as well as Francesco Redi) was one of its fiercest opponents. Thus, opened and dissected molluscs became a battleground for conflicting theories of biological propagation. 'However, I will resume my investigations about it. I was pleased to see that those who consider that some generations take place spontaneously are daily receiving less credit and are confirming more and more the reputation of those who are more certain of their ground in their investigation of nature. I have had sent to me the summary of a book printed in Rome and edited by Father Philippus Bonnanius S.J.; in which the latter maintains that animate beings spring from inanimate beings, such as Shellfish from mud [...] I believe this Gentleman lacks correct observations'.[38]

Martin Lister: Historia Conchyliorum, *1685–1692*

Four years after the publication of Buonanni's compendium, the first volume of Martin Lister's encyclopedic large-format *Historia Conchyliorum* was published. The work, which appeared in four parts between 1685 and 1692 and contained over 1000 plates, was a systematic study of all of the mollusc groups and families known to exist at the time in native and foreign coastal regions.[39] Already in 1669 the avowed Baroque fascination with whorls or spiral movements that would later preoccupy Buonanni had led him to the publication of a first paper on molluscs that dealt with the 'odd turn of some shell-snails'.[40] It was 'followed, in 1678, by a more ambitious work on the spiders and molluscs of the British Isles, illustrated by accurate figures of shells,

[38] Leeuwenhoek refers here to *Observationes circa viventia, quae in rebus non viventibus reperiuntur* which was published by Filippo Buonanni in Rome in 1691.

[39] For a bibliography of Lister's work see: Kenyes K.G., *Dr Martin Lister. A Bibliography* (Godalming: 1981).

[40] Martin Lister, *Concerning the odd turn of some shell-snailes, and the darting of spiders, made by an ingenious Cantabrigian and by way of letter communicated to Mr. I. Wray, who was transmitted them to the publisher for R.p. Phil. Trans.* 4: 1011–6 (1669).

including fossil forms (*Historiae animalium Angliae tres tractatus*).[41] In 1685 [Lister] distributed among his scientific friends a thin volume of copper-plate engravings of land shells mostly foreign to the British Isles and entitled *De Cochleis*.[42] Realizing what a large amount of material was available, he embarked on a much larger work, along the lines of the *De Cochleis*, which was to include figures of all the recent and fossil shells then known. Thus the first of the major shell iconographies was born'.[43] The work referred to is Lister's monumental *Historia*, which his friends and family also worked on. In 1696 Lister published another treatise regarding (and entitled) *Conchyliorum Bivalvium*.[44]

The value of Lister's *Historia* is almost exclusively a function of the quality of the illustrations and the taxonomical properties that they depict. 'There is no actual text [...]. The species figures are nearly all recognizable and none are reversed; and their scientific value is enhanced still further by numerous indications of locality. [...] He had intended to follow the plates of the *Historia* with anatomical descriptions of every 'family' in a systematic order and, undoubtedly, this would have improved its arrangement. Unfortunately the *Historia* was not completed and most of the illustrations of molluscan anatomy added to the second and subsequent editions were those previously published in the *Exercitationes*.[45] The *Historia Conchyliorum* represents, nevertheless, the first real attempt at a systematic arrangement of the Mollusca; Linnaeus cited figures from it and collectors and students used it as a reference work for many years'.[46] Lister's taxonomical investigations were based on a formalistic approach as well as on anatomical studies. For example, Lister's classification of the Venus shell (*Concha Veneris*) – 'unicolourous, black streaked, transversely waved, ringed or banded, black spotted, white spotted, transversely ribbed, granulated, smooth-mouthed or not toothed, umbilicated'[47] – embodies a complex formalistic strategy that refers to the sculptural and colouristic features

[41] Martin Lister, *Historiae animalium Angliae tres tractatus* [...] (London: 1678).
[42] Dance, *A History of Shell Collecting* 23.
[43] Dance, *Shell collecting* 44.
[44] Martin Lister, *Conchyliorum bivalvium* [...] *exercitatio anatomica tertia* [...] (London: 1696).
[45] Martin Lister, *Exercitatio anatomica* [...] (London: 1694); idem, *Exercitatio anatomica altera* [...] (London: 1695); idem, *Conchyliorum bivalvium* [...] *exercitatio anatomica termia.*
[46] Dance, *Shell collecting* 46.
[47] Martin Lister, *Historia Conchyliorum* (London: 1685–92), section 9.

of the shells. On the other hand, comparative anatomical studies – laid down in the *Exercitationes* – helped Lister to devise a well founded mollusc classification system.

Martin Lister's *Historia Conchyliorum* successfully united art and science. It is due to the accuracy and variability of his illustrations that they are more impressive than Buonanni's. The thorough and systematic comparisons of forms enabled him to identify subspecies, define characteristics and work out the slightest differences and similarities. The taxonomical characteristics were traced in a purely visual process. Unlike the proprietors of curiosity cabinets, Lister was not particularly interested in exotic specimens and equally studied native specimens such as the small and unassuming limpets (*Patella)* whose varying sizes and sculptured forms are depicted with true accuracy; the common tower shell (*Turritella*) whose dextral spirals Lister printed correctly; as well as the hard, ribbed cockles (*Cardium*) that were regularly washed ashore on the English coastline. His friends sent him beach finds or small sketches which were later mounted in a scrapbook along with the watercolours that Lister's daughter Anna painted to prepare the copper plate engravings. Figure 10 shows a representative page that is revealing, because, in addition to the sketches realized by Lister's daughter for the *Historia*, it also contains a fresh print of Rembrandt's well known etching of a *Conus marmoreus* [Fig. 10]. Later Lister had this image copied and incorporated it into the *Historia*, and in so doing corrected the direction of the spiral [Fig. 11].

We can see that in addition to illustrations based on natural objects which reveal Lister's empirical interest in the exact morphology of molluscs, the *Historia* also contains copies of older shell illustrations that are interpolated into the book at various points as if they were quotes. These drawings constitute a thumbnail early history of conchology, in which Lister primarily cites illustrations by Colonna[48] and Buonanni, as well as the aforementioned Rembrandt engraving, all of which he simply intersperses among his own plates with no or with short comment. These indirect references enable us to identify the literature that formed the basis for his investigations. Besides that, Lister's sketchbook is a useful source of information. The sketches often bear Latin inscriptions not found on the plates and show that Lister had read the

[48] Colonna's outstanding illustrations of *Cardium costatum*, in Fabio Colonna, *Purpura*.

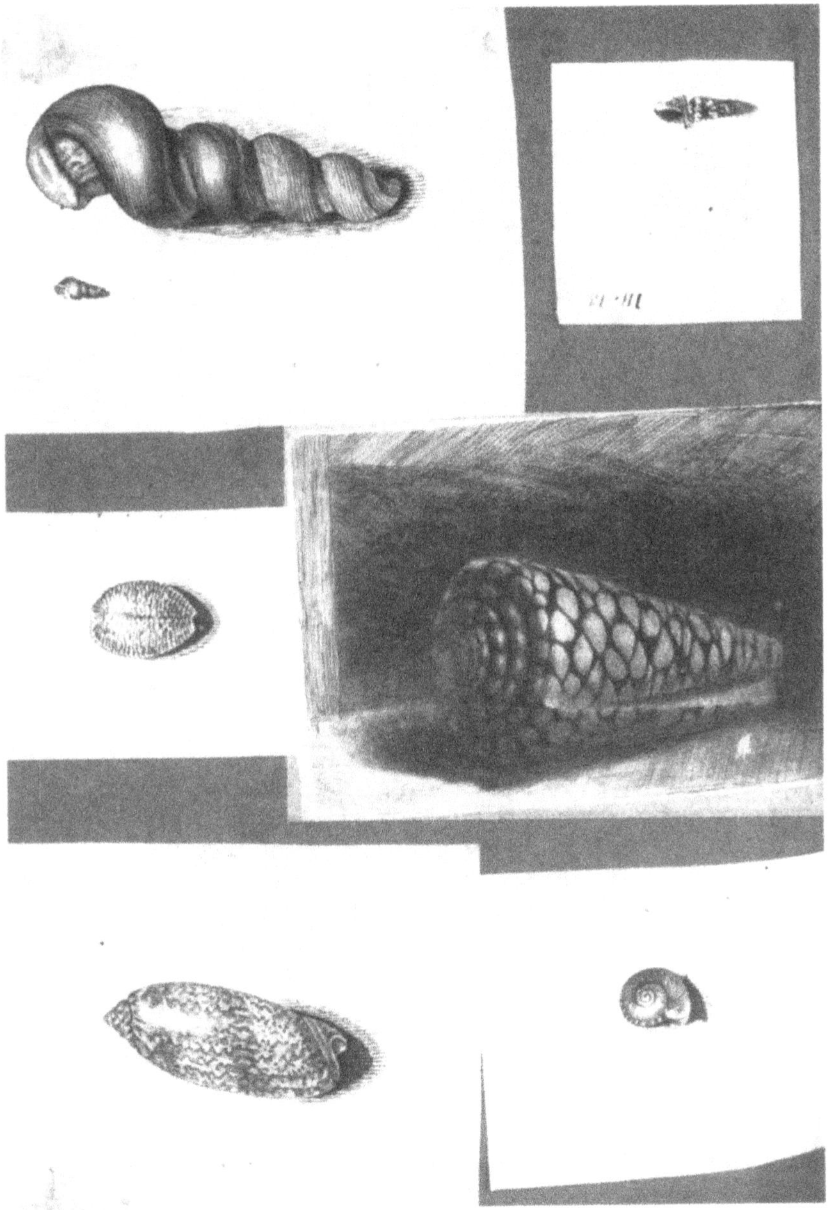

Fig. 10. Drawings of shells, along with an early print of Rembrandt's *Conus marmoreus* from Martin and Anna Lister, *Scrap-Book*, compiled in about 1680. MS Lister 9, fol. 29v. Bodleian Library, Oxford.

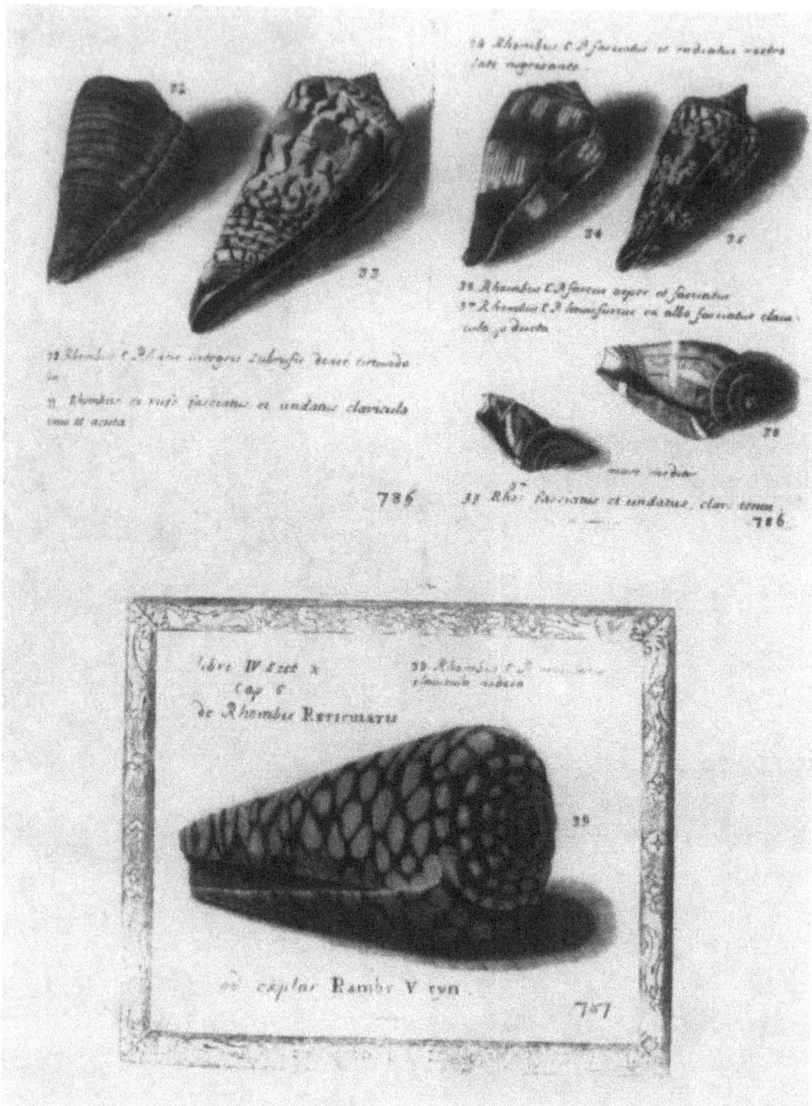

Fig. 11. Rembrandt's *Conus marmoreus*, here reprinted and reversed in Martin Lister, *Historia Conchyliorum* (London: 1685–92), Fig. 787.

works of Colonna and Buonanni, and above all those of Aldrovandi, Gesner[49] and Besler.[50]

But the sketches are worthy of attention in their own right. The watercolours were executed in soft somewhat cautious brush strokes, and some are signed A.L. (Anna Lister). Then the scrapbook contains a few delicate pen and brown India ink drawings that were realized by another artist and are notable for their firm and short strokes. Other drawings, finely done in India ink, a few of them tinted, were realized with a fine brush. These drawings, which again can be traced back to Anna Lister, provide a fascinating insight into the phased process of the book illustrations. For example, before the watercolours could be sent to the printer, they had to be "translated" into black and white tones. However, in this case the conventional cross hatching was dismissed and instead the entire shell was translated into a series of parallel lines. Thus the brush drawings allowed for the reproduction of the minutest details in the *Historia*'s engravings, and the graphic abstraction process was realized by the Lister family itself rather than by the engraver.

As for the engravings themselves, again an unorthodox method was used. In contrast to what had almost universally been done up to that time, the shells were not rendered beginning from their outline, but were instead "palpated" line by line, as if a finger had been run over the irregular surface of the shell and had recorded the various rises and falls. The procedure is reminiscent of laser scanning, except that each line has a variable starting point rather than a fixed one. Concave and convex forms were reproduced by following the curves of the shells rather than modeling their light and dark values [Fig. 12]. A comparison with Rembrandt's *Conus* reveals that Rembrandt renders plasticity by means of dense cross hatching, whereas Lister conveys the same quality using a few lines, which are deliberately used as an investigative tool. But Lister's blue scrap-book – testimony to a process of conchological brainstorming – also shows us other methods of describing nature in pictorial terms. Within the scrapbook watercolours alternate with hard pen and ink drawings, their lines are parallel or hatched, bunched together or scattered; unfinished sketches rub elbows with carefully elaborated studies, rapidly executed sketches show Latin notes to consult

[49] Conrad Gesner, *Icones* and *Historia*; Ulisse Aldrovandi, *De reliquis animalibus exanguibus*.

[50] Basil Besler, *Fasciculus rariorum et aspectu dignorum* and Michael Rupert Besler, *Gazophylacium rerum naturalium*.

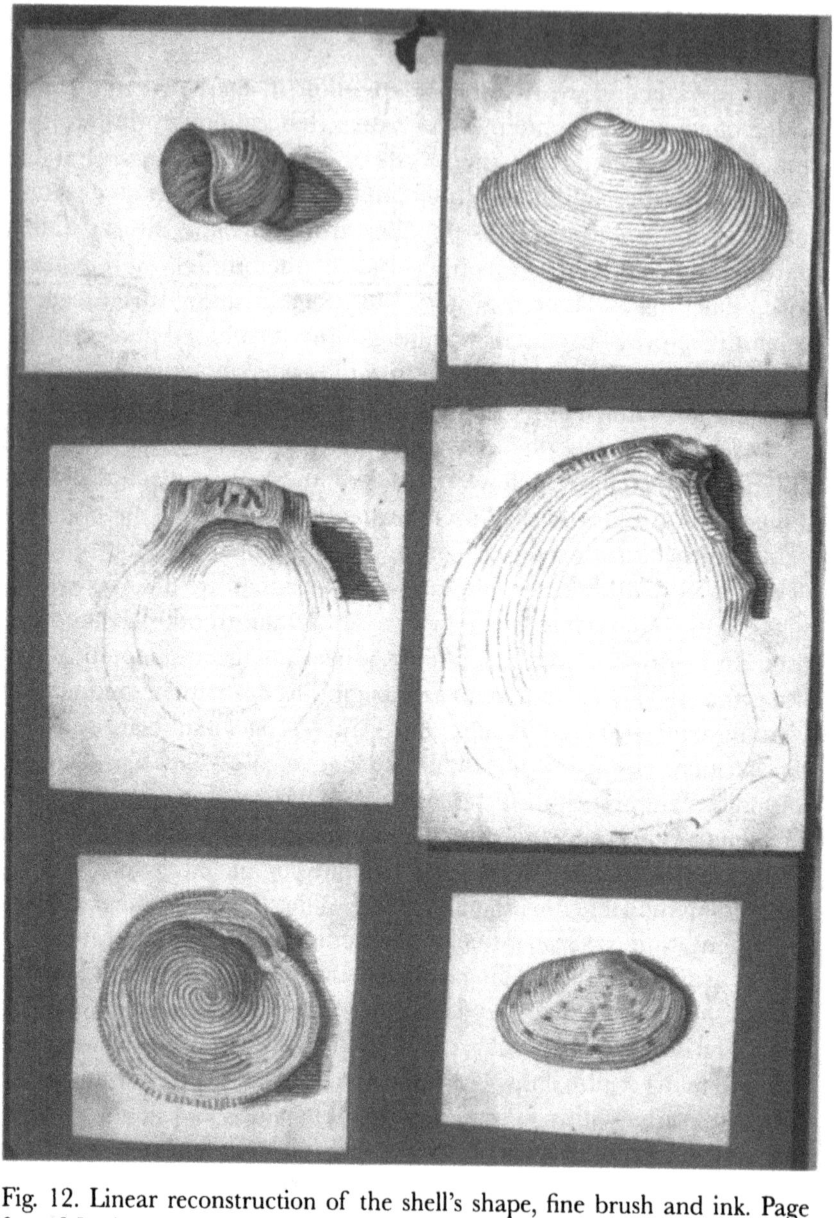

Fig. 12. Linear reconstruction of the shell's shape, fine brush and ink. Page from Martin and Anna Lister, *Scrap-Book*, written in about 1680, MS Lister 9, fol. 47v. Bodleian Library, Oxford. This work anticipated the process of engraving.

certain reference books; completed engravings follow hard upon monochrome studies realized with a fine brush. Some sketches were pinned to the pages with thin needles, and a crudely hand-coloured print of Rembrandt's *Conus* also appears.

Georg Eberhard Rumphius: D'Amboinsche Rariteitkamer *(1705)*

Georg Eberhard Rumphius (1627–1702) was a native of Hessen, Germany where his father was a construction engineer in the service of the prominent Solms family; during the Thirty Years War he mainly worked for members of the local aristocracy.[51] 'Burning with an insatiable desire to know foreign lands',[52] in 1652 Rumphius signed on with the East India Company and sailed off to Java. The East India Company had built a major trade center at Batavia, which was the name the Dutch had given to Java's capital city after conquering it in 1619. Rumphius's initial position as a *fabryck* or engineer, was mainly of a military nature, and before long he asked to be transferred to a new position, whereupon he became a trade overseer on a small island called Ambon. There, Rumphius soon began realizing botanical and zoological studies and sketches for an ambitious magnus opus entitled the *Amboinsche Kruid-boek*,[53] which he worked on concurrently with his conchological *Rariteitkamer*. During this period, Rumphius's own *curiosity cabinet* also grew and eventually contained 360 specimens of native shells and fossils. For financial reasons, Rumphius sold the collection to the equally avid patron of the arts and collector Cosimo III de Medici, Grand Duke of Tuscany in 1682. In the final chapter of his *Amboinsche Rariteitkamer*, Rumphius recalls the enormous effort that was involved in building up his collection. He then describes the animal habitats at length, a knowledge that he needed in order to acquire the various species, and points out the rarity of some of the specimens and the fact that they were often damaged or soiled. This account contains the parallel story of Baroque collection practices – which was

---

[51] Georg Eberhard Rumphius, *The Ambonese Curiosity Cabinet* (*D'Amboinsche Rariteitkamer*, Amsterdam 1705), translated, edited, annotated, and with an introduction by E.M. Beekmann (Ann Arbor: 1999) XLII. For literature on Rumphius see the selective bibliography at the end of this paper.

[52] Ibid. XLIII.

[53] Georg Eberhard Rumphius, *Het Amboinsche Kruid-boek*, 2 vols. (Amsterdam et al.: 1741).

from the standpoint of the colonies. In fact it constitutes an implicit contemporary critique of the high aesthetic demands of Dutch shell collectors and proprietors of *curiosity cabinets*:

> Our Compatriots and Friends in the Fatherland are commonly of the opinion that we find the Whelks and Curiosities on the beach, or haul them from the Sea, just as clean and pure as when we send them to those folks, and that consequently it requires little more than picking them up. It is almost not worth refuting this misconception at length; except that it is necessary to purge us, who live in India, from the suspicion that it is impolite or unreasonable of us, when we sometimes cannot fulfill our friends' demands nor answer to their cleanliness.[54]

Such lines reveal Rumphius's sceptical attitude toward *art connoisseurs* and *liefhebbers* who regarded exotic animals and other natural wonders solely as showpieces or refined consumer goods. Rumphius therefore discusses issues such as the uniqueness of some specimen, resource limits, and the rarity of certain species to distance himself from this world of pure *konst-liefhebbery*: 'Besides the length of time that is required for this, I will also indicate all the trouble and tedium one has to endure in order to clean them and make them presentable'.[55] On the one hand he provides exact instructions for preparing, presenting and conserving them properly. But in many places in his book, Rumphius criticizes a consumptuous attitude, e.g. when writing about strange small stones known as *Mesticae* that are sometimes formed in plants and animals and are said to bring the finder good luck: 'They will only be lucky for the one who found them, or the person who received one as a gift, but not for someone who buys them'.[56] Some chapters before he had reported about his own lucky find: a large nautilus shell had practically fallen from the sky:

> I should mention a rare event here. A Sea Eagle, being a bird that constantly hunts at sea, took such a Nautilus, while it was floating in the Sea, and bore it aloft, but while his business was with the fish, and since he did not care about it as a curiosity, he struck his claws mostly into the fish, wherefore the shell came to fall out of his claws and, by rare fortune, it fell on a small spot of sand between rocks in such a way, that nothing was broken off, except for a small corner of the foremost

[54] Rumphius, *The Ambonese Curiosity Cabinet* 223.
[55] Ibid. 224.
[56] Ibid. 328.

> edge; and a fisherman who was wandering thereabouts, quickly picked it up and brought it to me; and since this fell out of the sky like another Palladium, as if a small likeness of the fabled ship Argos, I sent it in the year 1683 as a memento to Mr. Johan Michael Fehr <and> the Collegium in which I was accepted as a Member in the same year.[57]

Rumphius thanked the renowned Collegium for having accepted him as a member by sending them the same *Carina nautili* that good fortune had caused to fall at his feet. We know that the reference to the fabled ship Argos pertains to a very common element of Baroque iconography. For example, an early emblem book by Joachim Camerarius (1604) contains an illustration of an *Argonauta argo* (paper boat) whose wafer-thin shell was able to remain afloat on the surface of the sea. This was regarded as a symbol of good fortune and of a ship's ability to survive ocean voyages in both fair and foul weather.[58]

Another specimen of *Nautilidae*, the more robust *Nautilus pompilius* (pearly nautilus) with its pearl-like shell served as a baroque dining table showpiece and symbolized the economic power of a seafaring nation. Fig. 13 shows a nautilus shell of this type as depicted in a still life painting by Willem Claesz Heda. But for the 17th century it also represented a masterpiece of hydraulic engineering: the whelk contains separate flotation chambers that are interconnected by siphon-like channels. A cross-section reveals its main feature: the exponentially growing spiral whose plane Archimedes had attempted to calculate using a specific method which came close to the differential calculus of the Baroque period. As a non-platonic figure, the spiral corresponded to the Baroque taste for dynamic forms that required physical force and symbolized movement, growth or expansion [Fig. 14].[59]

In 1670 Rumphius went blind. The expressive portrait at the beginning of the *Amboinsche Rariteitkamer* (after a drawing by his son Paulus Augustus) shows the author in old age with fingers bent into a claw trying to touch some shells on a table[60] – the folded whelk of an *Argonauta argo* is clearly distinguishable [Fig. 15].[61] It was such a *paper boat*

---

[57] Ibid. 94.

[58] Joachim Camerarius, "Tutus per summa, per ima", in idem, *Symbolorum et emblematum ex aquatilibus et reptilibus desumtorum* [...] (Nuremberg: 1605 [1604]).

[59] Cf. note 7. The suggestive cross-section in Rumphius's *Rariteitkamer* harks back to Frans Halma and Simon Schijnvoet.

[60] Rumphius, *The Ambonese Curiosity Cabinet* IV.

[61] Ibid. (*Nautilus Tenuis*).

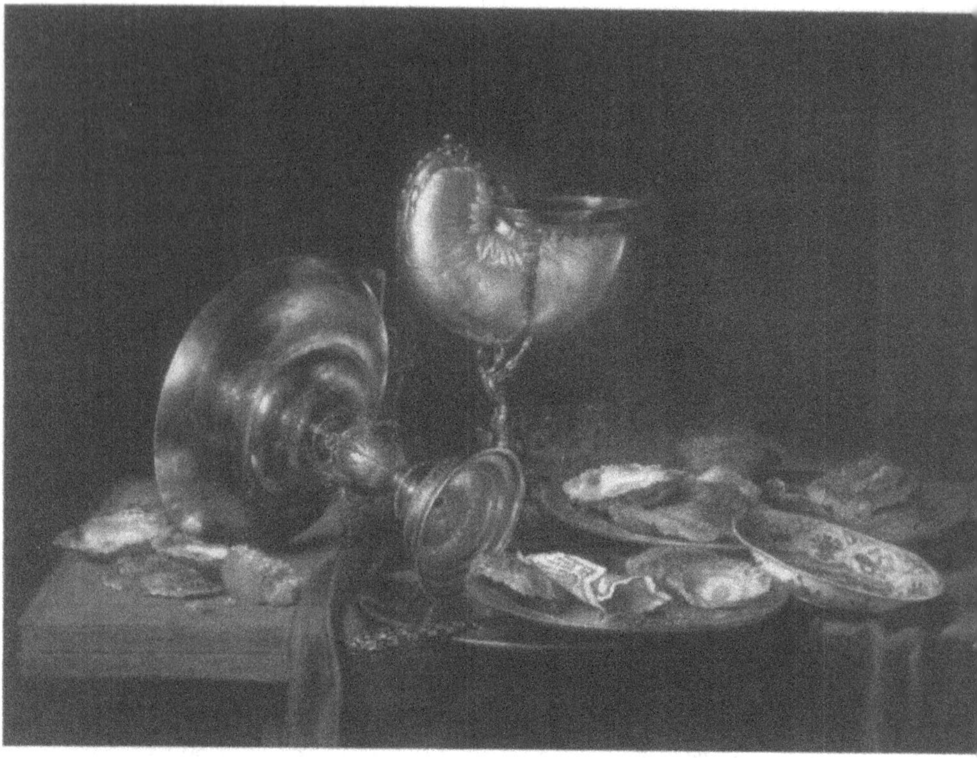

Fig. 13 [Col. pl. VI]. Still life painting with Nautilus goblet, showing its mother-of-pearl surface. Willem Claesz. Heda, Still life with Nautilus cup, 1649. Oil on wood, 49.5 × 68.2 cm Schwerin, Staatliches Museum.

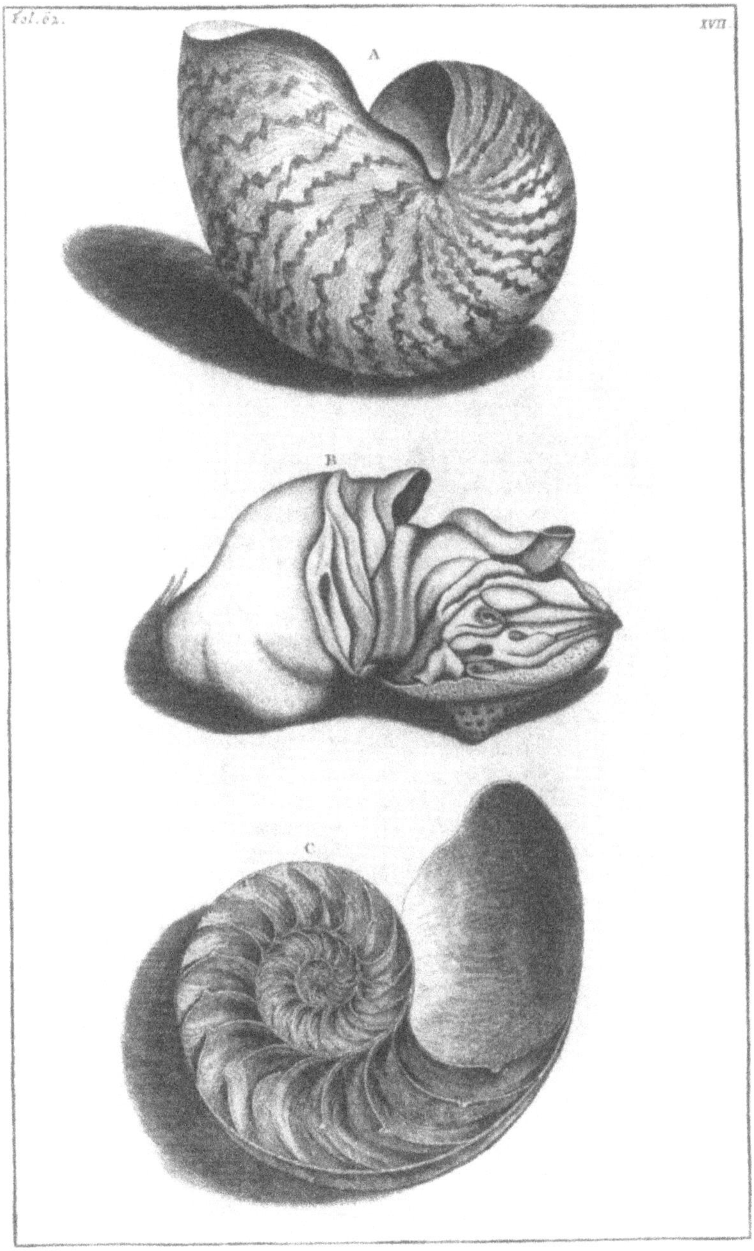

Fig. 14. The popular *Nautilus pompilius* and the spiral arrangement of the internal chambers in Georg Eberhard Rumphius, *The Ambonese Curiosity Cabinet (D'Amboinsche Rariteitkamer, Amsterdam 1705)*, Fig. 62. This object was inherited by Henri D'Acquet, the Delft Mayor, whose shell collection was used to complete the illustrations.

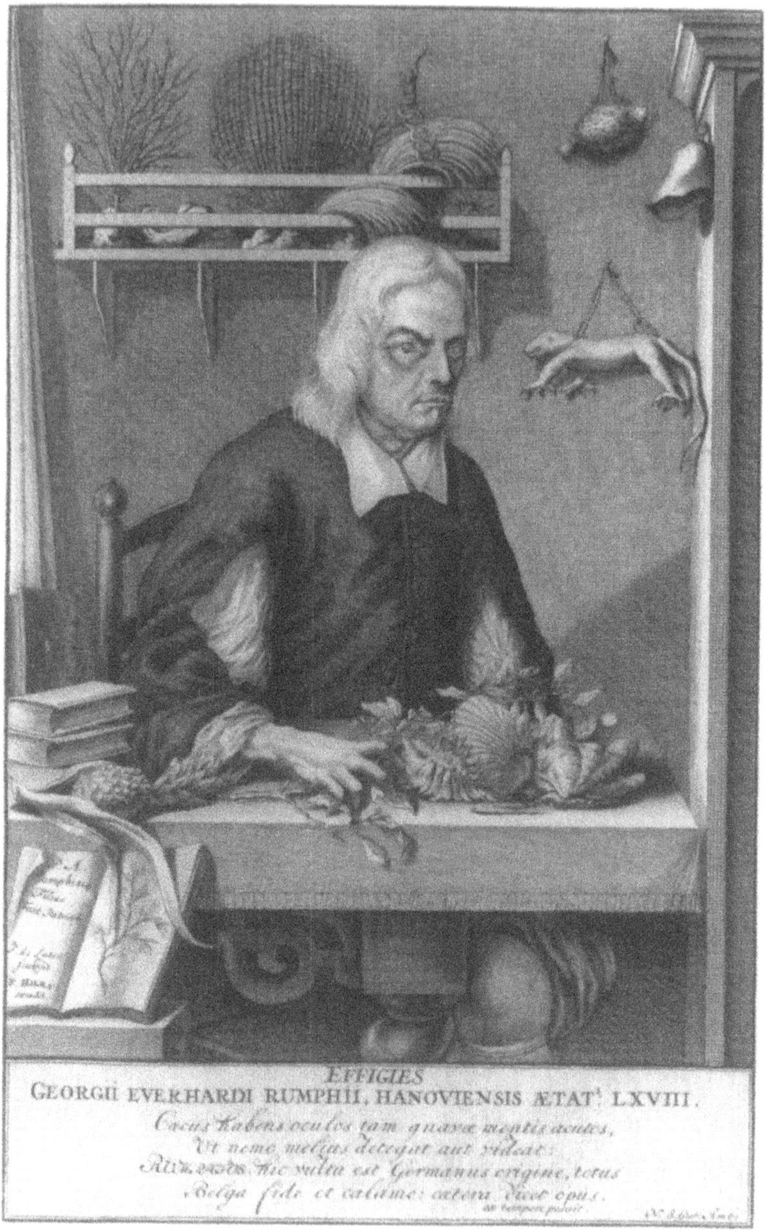

Fig. 15. Portrait of the blind malacologist Georg Eberhard Rumphius after a drawing by his son Paulus Augustus (1696): 'Though he is blind, his mental eyes are so sharp that no one can beat him in inquiry or discernment'. From Georg Eberhard Rumphius, *The Ambonese Curiosity Cabinet (D'Amboinsche Rariteitkamer, Amsterdam 1705)*, title page.

that had been carried up to the sky by a sea-eagle till its shell came off and fell on the ground. He also mentions it in the first chapters of his second book which is devoted solely to hard shellfish and was therefore most widely used by shell collectors:

> Now this is the true Nautilus which we [...] call the thin or fine one, and which seems to have been the only one known to the Ancients, because it has the right shape of a galley, with a narrow keel and two high sides [...], fashioned from a single thin shell that has the thickness of parchment, partially transparent, exceedingly white, like paper [...]. The entire whelk is covered with folds. [...]The Fish that lives there is notably similar to a Polypus or Many-Feet, [...]its flesh entirely soft, [...]white and covered with warts, and which it spreads out like a rose when swimming [...].[62]

This is followed by a magnificent description of the behaviour of these unusual creatures. For several centuries, this passage was to remain the sole account of the living nautilus and is yet another testimony to Rumphius's fascination with shells that shows how very different he was from his "curious" contemporaries.[63] Unlike them, Rumphius emphasized the importance of functional morphology and in so doing emulated Aristotle's emphasis on the priority of field observation. Beekman correctly observed that what made Rumphius remarkable was his precise notation of habitats and sites and his talent for classification.[64]

However, Rumphius's book also contains detailed descriptions of shells' shapes and patterns, and he draws numerous comparisons with similar forms in calligraphy, painting, sculpture and architecture. For example, Rumphius identifies small letters or images on shell surfaces, some of which remind him of the bluish perspectives that were common in Dutch landscape painting:

> *Chama optica*, Little Perspectives, are roundish and bossed little Shells, [...] decorated with blackish designs, which represent hills, little houses and peaks, in such a way, that the ones nearest to the edge are the largest and blackest ones; if one follows the other ones towards the back, one will see that they gradually become smaller and bluer, and fainter, just as if ones saw a drawing of a distant Landscape or one in perspective.[65]

---

[62] Ibid. 93.
[63] Ibid.
[64] Ibid. XCVIII.
[65] Ibid. 193.

The patterning on the *Chama ciricinata* reminded Rumphius of 'small towers or houses',[66] while another shell evoked 'certain tents with tiny flags on top, like a Turkish army in the field'.[67] *Letter Shells*, as Rumphius termed them, have 'black stripes or marks, which make for the letter W, and the more these are crowded together, the better the shell'.[68] Shells known as *Voluta* or (in Dutch) *whirlpools* are named after architectural volutes: '*Voluta*, is an art word, from Architecture, by which is meant the curls, which one sees on Ionian and Corinthian pillars, and these Whelks have been called after them because of their resemblance to this: because they have a flat head, made from many whorls that flow into each other, in the shape of a *linea spiralis*, or snake line'.[69] Known as *Conidae* today, these shells were favorite collector's items in the 17th century owing to its elegantly curved forms and elaborate patterns. Rumphius could identify 'rough, black, and square Characters' on their surfaces, as well as similarities to 'Music notes; but the Characters are long on some, and those next to them, are a little shorter, like the letters one sees in the little ABC books of children' [Fig. 16].[70] Other patterns evoked 'white church windows',[71] while others were 'marked with little gold stars or flowers'[72] or 'yellow feathers',[73] or had 'russet, crooked veins'[74] or 'thin and hacked Characters, that look quite like the ghosts, which the Map makers paint in the big and dreadful desert called Lob, to the west of Sina'.[75] 'These spots resemble various shapes, such as Clouds, Animals, or People carrying a large box, or whatever a clever person wants to contemplate therein".[76] Although such natural structures were regarded as a morphological phenomenon, the approach was not exclusively formalistic. Indeed, the observer was stimulated to use his semiotic capabilities to fill these forms and patterns with meaning.

Behind the Baroque interest in so-called 'surfaces painted by nature' lay the question as to the origin of these signs and symbols. If *Natura* herself had inscribed series of signs in the shells, these signs meant that

---

66 Ibid.
67 Ibid.
68 Ibid.
69 Ibid. 144.
70 Ibid. 146.
71 Ibid. 148.
72 Ibid.
73 Ibid. 149.
74 Ibid. 150.
75 Ibid.
76 Ibid. 146.

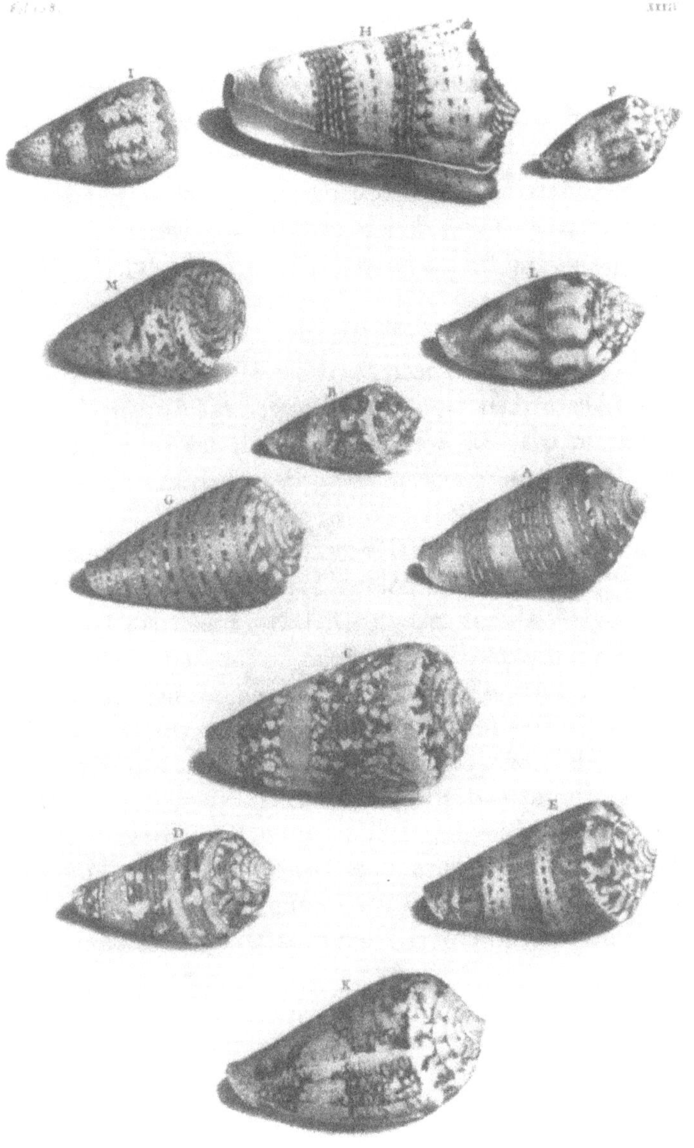

Fig. 16. Richly patterned species of *Conidae* as described in Georg Eberhard Rumphius's *D'Amboinsche Rariteitkamer*. The characters and signs on the surface seem to be directly 'painted by Nature'. The *Conida* injects poisonous barbs into its prey before devouring it. Georg Eberhard Rumphius, *The Ambonese Curiosity Cabinet (D'Amboinsche Rariteitkamer, Amsterdam 1705)*, Fig. 104.

an alternative to the man-made writing systems of western cultures must exist. It was based first and foremost on visual phenomena – to read nature's writing meant to be able to interpret natural traces, e.g. cracks in surfaces or colour differences. This may explain why specimens from the *Conidae* family were among the favourite subjects of Dutch shell still life paintings: in fact their patterns looked very much like some mysterious scripture or scribble. The possibility that both image and text had a common origin in a natural iconography was widely debated during the 17th century and both explored by means of linguistic models and by theory of art.

But, on the other hand, baroque observers may also have been so captivated by the *Conidae* species because of the fact that behind their alluring surface lurked the hidden danger of a highly toxic poison, a phenomenon that was well known to fishermen and collectors at that time. It is again Rumphius who described the deadly sting that the animals of this species deploy in self-defense if they are touched. For example, *Mitra papalis* or *Pope's Crown* 'is an oblong smooth little Whelk. [...] It can deliver a poisonous sting, when you hold it in your hand, and has killed some people: [...] who could tell, that such a holy *Pope's Crown* would conceal such deadly venomen?'.[77] In the annotation it is then laconically stated: 'The venomous little sting, which the Author noted in this fish, should be ignored by the Roman Catholic collectors'.[78] The obvious beauty – danger and evanescence – mortality interconnections evoked by these shells opened up a whole new realm of possible iconographical interpretation. However, the radical distinction between the alluring outer shell (conchology) and the living creature within (malacology) was becoming increasingly apparent, and with it the distinction between the formal and functional morphology of molluscs.

[77] Ibid. 137.
[78] Ibid. 140.

Selective Bibliography

Rumphius

BALLANTIJN G., *Rumphius. De blinde ziener van Ambon* (Utrecht: 1944).
BEEKMAN E.M., *The Poison Tree. Selected Writings of Rumphius on the Natural History of the Indies* (Amherst: 1981).
GRESHOFF M. (ed.), *Rumphius-Gedenkboek, 1702–1902* (Haarlem: 1902).
HUET C.B., *Het land van Rembrand. Studies over de Noordnederlandse beschaving in de zeventiende eeuw (1883–84)* (Amsterdam: 1987).
MARTELLI U. (ed.), *Le Collezioni di Giorgio Everardo Rumpf, acquistate dal Granduca Cosimo III de Medici* (Florence: 1903).
MARTENS E. von, "Die Mollusken (Conchylien) und die übrigen wirbellosen Thiere in Rumpf's Rariteitkamer", in Greshoff (ed.), *Rumphius Gedenkboek* 109–136.
RUMPHIUS GEORG EBERHARD, *The Ambonese Curiosity Cabinet* (*D'Amboinsche Rariteitkamer*, Amsterdam 1705), translated, edited, annotated, and with an introduction by E.M. Beekmann (Ann Arbor: 1999).
SIRKS M.J., "Rumphius, the Blind Seer of Amboina" in Honig P. – Verdoorn F. (eds.), *Science and Scientists in the Netherlands Indies* (New York: 1945) 295–308.
STRACK H.L. – Goud J., "Rumphius and the 'Amboinsche Rariteitkamer'", *Vita Marina* 44,1–2 (1996) 29–39.
WIT H.C.D. DE (ed.), *Rumphius Memorial Volume* (Baarn: 1959).
ZAUNICK R., "Georg Eberhard Rumphius und Cosimo III von Medici", *Physis. Rivista di storia della scienza* 3,2 (1961) 125–136.

Conchology, Palaentology and Geology

ALBRITTON C.C., *The Abyss of Time: Changing Conceptions of the Earth's Antiquity after the Sixteenth Century* (San Francisco: 1980).
ARIEW R., "A New Science of Geology in the Seventeenth Century?", in Barker P. – Ariew R. (eds.), *Revolution and Continuity: Essays in the History and Philosophy of Early Modern Science* (Washington D.C.: 1991).
COHN N., *Noah's Flood: The Genesis Story in Western Thought* (New Haven: 1996).
CUTLER A., *The Seashell on the Mountaintop* (New York: 2004).
GOULD S.J., *Time's Arrow and Time's Cycle: Myth and Metaphor in the Discovery of Geological Time* (Cambridge: 1987).
LINDNER G., *Muscheln und Schnecken der Weltmeere* (Munich et al.: 1999).

Dutch shell collections

JUTTING W.S.S. VAN BENTHEIM, "A brief history of the conchological collections at the zoological museum of Amsterdam", *Bijdragen tot de Dierkunde* (1938) 167–245.
BERGVELT E. – KISTEMAKER R. (eds.), *De wereld binnen handbereik. Nederlandse kunst- en rariteitenverzamelingen, 1585–1735* (Zwolle – Amsterdam: 1992).
BERGVELT E. – JONKER M. – WIECHMANN A. (eds.), *Schatten in Delft, burgers verzamelen 1600–1750* (Zwolle: 2002).
DANCE S.P., *A History of Shell Collecting* (Leiden: 1986).
——, *Shell Collecting. An Illustrated History* (London: 1966).
DIETZ B., "Mobile objects: the space of shells in eighteenth-century France", *The British journal for the history of science* 39,3 (2006), 363–382.
ENGEL H., "Alphabetical list of dutch zoological cabinets and menageries", *Bijdragen tot de Dierkunde* (1938) 247–346.

Kistemaker R., „Seba und Ruysch: Die Anfänge der petrinischen Kunstkammer; Peters Besuche in Holland", in *Palast des Wissens: Die Kunst- und Wunderkammer Zar Peters des Großen* (Munich: 2003) 54–66.
Roemer B. van de, "Neat nature: The relations between nature and art in a Dutch cabinet of curiosities from the early eighteenth century", *History of Science* (2004), 47–84.

Shells in art

Barten S. (ed.), *Die Muschel in der Kunst: von der Renaissance bis zur Gegenwart* (Zurich: 1985).
Hogendorp Prosperetti L. van, "'Conchas legere': Shells as trophies of repose in Northern Europe Humanism", *Art History* 29,3 (2006) 387–413.
Kemp M., "Palissy's Philosophical Pots: ceramics, grottos and the 'Matrice' of the earth", in Tega W. (ed.), *Le origini della Modernità* (Florence: 1999) 69–88.
Leonhard K., "Ueber Links und Rechts und Symmetrie im Barock", in Guenzel S. (ed.), *Topologie* (Weimar: 2007) 135–151.
Mauries P. (ed.), *Shell Shock: Conchological curiosities* (London: 1994).
Mette H.-U., *Der Nautiluspokal: Wie Kunst und Natur miteinander spielen* (Munich et al.: 1995).
Shell H.R., "Casting Life, Recasting Experience: Bernard Palissy's Occupation between Maker and Nature", *Configurations* 12,1 (2004) 1–40.

Shell still life painting

Atwater G.D., *The impact of trade by the Dutch East India Company on seventeenth-century Netherlandish art* (Kansas: 1992).
Bergström I., *Dutch still-life painting in the seventeenth century* (London: 1956).
Bol L. (ed.), *Adriaen Coorte, stillevenschilder* ex. cat. Dordrecht 1958 (Dordrecht: 1958).
—— *Adriaen Coorte: A unique late seventeenth century Dutch still-life painter* (Assen: 1977).
Grisebach L., *Willem Kalf 1619–1693* (Berlin: 1974).
Grohé S., *Stillleben: Meisterwerke der holländischen Malerei* (Munich et al.: 2004).
Gemar-Koeltzsch E., *Holländische Stillebenmaler im 17. Jahrhundert*, ed. by K. Ertz (Lingen: 1995).
Lowenthal A., *The Object as Subject. Studies in the Interpretations of Still Life* (Princeton: 1996).
Luther G., „Stilleben als Bilder der Sammelleidenschaft", in Langemeyer G. – Peters H.-A. (eds.), *Stilleben in Europa* (Münster: 1979) 88–128.
Segal S., *A fruitful past. A survey of the fruit still lives of the Northern and Southern Netherlands from Brueghel till van Gogh* (Brunswick: 1983).
—— *A prosperous past. The Sumptuous Still-Life in the Netherlands 1600–1700* (The Hague: 1989).
—— *Flowers and Nature. Netherlandish flower paintings of four centuries* (The Hague: 1990).

THE CONSTRUCTION OF THE ANIMAL IN ZOOLOGICAL ILLUSTRATIONS

IN MINIMIS MAXIME CONSPICUA. INSEKTENDARSTELLUNGEN UM 1600 UND DIE ANFÄNGE DER ENTOMOLOGIE

Thea Vignau-Wilberg

Als eines der seltsamsten, mit Sicherheit aber als eines der kostbarsten Kompendien von Tierdarstellungen haben sich vier Alben mit Miniaturen erhalten, die der flämische Miniaturist Joris Hoefnagel (1542–1600) am Ende des 16. Jahrhunderts schuf.[1] Nach van Mander zahlte Kaiser Rudolf II. die exorbitant hohe Summe von 1000 Goldkronen, um sie seiner Kunstkammer einverleiben zu können.[2] Die vier kleinen, querrechteckigen Bändchen enthalten naturalistische und kunstvoll auf Pergament gemalte Darstellungen verschiedener Tiere. Die Tierbilder sind auf dem Blatt jeweils von einem schmalen ovalen Goldrahmen eingefaßt. Über und unter dem Oval steht häufig ein (meist lateinischer) Sinnspruch oder eine längere Sentenz klassischer oder biblischer Provenienz, wodurch sich die Blätter in der Struktur dem klassischen Emblem nähern. Die Tiere sind jeweils einem der Vier Elemente der antiken Elementenlehre zugeordnet: der Erde (terra), dem Wasser (aqua), der Luft (aier) und dem Feuer (ignis). Die Miniaturenfolge wird gewöhnlich als *Die vier Elemente* bezeichnet,[3] die Titelblätter der Alben tragen entsprechende Aufschriften: „Terra: animalia quadrupedia et reptilia"; „Aqua: animalia aquatilia et conchiliata"; „Aier: animalia volatilia et amphibia"; „Ignis: animalia rationalia et insecta".

Diese augenscheinlich ausgefallene Anordnung der Tiere nach den Vier Elementen entspricht größtenteils der traditionellen Klassifizierung, wie man sie beispielsweise in Conrad Gesners *Historia animalium* findet:

---

1 Washington D.C., National Gallery of Art, Gift of Mrs. Lessing J. Rosenwald, Inv. Nr. 1987.205–20.8. Dazu in erster Linie Hendrix M.L., *Joris Hoefnagel and the „Four Elements": a study in Sixteenth-Century Nature Painting* (Ann Arbor: 1985) (Diss. Princeton University: 1984).

2 Karel van Mander, *Het Schilder-boeck* (Haarlem: 1604) fol. 263r.

3 Zu den einzelnen Miniaturen in verschiedenen Sammlungen siehe Hendrix, *Four Elements* 8 und 126, Anm. 10; Vignau-Wilberg Th., „Naturemblematik am Ende des 16. Jahrhunderts", *Jahrbuch der Kunsthistorischen Sammlungen in Wien* 82/83 (1986/87) (145–156) 145, Anm. 3.

Vierfüßer (terra), Fische (aqua) und Vögel (aier).[4] Dem Element Feuer wird gewöhnlich kein Tier zugeordnet, außer manchmal der Salamander. Als erste Darstellung in Hoefnagels Band *Ignis*, in dem „animalia rationalia" und Insekten zusammengefasst sind, ist eine Miniatur des so genannten Haarmannes Petrus Gonsalus (Pedro Gonzales) und seiner (unbehaarten) Frau aufgenommen, die nächste zeigt ihre beiden behaarten Kinder.[5] Sie galten in dieser Zeit als medizinische Wunder: Wesen mit tierischer Behaarung und mit menschlichem Geist begabt. Gonsalus, gebürtig aus Teneriffa, wurde an den europäischen Höfen vorgeführt und – laut dem Text von Hoefnagel in den *Vier Elementen* „comparuit Monachii boiorum anno 1582" – hielt er sich einige Zeit in München auf (im Kunstkammerinventar von Schloss Ambras, 1621, wird er als „rauch man zu Münichen" bezeichnet).[6]

Nach den zwei Miniaturen der Haarfamilie folgen in *Ignis* fast ausschließlich Abbildungen von Insekten oder von Kleintieren wie Spinnen, die damals zu den Insekten gerechnet wurden. Sie sind meist zu mehreren und zusammen mit Pflanzen im golden umrahmten Oval angeordnet. Ihre Wiedergabe ist naturgetreu und meist lebensgroß; die einzelnen Tiere sind jeweils durch kleine farbige Zahlen nummeriert. Seitlich des Ovals erscheint eine römische Nummerierung. Zwischen den bemalten Pergamentblättern wurden gleich bei der Entstehung dünne Papierblätter eingebunden, die einerseits zum Schutz der Miniaturen dienten, andererseits durch den Künstler als Träger weiterer Kommentare verwendet wurden. Sie enthalten Zitate aus der Bibel, aus antiken und humanistischen Schriften. Die arabischen Zahlen neben den einzelnen Tieren, die es auch in den drei anderen Bändchen gibt, legen nahe, daß ursprünglich ein separates Verzeichnis existierte, in

[4] Conrad Gesner, *Historia animalium*, 4 Bde. (Zürich, Christoffel Froschauer: 1551–1558).

[5] Hendrix, *Four Elements* 266–267; *Bestiaire*, Cod. Min. 129, fol. 1 und 2. Siehe Haupt H. – Vignau-Wilberg Th. – Irblich E. – Staudinger M., *Le Bestiaire de Rodolphe II. Cod. Min. 129 et 130 de la Bibliothèque Nationale d'Autriche* (Paris: 1990), Text 46–47 und 92–97.

[6] So im Porträtinventar 1621 Schloss Ambras, in dem vier Gemälde und zwei Kopien verzeichnet sind. Siehe Kenner F., „Die Porträtsammlung des Erzherzogs Ferdinand von Tirol", *Jahrbuch der Kunsthistorischen Sammlungen des Allerhöchsten Kaiserhauses* 15 (1894) (147–259) 250–251; Haupt et alii, *Bestiaire* 46; Scheicher E., *Die Kunst- und Wunderkammern der Habsburger* (Wien: 1979) 134–135. Nach Hendrix, *Four Elements* 100–101, sind die Miniaturen Kopien nach den vier lebensgroßen Ölbildern in Ambras. Nach Kenner wurden sie in München im Auftrag von Herzog Wilhelm V. gemalt und gelangten als Geschenke in die Kunstkammer von Ferdinand von Tirol.

dem jedes Tier namentlich aufgeführt wurde. Hier waren wohl weitere Texte aufgenommen, die jenen in den gedruckten zoologischen Werken entsprechen, in denen man sich auf die Forschungsergebnisse älterer Autoren (Aristoteles, Plinius, Gesner u. a.) berief.

Während die Tiere der Elemente terra, aqua und aier größtenteils nach gemalten, in Kupfer gestochenen oder in Holz geschnittenen Vorlagen zeitgenössischer Künstler (Hans Bol, Adriaen Collaert, Nicolaes de Bruyn u. a.)[7] gemalt wurden – eine gängige und keineswegs abwegige Vorgehensweise –, ist das bei den Tieren von *Ignis* nicht der Fall. Es existierte noch keine illustrierte Publikation über die Welt der Insekten, auf die Hoefnagel hätte zurückgreifen können. Das erste neuzeitliche Werk über die Insekten,[8] *De animalibus insectis libri septem cum singulorum iconibus ad vivum expressis* von Ulisse Aldrovandi, erschien erst 1602, zwei Jahre nach Hoefnagels Tod.[9] Trotz des vielsprechenden Titels (*cum singulorum iconibus ad vivum expressis*) ist Aldrovandis Buch spärlich und noch dazu kümmerlich illustriert. Als nächstes und künstlerisch weitaus wichtigeres Insektenbuch erschien das *Insectorum sive minimorum animalium theatrum* des Engländers Thomas Mouffet (1553–1604). Es kam 1634, dreißig Jahre nach Mouffets Tod, heraus.[10]

Im gelehrten Europa war um 1600 das Interesse für die Insektenwelt groß. Man entdeckte in diesen winzigen Kreaturen die in Bezug auf die Artenzahl weitaus größte Tierklasse, ihre skurille Schönheit und die schnelle Abfolge der Generationen. Gelehrte wie Mouffet und Aldrovandi korrespondierten um 1590 über diese Tierchen ebenso wie Clusius (Vermerk im Buch von Mouffet, siehe unten) und Joachim Camerarius der Jüngere (1534–1598). Der gelehrte humanistische Künstler Hoefnagel, war, wie seine Korrespondenz zeigt, ein

---

[7] Siehe Hendrix, *Four Elements* passim.

[8] Plinius hatte mit seinem elften Buch der *Naturalis historia* bereits ein Buch über die Insekten geschrieben. Er konzentrierte sich auf wenige Arten: Biene, Seidenraupe, Spinne, Skorpion, Zikade, Käfer, Heuschrecke, Ameise, Schmetterling.

[9] Ulisse Aldrovandi, *De animalibus insectis libri septem cum singulorum iconibus ad vivum expressis* [...] (Bologna: 1602). Im Vorwort vom 1. April 1599 der Ausgabe 1602 wird das kaiserliche Privileg für zehn Jahre verbrieft.

[10] Thomas Mouffet, *Insectorum sive minimorum animalium theatrum. olim ab Edoardo Wottono, Conrado Gesnero, Thomaque Pennio inchoatum: tandem Thomae Moufeti Londinatis opera sumptibusque maximis concinnatum, auctum, perfectum et ad vivum expressis iconibus supra quingentis illustratum* (London: 1634); englische Zitate nach der Facsimile-Ausgabe *The Theater of Insects: or Lesser living Creatures, as Bees, Flies, Caterpillars, Spid[e]rs, Worms* [...] (London: 1658; Facsimile-Ausgabe Edward Topsell, *The history of Four-Footed Beasts and Serpents and Insects*, vol. III. *The Theater of Insects* by T. Muffet (New York: 1967)).

hochgeschätztes Mitglied desselben Kreises. Camerarius starb 1598, Hoefnagel 1600, Mouffet 1604, Clusius im Jahr 1609. Sie gehören derselben Generation an.

In der flämischen Buchmalerei, besonders während ihrer späten Blüte in den Gent-Brügger Streublumenhandschriften, finden sich bereits vereinzelt Insekten, allerdings häufig kombiniert mit einer Pflanzendarstellung. Es sind bunte Schmetterlinge, Raupen und Libellen, sie werden – zusammen mit der Blume –häufig im Rahmen der marianischen Symbolik verwendet. Der in Antwerpen gebürtige Hoefnagel, der als Kaufmann ausgebildet wurde, übte die Miniaturmalerei ursprünglich nur als Freizeitbeschäftigung aus. Obwohl Außenseiter und kein zunftverbundener Maler, sollte er sich zum virtuosesten Miniaturisten der zweiten Hälfte des 16. Jahrhunderts nördlich der Alpen entwickeln (die spezialisierten englischen Porträtminiaturisten Nicholas Hilliard und Isaac Olivier seien bei dieser Wertung außer Betracht gelassen). Hoefnagel schließt in einigen seiner Werke gezielt an die Tradition der Gent-Brügger-Handschriften an, so auch bei manchen Illuminationen des Missales für Erzherzog Ferdinand II. (Ferdinand von Tirol)[11] und vor allem bei seinen in Miniatur gemalten Ergänzungen im Stundenbuch für Philipp von Kleve in Brüssel, das bereits Ende des 15. Jahrhunderts illuminiert worden war.[12]

Ebenso große Wirkung ging von den Aquarellen mit Blumen und kleinen Tieren von Albrecht Dürer aus. Sie waren schon zu Hoefnagels Zeiten berühmt und dieser Ruhm sollte nie verblassen. Kaiser Rudolf II. war bemüht, alles, was sich aus Privatbesitz erwerben ließ, in seinen Besitz zu bringen: den Liegenden Feldhasen (1502) und das Große Rasenstück (1503). Als weiteres bedeutendes Blatt ist der Hirschkäfer (1505) zu nennen [Abb. 1].[13] Dürers Zeichnung charakterisiert zum ersten Mal ein Insekt in seiner *malerischen* Gestalt – sein

[11] Wien, Österreichische Nationalbibliothek, Cod. 1784, fol. 83r; 347v; 403r etc.

[12] Brüssel, Bibliothèque Royale de Belgique, Ms. IV, 40; Hendrix, *Four Elements* 37; Thoss D., „Georg Hoefnagel und seine Beziehung zur Gent-Brügger Buchmalerei", *Jahrbuch der Kunsthistorischen Sammlungen in Wien* 82/83 (1986/97) 199–212; Vignau-Wilberg Th., „Die Randilluminationen und Initialen", in *Das Gebetbuch Kurfürst Maximilians I. von Bayern* (Frankfurt-Stuttgart: 1986) (65–118) 85–86.

[13] Wien, Albertina, Inv. Nr. 3073 (Hase) und Inv. Nr. 3075 (Großes Rasenstück). Los Angeles, J. Paul Getty Museum, Inv. Nr. 84 GC.214 (Hirschkäfer). Koreny F., *Albrecht Dürer und die Tier- und Pflanzenstudien der Renaissance* (München: 1984; Ausstellungskatalog Wien, Albertina 1984/85) Nr. 43, 61 und 36.

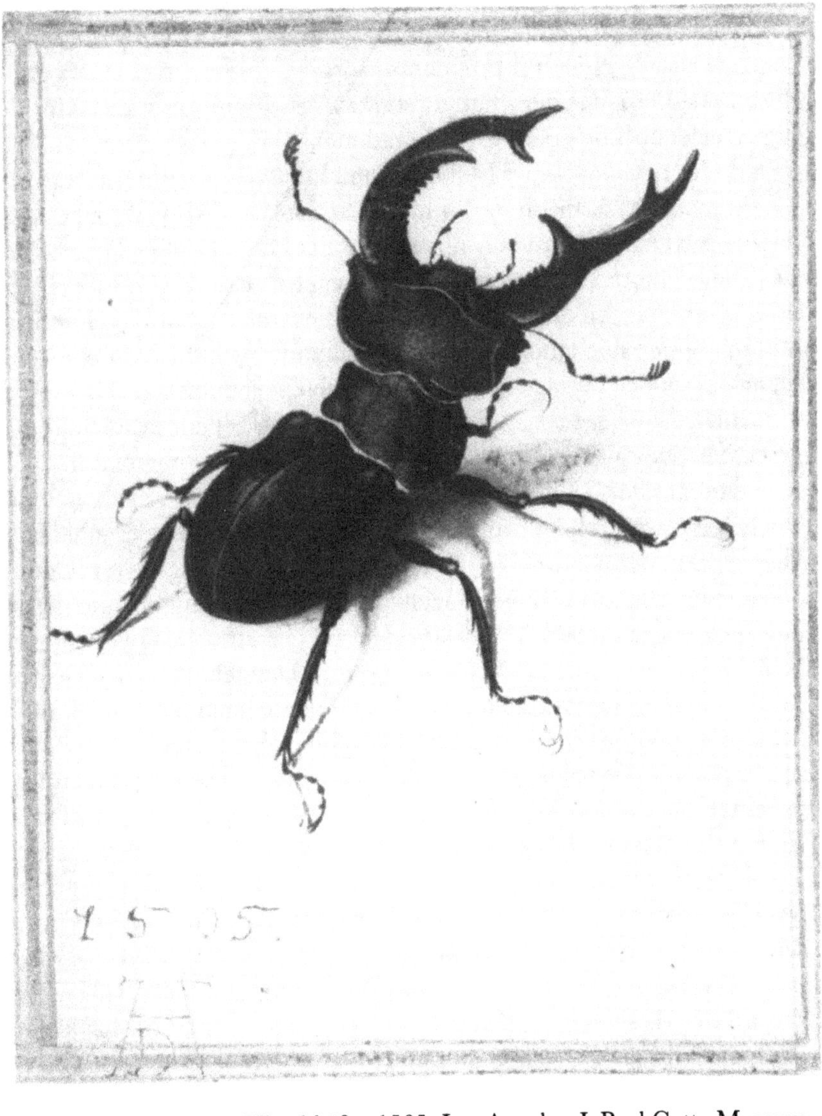

Fig. 1. Albrecht Dürer, Hirschkäfer, 1505. Los Angeles, J. Paul Getty Museum.

glänzender, knackiger Panzer, sein skurilles „Geweih" – und zeichnet es damit zugleich wissenschaftlich-determinierend auf: Es bleibt nichts im Unklaren. Dürers Großes Rasenstück, in dem sich der Bezug der kleinen Natur, des Mikrokosmos, zum All, zum Makrokosmos, in einer ansprechenden Form und einer für die Humanisten des 16. Jahrhunderts völlig verständlichen Bildsprache ausdrückt,[14] sowie das lebensgroße Bildnis des Hirschkäfers hatten starke Auswirkung auf die Künstler seiner eigenen und späterer Generationen.[15]

Als Georg Flegel in Frankfurt den Hirschkäfer am Anfang des 17. Jahrhunderts in einem Aquarell darstellte [Abb. 2],[16] geschah dies selbstverständlich auch mit einem Blick auf Dürers Hirschkäfer. Vor Flegel hatte es bereits Joris Hoefnagel in den *Vier Elementen* (vgl. *Ignis* V) und in einem dazugehörigen Einzelblatt getan.[17] Hoefnagel ahmt in seiner Wiedergabe der ersten Miniatur nicht nur Dürers Aquarell nach, sondern versucht, es zu übertreffen. Neben der *imitatio* strebten die Künstler die *aemulatio*, die das Original übertreffende Nachbildung, an. Hoefnagel gab das Insekt nicht nur, wie Dürer, diagonal in leichter Seitenansicht wieder, sondern auch in Aufsicht, mit Blick auf den Rücken, wie aufgespießt mit ausgebreiteten Flügeln, in der stofflichen Wiedergabe dennoch völlig lebendig [Abb. 3]. Neben vielen anderen Insekten erscheint in einem Einzelblatt zu *Ignis* ein weiterer und ebenso eindrucksvoller Käfer, der Nashornkäfer (korrekter: Elefantenkäfer).

Zahlreich sind in den *Vier Elementen* die Darstellungen von Faltern und Schmetterlingen. Es fehlt dabei auch die sogenannte vollkommene Metamorphose nicht, die in ihren drei Stufen abgebildet ist: die Raupe, die Puppe und der Schmetterling. Wie nahe Hoefnagel bei seinen Darstellungen der Natur folgt, lässt sich an der Miniatur der drei Libellen (*Ignis* LIV) ablesen; bei zwei Exemplaren sind die Flügel wahrhaftiger Libellen an der entsprechenden Stelle auf das Pergament geklebt [Abb. 4]. Sicher ist das auch eine Spielerei. Man darf es sich wohl nicht so vorstellen, dass Hoefnagel vor einem (toten) Insekt oder vor einem Präparat (vgl. den Hirschkäfer mit den ausgebreiteten Flügeln) saß und es direkt auf das Pergament porträtierte. Die Möglichkeiten zur Korrektur der Darstellung bei dem hochempfindlichen, präparierten

---

14 Vgl. Vignau-Wilberg, „Naturemblematik" 145–156.

15 Siehe Koreny, *Tier- und Pflanzenstudien* zu Nr. 61 und 36.

16 Ehem. Berlin, Kupferstichkabinett, Kriegsverlust.

17 Berlin, Staatliche Museen zu Berlin, Kupferstichkabinett, KdZ 4807. Koreny, *Tier- und Pflanzenstudien* Nr. 39.

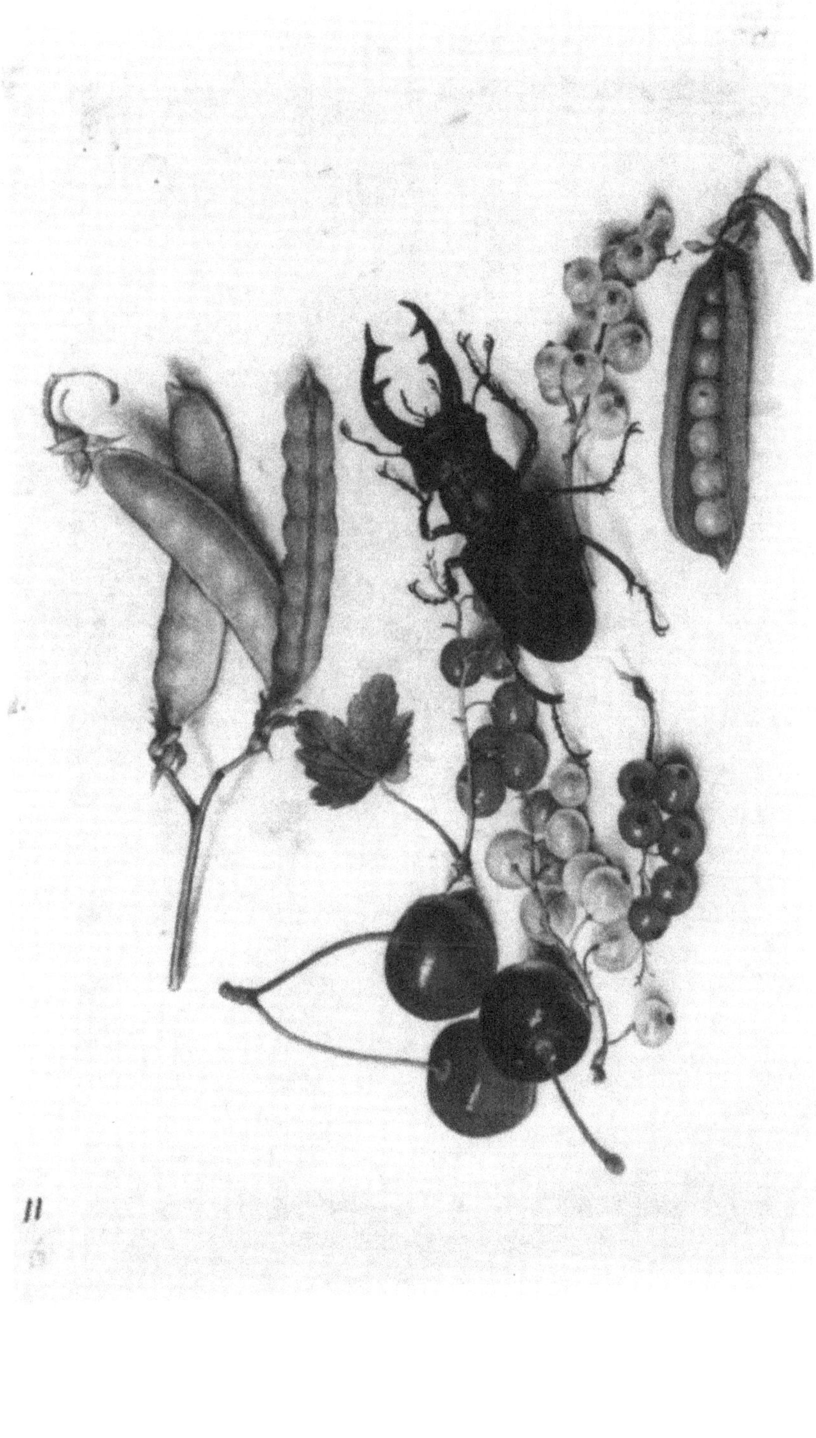

Fig. 2. Georg Flegel, Stilleben mit Hirschkäfer, um 1630. Ehem. Berlin, Kupferstichkabinett (Kriegsverlust).

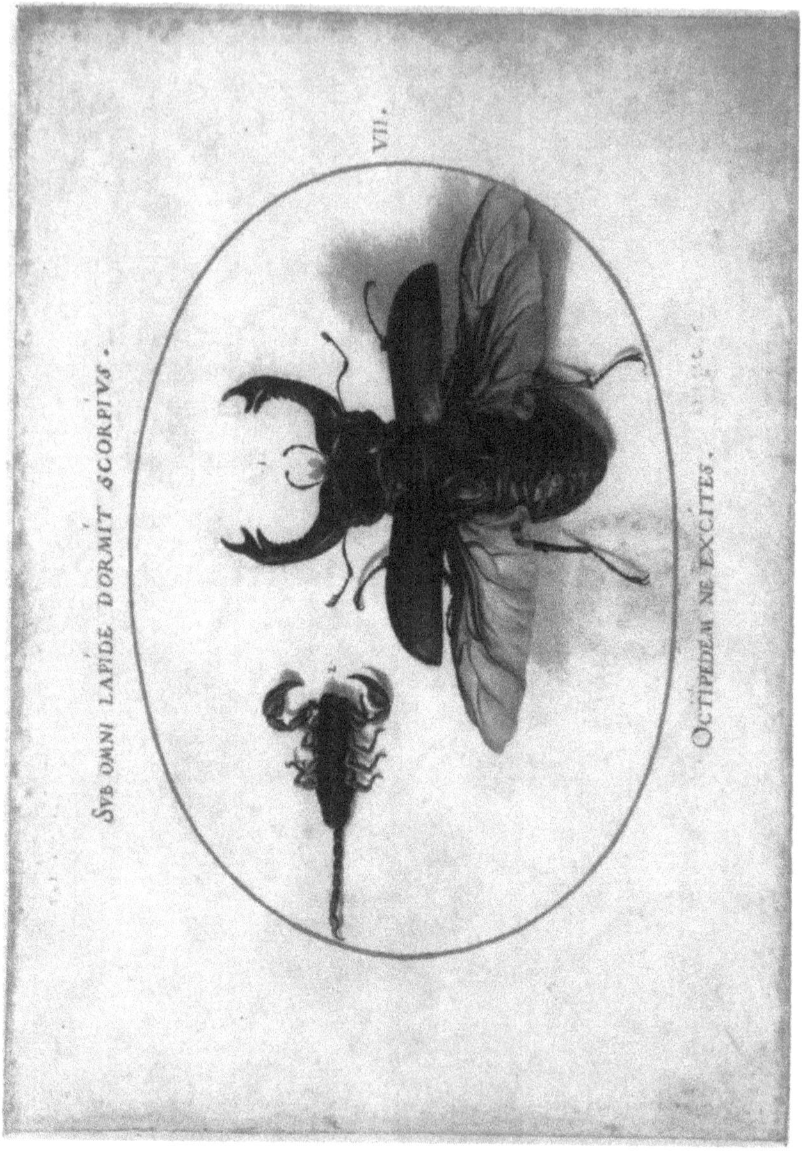

Fig. 3 [Col. pl. VIII]. Joris Hoefnagel, Hirschkäfer. Berlin, Staatliche Museen zu Berlin, Kupferstichkabinett.

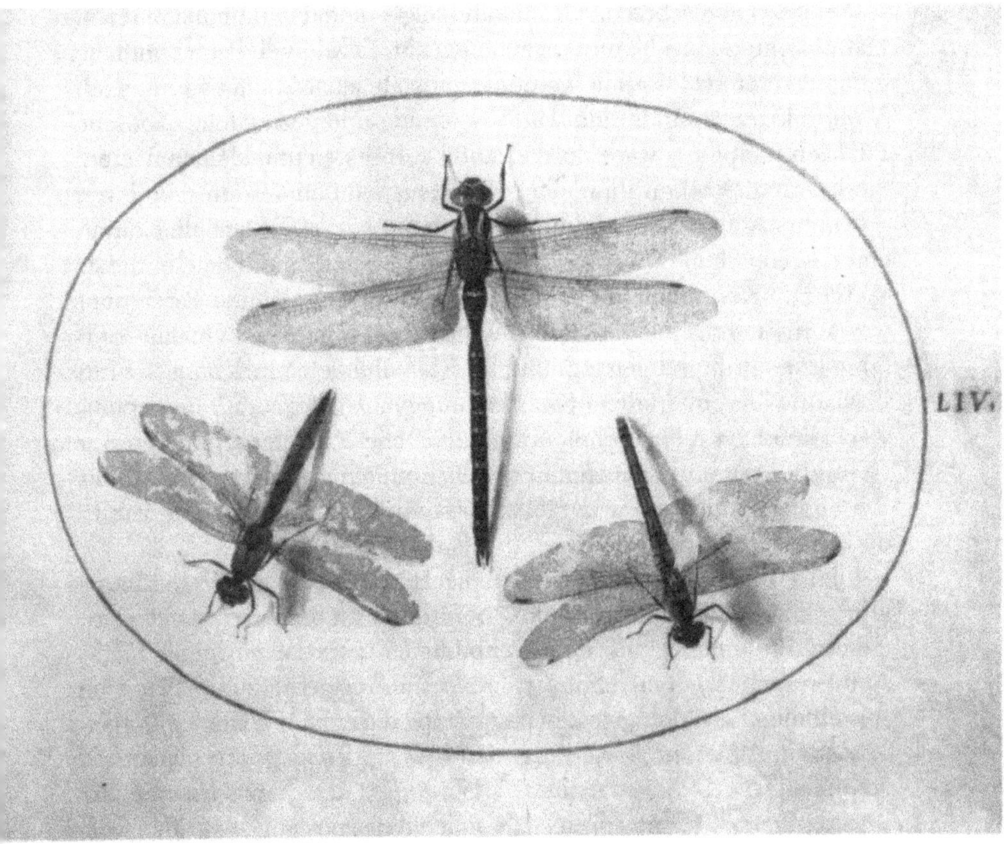

ig. 4. Joris Hoefnagel, *Die Vier Elemente – Ignis* LIV. Washington D.C., National Gallery of Art.

Pergament wären dafür zu gering, das Material Pergament zu kostbar gewesen. Hoefnagel arbeitete, wie andere Maler auch, bei den *Vier Elementen* mit Hilfe einer Vorlagesammlung, eines Musterbuchs, in dem er seine Motive zeit seines Lebens sammelte und aus dem er bei Bedarf kopierte. Nur so lässt sich erklären, dass sich bestimmte Naturmotive in Werken, die in Abstand von bis zu zwanzig Jahren gefertigt wurden, in derselben Form wiederholen.

Hoefnagel malte neben den Miniaturenfolgen und den Illuminationen in Handschriften auch Naturdarstellungen auf einzelnen Pergamentfolien, die in einer geschlossenen Komposition stillebenartig angeordnet sind. Seine Bildgegenstände sind wild wachsende und gezüchtete, exotische Pflanzen (Tulpen) sowie eine Vielzahl von Insekten und kleinen Tieren. Sie liegen gleichsam auf dem Pergament auf und werfen dort ihren Schatten [Abb. 5]. Fast immer wurde eine solche Darstellung durch einen lateinischen Sinnspruch inhaltlich ausgelegt und leitete in dieser Weise zur Kontemplation und Meditation an. Das feine Pergament wurde meistens auf eine Holztafel aufgeklebt und das Gemälde sorgfältig in ein Futteral gelegt und in der Schublade eines Kunstschranks verwahrt. Es sind erlesene Kunstkammerstücke, die zum kontemplativen Betrachten hervorgeholt wurden.[18] Diese Kabinettminiaturen mit Naturobjekten und emblematischen Deutungen waren am Hof Kaiser Rudolfs II. und unter der Hocharistokratie sehr gefragt und nur in diesem Kreis bekannt.

Joris Hoefnagel selbst ergriff die Initiative, aus der Motivsammlung seines Musterbuchs Bildkompositionen zusammenzustellen und diese durch einen Sinnspruch sinnbildlich zu vertiefen. Sein 1573 in Antwerpen geborener Sohn Jacob, der im Gegensatz zum Vater eine Ausbildung als Maler genossen hatte, stach die achtundvierzig Vorlagen in Kupfer. Als *Archetypa studiaque patris Georgii Hoefnagelii* erschienen die Kupferstiche 1592 in Frankfurt.[19] Die *Archetypa*, d. h. der Formenschatz seines Vaters, wurden ihrerseits von Künstlern als Musterbuch verwendet, die Hoefnagel-Motive wurden bis ins 19. Jahrhundert hinein kopiert. Einzelne Blätter der *Archeypa*-Folge wurden von Künstlern und Liebhabern liebevoll koloriert und wie profane Andachtsbildchen behandelt. Auch in den *Archetypa* finden sich der Nashorn- (Elefanten-) Käfer

[18] Über Hoefnagels Kabinettminiaturen generell siehe Vignau-Wilberg Th., „Unbekannte Kabinettminiaturen von Joris Hoefnagel", *Jahrbuch der Kunsthistorischen Sammlungen in Wien* 85/86 (1989/90) 67–77.

[19] Kommentierte Facsimile-Ausgabe (München: 1994).

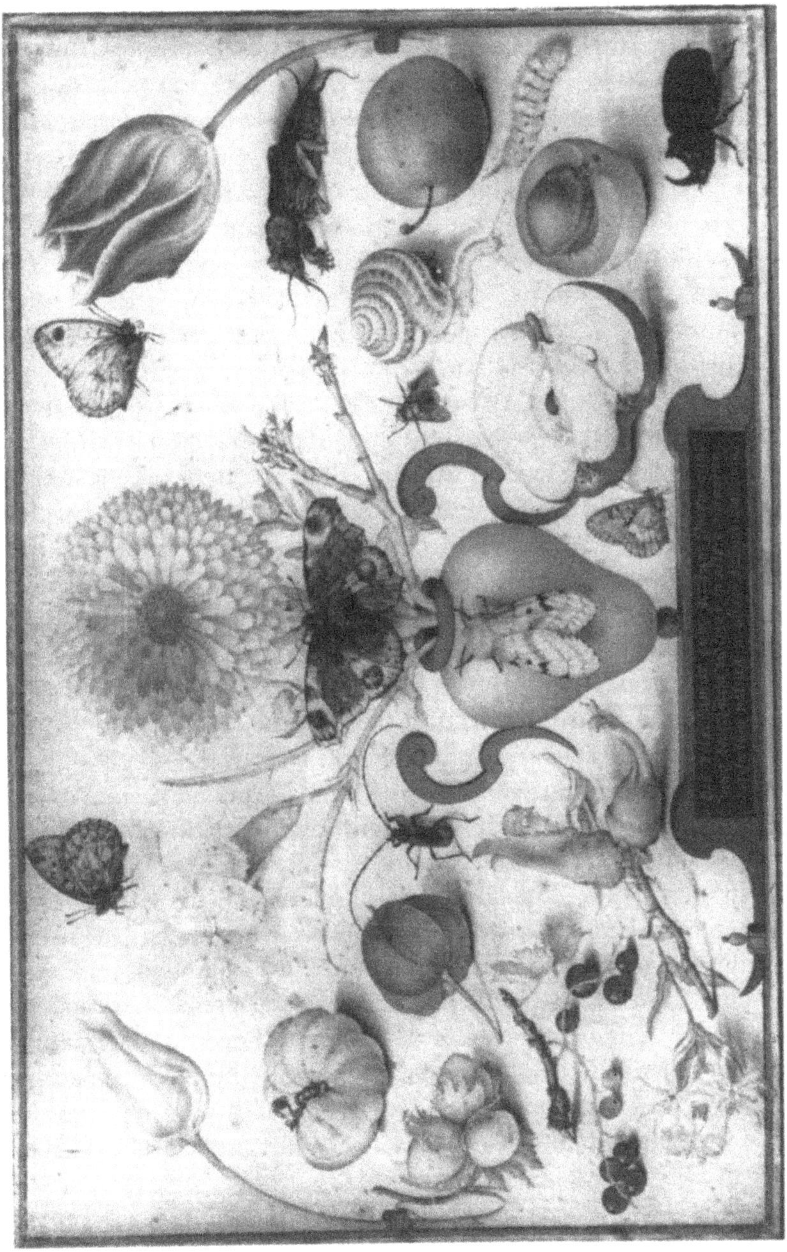

Fig. 5. Joris Hoefnagel, Kabinettminiatur, 1597. Sibiù, Muzeul Brukenthal.

(I, 1) und der Hirschkäfer (I, 6), durch das Stechverfahren seitenverkehrt zum gemalten Bild. Der diagonale Hirschkäfer in leichter Seitenansicht (*Archetypa* II, 1) [Abb. 6] ist Mittelpunkt einer Komposition, bei der Hoefnagel im Text auf das einmalige Werden und Vergehen, auf die Vergänglichkeit alles Irdischen anspielt: Dafür verwendet er die von ihm häufig benutzte symbolische Gegenüberstellung der knospenden und der verblühten Rose mit dem Spruch „Rosam quae praeterierit ne quaeras iterum"[20] („Frage die verblühte Rose nicht um eine neue Blüte"). Auf den Hirschkäfer selbst bezieht sich ein nicht ganz passendes Zitat (in *Ignis* V, erhielt er denselben Kommentar), wie viele andere den *Adagia* von Erasmus entnommen.[21] Er lautet: „Scarabei umbra" („der Schatten eines Skarabäus"), und macht das Tier (Erasmus folgend) zum Sinnbild unbegründeter Furcht, weil es mit schrecklichem Getöse im Nachtflug zu imponieren versucht. Die Wiedergabe des Schattens bei den einzelnen Naturgegenständen suggeriert, dass sie frisch abgepflückt oder eingefangen auf das Pergament gelegt wurden. Die Schattengebung ist wie bei einem Stilleben konsequent durchgeführt, auch zum Beispiel bei den zarten Beinchen der Spinnenartigen, nicht aber bei den im seitlichen Rahmen „befestigten" Blumen. Bei der gemalten Darstellung kommt das Licht von links, im Stich durch die Umkehrung von rechts.

Der Hirschkäfer in Rückenansicht beherrscht das Blatt 6 der ersten Folge der *Archetypa* (I, 6) [Abb. 7]. Er folgt der gleichen Vorlage wie der Hirschkäfer auf dem Einzelblatt in Berlin (siehe oben). Während der untere Spruch, „Habet et Musca splenem et Formicae sua bilis inest" („Es hat auch eine Fliege eine Milz [d. h. zum Lachen] und eine Ameise eine Galle [d. h. Zorn]"), mit einem erneuten Zitat aus den *Adagia* des Erasmus nur lose auf die Darstellung der Fliege links und der Ameise anspielt[22] (das gleiche Zitat findet sich in *Ignis* LXXV), ist der Bezug des oberen Spruches zum Hirschkäfer von größerer Bedeutung: „Me neque mas gignit neque foemina concipit: autor // Ipse mihi solus seminiumque mihi" („Mich zeugt weder ein männliches Wesen, noch empfängt mich eine Frau. Für mich ist Gott allein der Schöpfer und der Samen zugleich"). Dem Hirschkäfer wurde die ungeschlechtliche Fortpflanzung nachgesagt, die Jungfernzeugung. Nur so ist es zu erklären, dass der Hirschkäfer zum Sinnbild Christi werden konnte, wie dies sichtlich der

---

[20] Desiderius Erasmus, *Adagia* (Basel: 1559) chil. II, cent. VI, XL.
[21] Desiderius Erasmus, *Adagia* (Basel: 1559) chil. III, cent. II, XLV.
[22] Desiderius Erasmus, *Adagia* (Basel: 1559) chil. III, cent. V, VII.

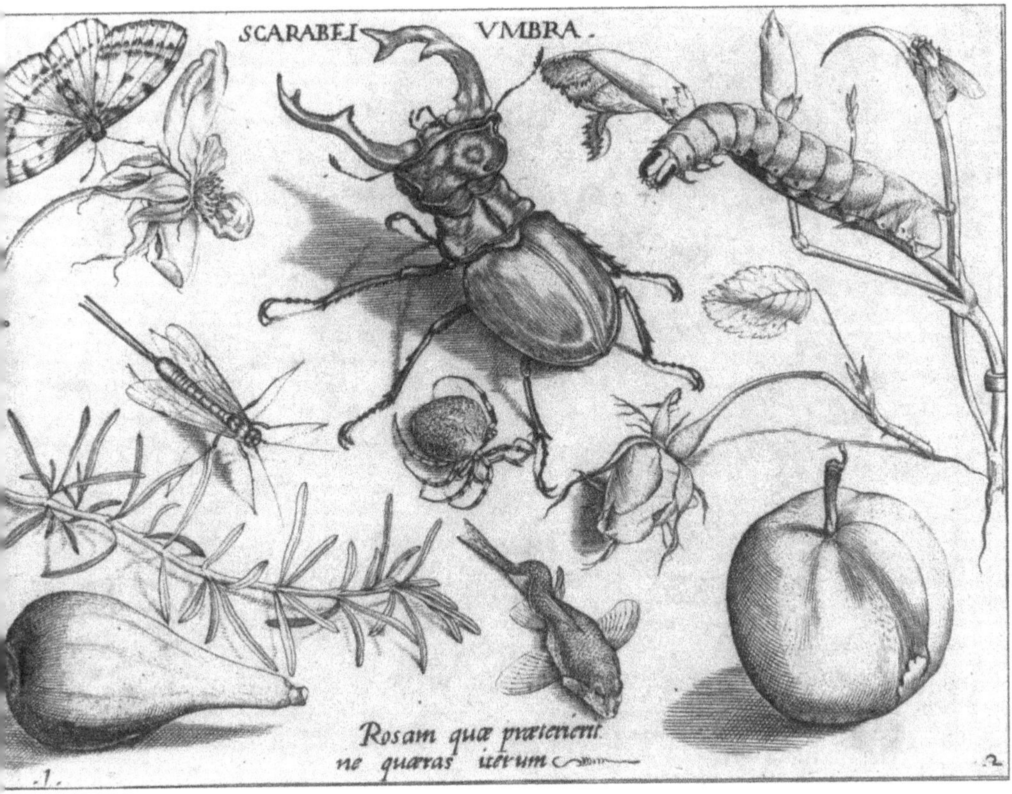

6. Jacob Hoefnagel nach Joris Hoefnagel, *Archetypa studiaque patris Georgii Hoefnagelii* (Frankfurt: 1592) II, 1, Kupferstich. München, Staatliche Graphische Sammlung.

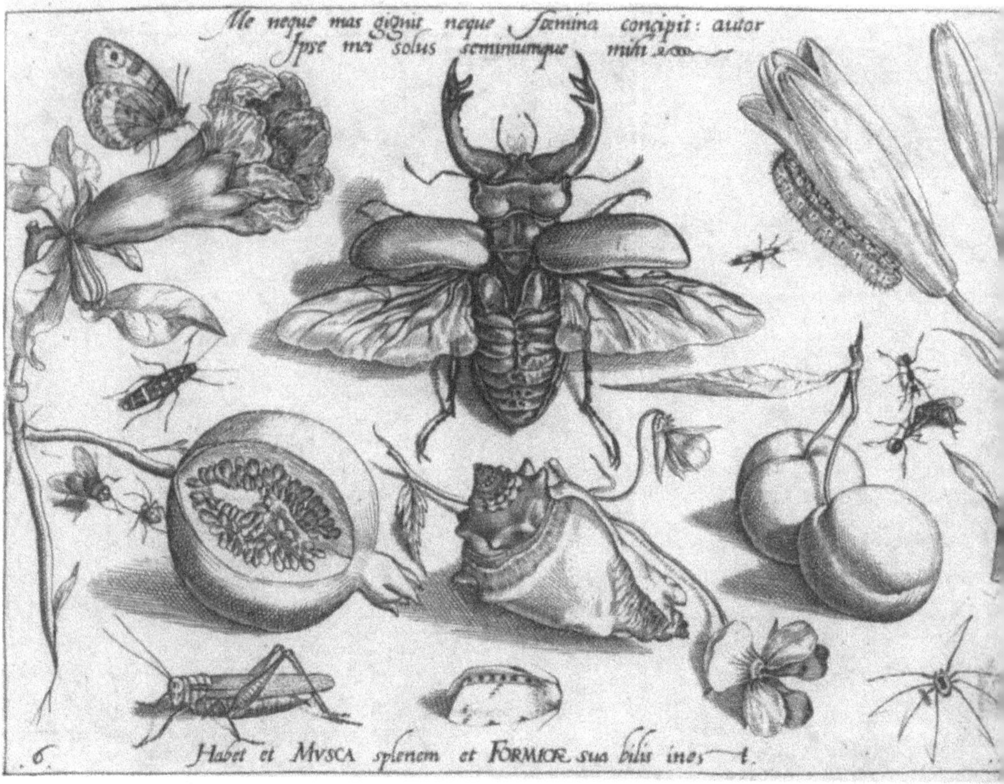

Fig. 7. Jacob Hoefnagel nach Joris Hoefnagel, *Archetypa studiaque patris Georgii Hoefnagelii* (Frankfu
1592) I, 6, Kupferstich. München, Staatliche Graphische Sammlung.

Fall ist auf Dürers Aquarell Maria mit den vielen Tieren.[23] Derselbe Spruch findet sich auch in dem Insektenwerk von Mouffet, allerdings nicht zum Hirschkäfer, sondern zum Großen Nashornkäfer (korrekter: Elefantenkäfer), den er ebenfalls abbildet [Abb. 8].[24] Der Spruch stammt nach Mouffet von Joachim Camerarius, aus dessen Sammlung, wie er schreibt, auch die Vorlage zum Großen Nashornkäfer stammt, die über Thomas Penny zu Mouffet gelangte. Sie porträtiert das Tier nach dem Exemplar aus der Kunstkammer, dem Naturalienkabinett des Herzogs von Sachsen.[25] Der Nashornkäfer[26] stimmt völlig überein mit jenem in den *Archetypa* I, 1 [Abb. 9]. Mouffet erwähnt auch einen Kleinen Nashornkäfer und bildet ihn ab; er ist identisch mit dem Kleinen Nashornkäfer in den *Archetypa* III, 1. Vom letzteren Käfer wurde ihm laut Text die Vorlage aus Wien von Carolus Clusius geliefert; er sei dort zu Lande häufig.[27] Der Große Nashornkäfer in den *Archetypa* ist ein angsteinflößendes Prachtexemplar, er ist würdig, die Folge der *Archetypa* bedeutungsvoll einzuleiten. Die Sprüche, die das Tier im Kupferstich begleiten, sind nicht so sehr symbolisch auf ihn bezogen als dass sie die demütige Ehrfurcht vor der Schöpfung und die Verehrung des christlichen Humanisten für die Vielfalt und Allmacht des Schöpfers zum Ausdruck bringen. Der obere Spruch ist ein Psalmzitat („Dicite Deo quam terribilia sunt opera tua, Domine, in multitudine virtutis tuae mentientur tibi inimici tui"; Ps. 65/66, 3: „Sprechet zu Gott: wie wunderbar sind Deine Werke. Es wird Deinen Feinden fehlen vor Deiner großen Macht"). In *Ignis* XXXV erscheint es bei demselber Tier. Der zweite Sinnspruch ist ein persönliches Motto des Künstlers Hoefnagel: „Danti mihi artem dabo gloriam" („Demjenigen, dem ich meine Kunstfertigkeit verdanke, werde ich durch sie Ehre erweisen").

---

[23] Wien, Albertina, Inv. Nr. 3066. Koreny, *Tier- und Pflanzenstudien* Nr. 35. Man hat versucht, die Symbolik vom Hirschen als Sinnbild des gläubigen Menschen des 41. (42.) Psalms herzuleiten.

[24] Thomas Mouffet, *Insectorum sive minimorum animalium theatrum* 150; *The Theatre of Insects* 1009.

[25] Thomas Mouffet, *Insectorum sive minimorum animalium theatrum* 150: „Cantharorum lege faeminam non habet, at ipse suae sibi faber est formae; faetum solo sibi genitum producit, quod Joachim Camerarius filius non inscite expressit, cum ad Pennium huius insecti iconem e Ducis Saxoniae rerum naturalium penu mitteret".

[26] Das Insekt wird in der heutigen Entomologie als Elefantenkäfer (*Coleoptera Dynastidae, Megasoma elephas*) bezeichnet.

[27] Thomas Mouffet, *Insectorum sive minimorum animalium theatrum* 151: „secundam Nasicornium speciem raram atque pulcherrimam, Mercurio sacram, Carolus Clusius Viennae depictam misit, in cuius agro frequens est."

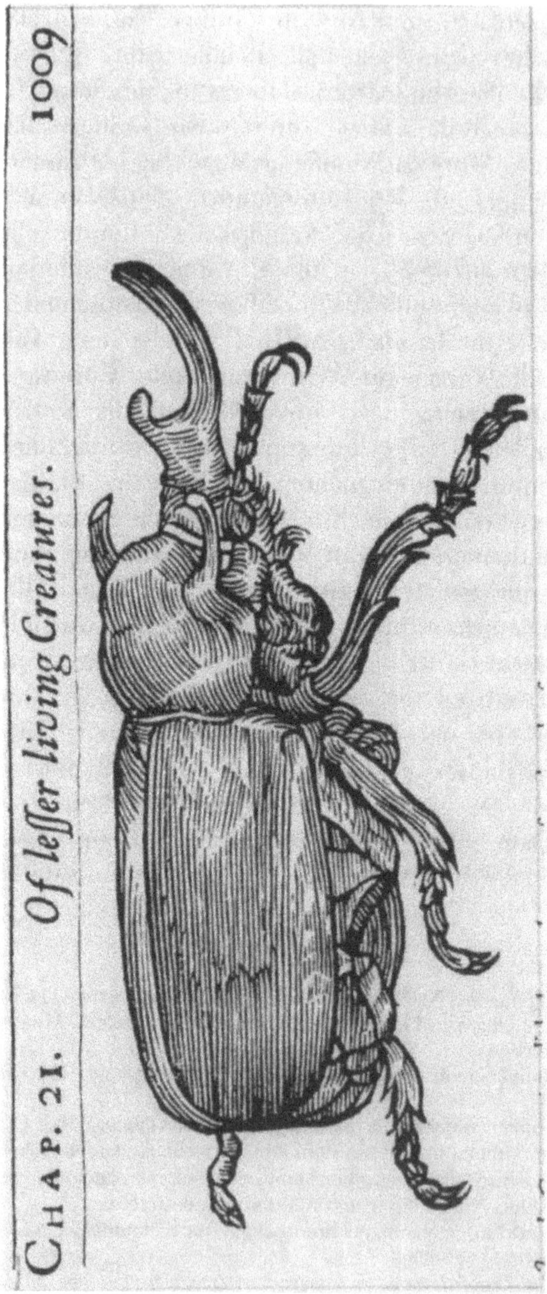

Fig. 8. Großer Nashornkäfer (Elefantenkäfer). Aus Thomas Mouffet, *The Theater of Insects: or Lesser living Creatures, as Bees, Flies, Caterpillars, Spiders, Worms etc. a most Elaborate Work by Thomas Mouffet, Doctor in Physick* (London: 1658) 1009.

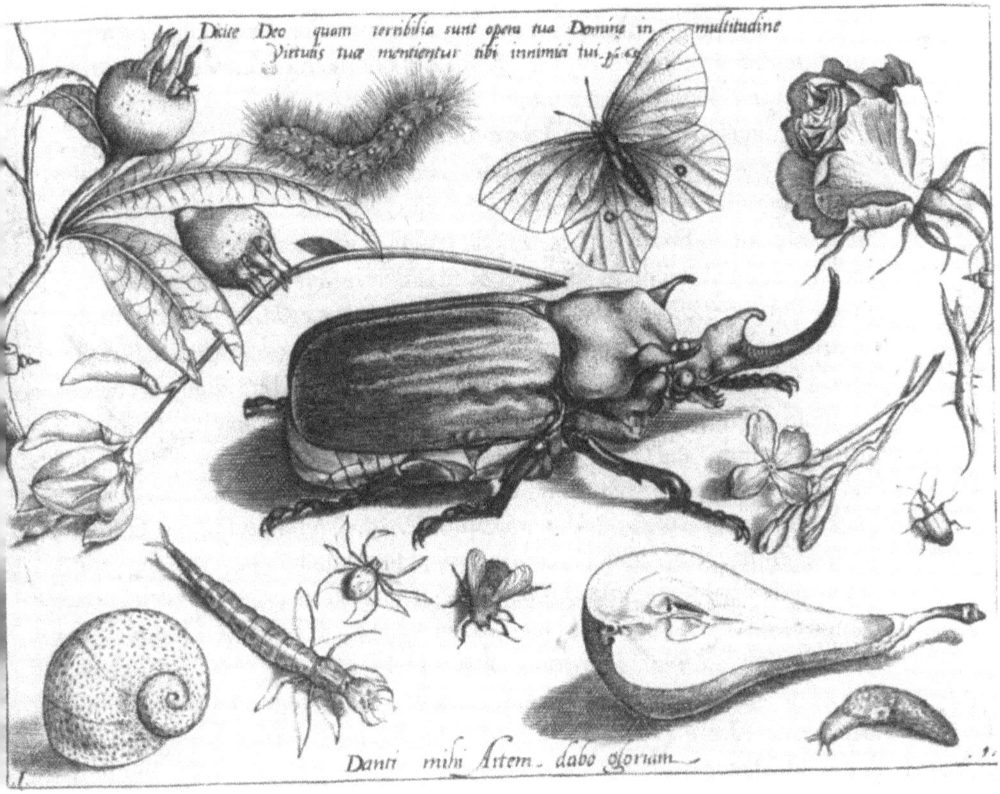

g. 9. Jacob Hoefnagel nach Joris Hoefnagel, *Archetypa studiaque patris Georgii Hoefnagelii* (Frankfurt: 1592) I, 1, Kupferstich. München, Staatliche Graphische Sammlung.

Man könnte sich vorstellen, dass mehrere Miniaturen von Insekten von Joris Hoefnagel den Ausgangspunkt von Mouffets Insektendarstellungen bildeten. Sie wurden jeweils kopiert, befanden sich unter den Naturalien in den Kunstkammern und kursierten, wie man sieht, beigelegt in der Korrespondenz der humanistischen Gelehrten. Aldrovandi nennt in seinem oben erwähnten Insektenwerk als Gewährsleute neben Carolus Clusius, Josephus und Julius Caesar Scaliger, Justus Lipsius und Lambertus Dodonaeus, auch den Maler „Georgius".[28] Damit lässt sich allerdings nicht unbedingt auf einen persönlichen Kontakt zwischen Aldrovandi und Hoefnagel schließen. Aldrovandis Bildquelle könnten auch die *Archetypa* gewesen sein, wobei zu bedenken ist, dass Aldrovandis Insektenbuch kaum Illustrationen aufweist.

Hoefnagels musterhafte Darstellung der Eintagsfliege (Ephemeroptera) in *Ignis* XXXXIV und in den *Archetypa* II, 10 [Abb. 10] war die Grundlage für eine Buchpublikation, die Outgert Cluyt, Augerius Clutius, Sohn des Dirck Outgertsz. Cluyt, Direktor des Hortus Botanicus der Leidener Universität und rechte Hand von Carolus Clusius, diesem Tier widmete: *De hemerobio sive ephemero insecto* (Amsterdam: 1634). Clutius schreibt in seiner Einleitung: „Accessit et altera occasio perscrutandi, quod Caesaris Rudolphi II pictor eximius Houfnagel bestiolae ad vivum depictae (Palingenia longicauda) imaginem mihi donavit, cum brevi historia natalis eius".[29]

Wie wir oben hinsichtlich des Hirschkäfers sahen, hat sich auch Joachim Camerarius mit dem Studium der Insekten beschäftigt. Camerarius gehörte zum gleichen Kreis wie Hoefnagel[30] und Mouffet und es existiert auch ein umfangreicher Briefwechsel zwischen Clusius und Camerarius.[31] Der Arzt und Naturwissenschaftler Camerarius hatte 1580 den schriftlichen Nachlass von Conrad Gesner (1516–1566) erworben. Er war selbst ein ausgezeichneter Latinist, sein Vater Joachim Camerarius der Ältere (1500–1574) hatte noch zum Humanistenkreis um Albrecht Dürer gehört.[32] Vom jüngeren Camerarius erschien zwischen

---

[28] In „Catalogus Authorum quibus in hoc opere de Insectis usus sum".

[29] Augerius Clutius, *De hemerobio sive ephemero insecto* (Amsterdam: 1634) 66–67.

[30] Siehe den Brief von Hoefnagel in München, Bayerische Staatsbibliothek, HS. Clm 10378, fol. 61r, publiziert in Vignau-Wilberg Th., „Joris Hoefnagels Tätigkeit in München", *Jahrbuch der Kunsthistorischen Sammlungen in Wien* 81 (1985) (103–167) 161.

[31] Briefe von Clusius in der Collectio Cameriana (München, Bayerische Staatsbibliothek, HS. Clm 10378); von Camerarius an Clusius in Leiden, Universiteitsbibliotheek, Cod. Vulc. 101.

[32] Siehe Vignau-Wilberg, „Naturemblematik".

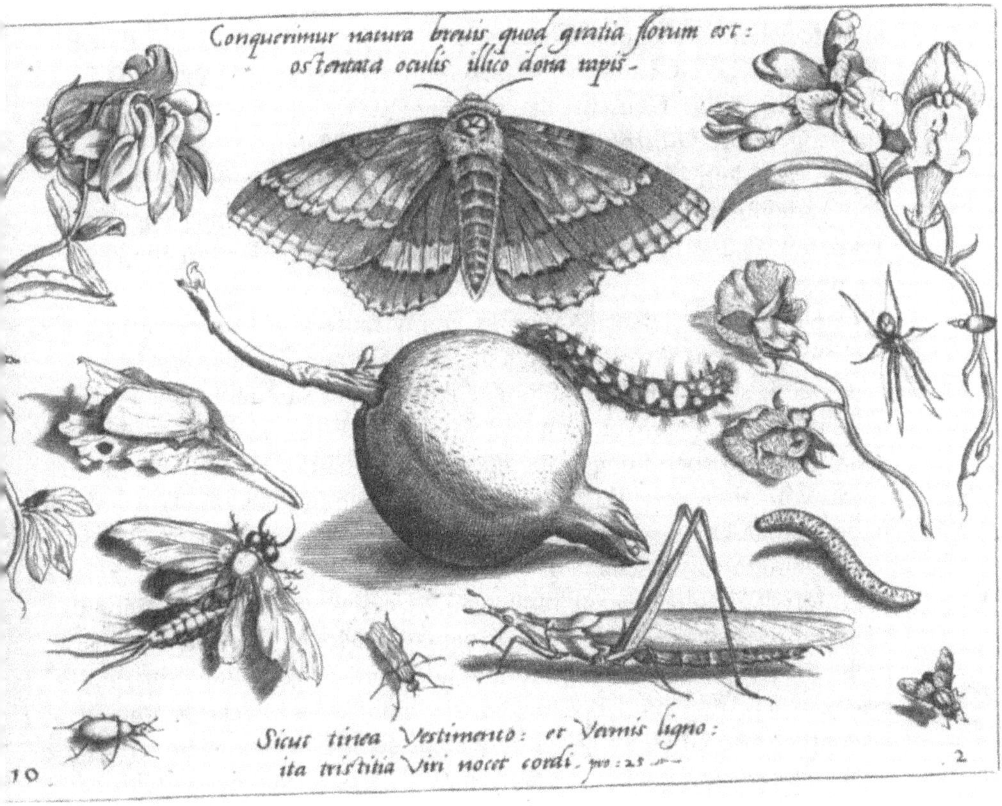

. 10. Jacob Hoefnagel nach Joris Hoefnagel, *Archetypa studiaque patris Georgii Hoefnagelii* (Frankfurt: 1592) II, 10, Kupferstich. München, Staatliche Graphische Sammlung.

1590 und 1604 in Nürnberg ein vierbändiges Emblembuch, dessen Embleme Naturobjekte zum Gegenstand haben. Der erste Band von 1590 (erschienen 1593) handelt von den Pflanzen, der zweite, 1595 publiziert, von den Vierfüßern, der dritte, 1596, von Vögeln und Insekten und der vierte, 1604 postum erschienen, von Wassertieren und Reptilien.[33] Diese Klassifizierung der Tiere ist noch konventionell. Die Insekten-Embleme erwecken die Neugierde, jedoch beschäftigen sich von den hundert Emblemen weniger als zehn mit Insekten (Embleme XC bis IC; C ist dem Phönix gewidmet). Es handelt sich größtenteils um die schon bei Plinius behandelten Insekten: die nützliche Biene (Embleme XC, XCI und XCIII), das Glühwürmchen (Emblem XCIII), die (biblische) Heuschrecke (Emblem XCVI), die von der Flamme angezogene und durch sie versengte Motte (Emblem XCVII), die Fliege (Emblem XCVIII), die Spinne im Gewebe (Emblem XCIX) und undefinierbare fliegende Insekten. Auch ihre Auslegung folgt der Tradition. Camerarius war offensichtlich nicht sonderlich an Insekten interessiert.

Der englische Gelehrte Thomas Penny hatte von Camerarius die entomologischen Aufzeichnungen aus dem Gesner-Nachlass erworben. Er verleibte sie seiner eigenen insektenkundlichen Sammlung, die größtenteils von Thomas Wotton stammte, ein. Als Penny 1589 starb, setzte Mouffet die Vorbereitungen für die Herausgabe des entomologischen Werks von Penny fort. Im Jahr 1590 betrachtete Mouffet die Arbeit als nahezu abgeschlossen.[34] Als Mouffet selbst 1604 starb, war das Werk jedoch noch nicht erschienen.

Unter den vielen Darstellungen in Mouffets Insektenbuch finden sich mehrere nach Vorlagen von Hoefnagel. Mouffet ließ sie offenbar nicht aus den *Archetypa* kopieren, sondern nach gemalten Vorlagen zeichnen. Die Insektendarstellungen, die ihm Clusius schickte, stammen vermutlich von Hoefnagel oder gehen auf seine Darstellungen zurück. In diesem Zusammenhang sei auf einen Brief von Mouffet an Clusius aus dem Jahre 1590 hingewiesen.[35]

---

[33] Joachim Camerarius, *Symbolorum et emblematum ex re herbaria desumtorum centuria una/ altera/ tertia/ quarta.*

[34] Siehe den Brief an Clusius, zitiert in Hoefnagel, *Archetypa* 39.

[35] Brief Leiden, Universiteitsbibliotheek, Cod. Vulc. 101, 10: London, Kal. Aprilis 1590 (teilweise zitiert in Hoefnagel, *Archetypa* 39).

Das vollendete Insektenbuch blieb nach Mouffets Tod vorerst unpubliziert liegen. Mouffets Widmung an Königin Elisabeth veralterte.[36] Nach der Zueignung heißt es: „He [Mouffet] thought it no indignity to Dedicate to the greatest Princess the miracles of Nature, which are most conspicious in the smallest things; which testifie the infinite power of the supreme Creator of all things, and raises the mindes of Princes who are the children of the most Highest, to the cause of all causes [...]". Nach einer allgemeinen Betrachtung über einzelne Insekten und kleine Tiere erhebt der Text die Insekten zu Objekten der Kontemplation und der christlichen Philosophie. Das Verhalten der Biene und der Ameise wird als Muster für die optimale Regierungsform bzw. für vernünftiges wirtschaftliches Denken gesehen („Monarchical government of Bees, the Democratical of Ants, and the oeconomical providence of them both"). Die sogenannte vollkommene Metamorphose von Raupe über Puppe zu Schmetterling spiegelt das irdische Leben, der Todesschlaf und die Auferstehung „crowns all the meditations of a Christian man".

In der Vorrede schreibt Mouffet, daß er zum ersten Mal 150 Illustrationen wiedergibt, die Gesner und Penny noch nicht kannten. Mouffet kommentiert im Vorwort seines Insektenwerkes, sich auf die Tiere beziehend, folgenderweise:

> [...] that in the universal world there is nothing more divine than these, exept Man. For however in shew [= show] they are most abject and sordid, yet if we look more nicely into them, they will appear far otherwise than they promise in the bare outside. [...] It will be easie to observe, that the divine force and power shew themselves more effectually in mean things, and they are far more miraculous, than those things the world with open mouth respects so much and admires [...]. And truly, if the fabrick of Insects were worthy of so great and divine Artificer, how can the contemplation of them be unworthy of the understandings of poor contemptible men?

Das erste Kapitel, über die Fliegen (Kap. X), ist eine allgemeine Betrachtung dieser Gattung, im nächsten Kapitel folgen die Beschreibung und die Darstellung zahlreicher Fliegenarten. Hier werden auch die Quellen der Darstellungen erwähnt: „The figure of this Fly did the abovesaid Joachim send to Pennius" und „The third of these Carolus Clusius sent from Vienna" (S. 938). Seite 941 besteht ausschließlich aus Abbildungen von geflügelten Insekten (hauptsächlich Libellen).

---

36 Der besseren Verständlichkeit wegen wird hier nicht aus dem lateinischen Erstdruck, sondern aus der englischen Übersetzung zitiert.

Der Titel von Kap. XII heißt „Of the use of Flyes" (Seite 944 ff.). An vielen Stellen ist die Rede von gezeichneten oder gemalten Insekten, die Clusius aus Wien an Mouffet schickte (Seite 948, 959, 963 und 983). Auch der Antwerpener Jacobus Quickelberg (Quiccheberg) und der Nürnberger Camerarius werden als Lieferanten von Bildquellen genannt (Seite 997, 1001, 1007 und 1015). Der Große und der Kleine Nashornkäfer wurde nach den Vorlagen von Joris Hoefnagel in Holz geschnitten [Abb. 11]. Die Beschreibung des Großen Nashornkäfers folgt auf Seite 1008–1009:

> Like to Beetles it hath no female, but shapes its own form it self. It produceth its young one from the ground by it self. Joachim Camerarius did elegantly express, when he sent to Pennius the shape of this Insect out of the storehouse of natural things of the Duke of Saxony, with these Verses:
>
> A Hee begat me not, nor yet did I proceed
> From any Female, but my self I breed.[37]

Das Tier stirbt einmal pro Jahr und wird, wie ein Phönix, nach dem Tod durch die Sonnenhitze wieder zum Leben erweckt. Durch die Wiedergeburt wird der symbolische Bezug des Tieres zu Christus noch evidenter:

> A thousand summers heat and winters cold
> When she hath felt, and that she doth grow old,
> Her life that seems a burden, in a tomb
> Of spices laid, comes younger in her room.

Es ist wohl kein Zufall, dass etwa zur gleichen Zeit, als Mouffets Insektenbuch auf Betreiben der Royal Society in London erschien, in Amsterdam bei dem rührigen Verleger Claes Jansz. Visscher, ein „neues" Stichwerk nach Werken des „berühmten" Joris Hoefnagel erschien, das ausschließlich Darstellungen von Insekten enthält: *Diversae insectarum volatilium icones* (Amsterdam: 1630).[38] Die Insekten sind dekorativ über das Blatt angeordnet, jedes Insekt ist ein Nachstich eines Tieres aus den *Archetypa* und erscheint seitenverkehrt. Die Wiedergabe ist schematischer und weniger subtil als sie es in den *Archetypa* war. Der breitflügelige Hirschkäfer erscheint als zentrale Bildfigur auf Blatt 4 [Abb. 12], der diagonale Hirschkäfer in Seitenansicht auf Blatt 10, der Nashornkäfer

---

[37] Siehe oben.

[38] Claes Jansz. Visscher, *Diversae insectarum volatilium icones.*

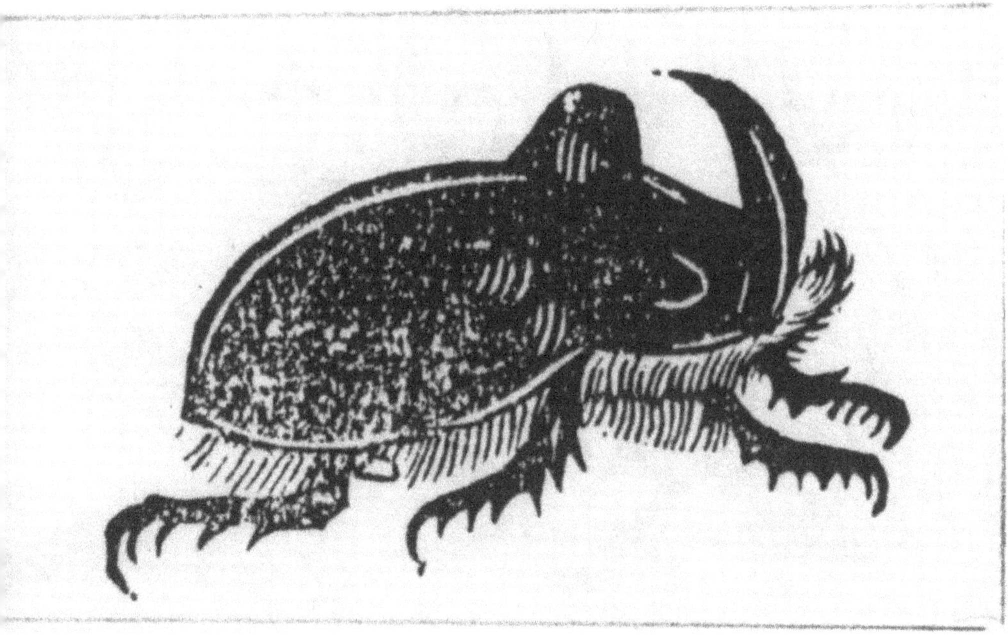

11. Kleiner Nashornkäfer, Aus Thomas Mouffet, *The Theater of Insects: or Lesser living Creatures, Bees, Flies, Caterpillars, Spiders, Worms etc. a most Elaborate Work by Thomas Mouffet, Doctor in Physick* (London: 1658) 1009.

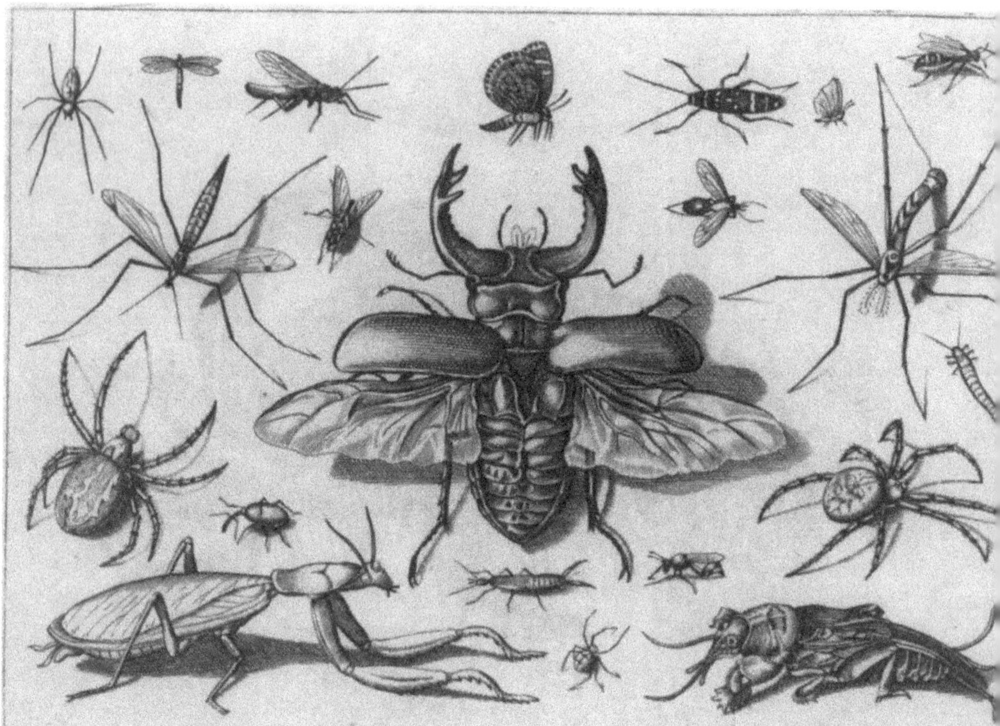

Fig. 12. Claes Jansz. Visscher, *Diversae insectarum volatilium icones ad vivum accuratissime depictae per cele mum pictorem D. I. Hoefnagel typisque mandatae a Nicolao Ioannis Vißcher* (Amsterdam: 1630) 4, Kupfers München, Staatliche Graphische Sammlung.

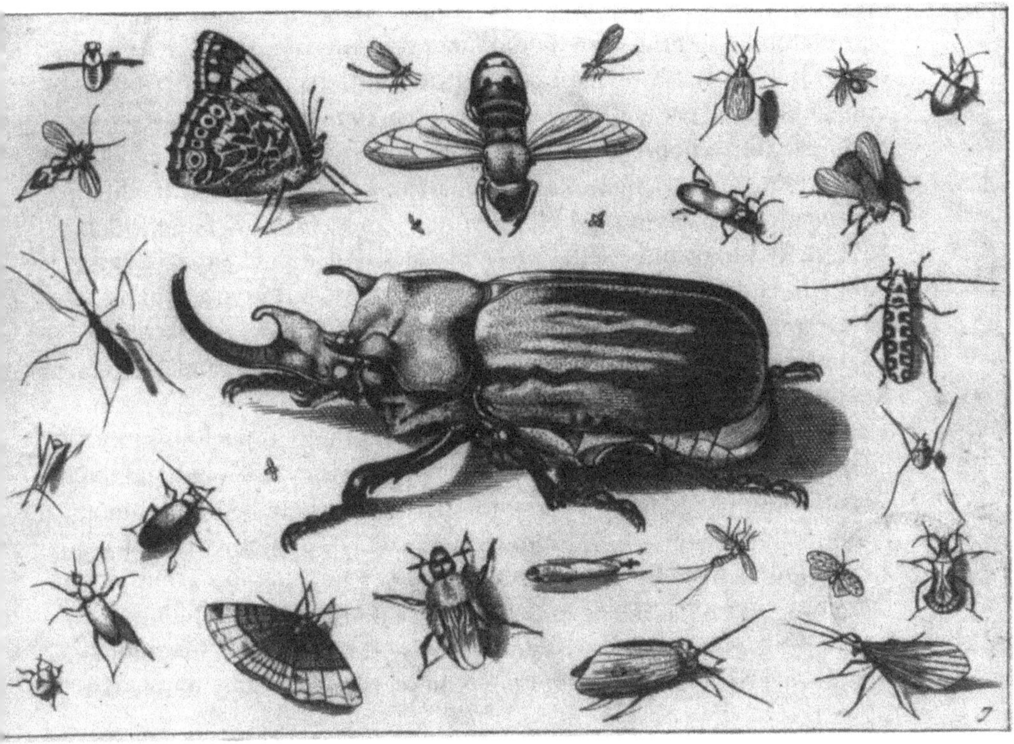

g. 13. Claes Jansz. Visscher, *Diversae insectarum volatilium icones ad vivum accuratissime depictae per eberrimum pictorem D. I. Hoefnagel typisque mandatae a Nicolao Ioannis Vißcher* (Amsterdam: 1630) 7, Kupferstich. München, Staatliche Graphische Sammlung.

(Elefantenkäfer) auf Blatt 7 [Abb. 13] und der Kleine Nashornkäfer auf Blatt 13, jeweils mit dem gleichen Schattenwurf, den er in den *Archetypa* hatte. Während die *Archetypa* bis ins 19. Jahrhundert immer wieder aufgelegt wurden, blieb es bei dem Insektenwerk nach Hoefnagel wohl nur bei einer einzigen Auflage.

Im letzten Viertel des 16. Jahrhunderts, als man bemüht war, die gesamte Tierwelt systematisch zu erfassen, wurden durch die Gelehrten des Humanismus die Insekten als eigene zoologische Tierklasse entdeckt, deren Vielfalt und skurille Erscheinungsform sie faszinierten. Unter den Naturalien der Kunst- und Wunderkammern dieser Zeit befanden sich auch Käfer, die von kundigen Künstlern porträtiert wurden und in Kopien für andere Sammler kursierten. Hinzu kam, dass es durch die fortschreitende Entwicklung verschiedenartiger geschliffener Linsen möglich war, auch sehr kleine Lebewesen zu betrachten und wiederzugeben (das Mikroskop aus dem frühen 17. Jahrhundert hatte ja Wegbereiter). Auch der Miniaturmaler Joris Hoefnagel kann nur mit vergrößernden Linsen gearbeitet haben. Kunstvolle Vorlagen bildeten die Basis der Illustrationen der Insektenwerke, die ab 1600 erschienen. Von ihnen ist Mouffets postum publiziertes Insektenbuch das bedeutendste.

Naturwissenschaft und Philosophie waren um 1600 noch in großer Selbstverständlichkeit verbunden. In der Naturwissenschaft bestand das Bedürfnis, überliefertes Wissen mit der auf empirische Wahrnehmung beruhenden neuesten Forschung in Einklang zu bringen oder mit ihr zu verbinden. Man suchte im Studium zu Erscheinungsform und Verhaltungsweise der Tiere Aufschluss über den Plan der Schöpfung und war der Meinung, dass sich gerade in den allerkleinsten Lebewesen Sinn und Ziel der Schöpfung und des Schöpfers am klarsten manifestiert.

Auswahlbibliographie

ALDROVANDI ULISSE, *De animalibus insectis libri septem cum singulorum iconibus ad vivum expressis* [...] (Bologna: 1602).
CAMERARIUS JOACHIM der Jüngere, *Symbolorum et emblematum ex re herbaria desumtorum centuria una collecta* [...] (Nürnberg: 1590).
——, *Symbolorum et emblematum ex animalibus quadrupedibus desumtorum centuria altera collecta* [...] (Nürnberg: 1595).
——, *Symbolorum et emblematum ex volatilibus et insectis desumtorum centuria tertia collecta* [...] (Nürnberg: 1596).
——, *Symbolorum et emblematum ex aquatilibus et reptilibus desumtorum centuria quarta a Ioachimo Camerario* [...] *coepta; absolutum post eius obitum a Ludovico Camerario* (o.O. [Nürnberg]: 1604; Nürnberg: 1605).
CLUTIUS AUGERIUS, *De hemerobio sive ephemero insecto* (Amsterdam: 1634).
ERASMUS DESIDERIUS, *Adagiorum chiliades quatuor* (Basel: 1559).
GESNER CONRAD, *Historiae animalium Liber I* (Zürich: 1551); *Icones animalium quadrupedum* (Zürich: 1553); *De avium natura Liber III* (Zürich: 1555); *Historiae animalium Liber IV qui est de piscium et aquatilium animantium natura* (Zürich: 1558).
HAUPT H. – VIGNAU-WILBERG TH. – IRBLICH E. – STAUDINGER M. (Hrsg.), *Le Bestiaire de Rodolphe II. Cod. Min. 129 et 130 de la Bibliothèque Nationale d'Autriche* (Paris: 1990).
HENDRIX M.L., *Joris Hoefnagel and the ‚Four Elements': a study in Sixteenth-Century Nature Painting* (Ann Arbor: 1985, diss. Phil. Princeton University: 1984).
HOEFNAGEL JORIS, *Archetypa studiaque Patris Georgii Hoefnagelii* (Frankfurt: 1592; Facsimile-Ausgabe Staatliche Graphische Sammlung München: 1994).
KENNER F., „Die Porträtsammlung des Erzherzogs Ferdinand von Tirol", *Jahrbuch der Kunsthistorischen Sammlungen des Allerhöchsten Kaiserhauses* 15 (1894) 147–259.
KORENY F., *Albrecht Dürer und die Tier- und Pflanzenstudien der Renaissance* (München: 1984; Ausstellungskatalog Wien, Albertina: 1984/85).
MANDER KAREL VAN, *Het Schilder-boeck* (Haarlem: 1604).
MOUFET THOMAS, *Insectorum sive minimorum animalium theatrum olim ab Edoardo Wottono, Conrado Gesnero, Thomaque Pennio inchoatum: tandem Thomae Moufeti Londinatis opera sumptibusque; maximis concinnatum, auctum, perfectum, et ad vivum expressis iconibus supra quingentis illustratum* (London: 1634).
——, *The Theater of Insects: or Lesser living Creatures, as Bees, Flies, Caterpillars, Spiders, Worms etc. a most Elaborate Work by Tho. Mouffet, Doctor in Physick* (London: 1658; Facsimile-Ausgabe Edward Topsell, *The history of Four-Footed Beasts and Serpents and Insects*, vol. III. *The Theater of Insects* by T. Muffet. New York: 1967).
ROTH M., unter Mitarbeit von PÉNOT S., *Georg Flegel. Die Aquarelle* (München: 2003).
SCHEICHER E., *Die Kunst- und Wunderkammern der Habsburger* (Wien: 1979).
THOSS D., „Georg Hoefnagel und seine Beziehung zur Gent-Brügger Buchmalerei", *Jahrbuch der Kunsthistorischen Sammlungen in Wien* 82/83 (1986/97) 199–212.
VIGNAU-WILBERG TH., „Joris Hoefnagels Tätigkeit in München", *Jahrbuch der Kunsthistorischen Sammlungen in Wien* 81 (1985) 103–167.
——, „Naturemblematik am Ende des 16. Jahrhunderts", *Jahrbuch der Kunsthistorischen Sammlungen in Wien* 82/83 (1986/87) 145–156.
——, „Die Randilluminationen und Initialen", in *Das Gebetbuch Kurfürst Maximilians I. von Bayern* (Frankfurt-Stuttgart:1986) 65–118.
——, „Unbekannte Kabinettminiaturen von Joris Hoefnagel", *Jahrbuch der Kunsthistorischen Sammlungen in Wien* 85/86 (1989/90) 67–77.
VISSCHER CL. J., *Diversae insectarum volatilium icones ad vivum accuratissime depictae per celeberrimum pictorem D.I. Hoefnagel typisque mandatae a Nicolao Ioannis Vißcher* (Amsterdam: 1630).

CURIOUS FISH: CONNECTIONS BETWEEN SOME SIXTEENTH-CENTURY WATERCOLOURS AND PRINTS

Florike Egmond

Two fishes

If the similarities between two sets of sixteenth-century watercolours depicting peculiar fish are so great that they can hardly be a coincidence, what should we conclude? This question was only the first of a series of queries that emerged while probing the possible reasons for (and implications of) these similarities. Was the one the model for the other? How could we know, and what were the implications? The following paragraphs, which also touch upon questions about the contexts in which these particular watercolours could be studied, and the problem of to which genre they actually belong, present our by no means final findings.

The two sets of remarkably similar pictures of fish belong to two very different sixteenth-century collections of watercolours depicting *naturalia*. The first is the collection known as *Libri Picturati* A. 16–31, held at the Bibliotheka Jagiellonska in Kraków and consisting of sixteen volumes of mainly botanical and some zoological watercolours. The core of this collection originated in the Southern Netherlands during the 1560s under the patronage of the aristocratic collector Charles de Saint Omer (1533–1569), lord of Moerkercke, who was assisted and probably inspired by his protégé, the famous botanist Carolus Clusius (1526–1609).[1] The circa 1500 watercolours contained in these volumes have been recognised at least since the 1930s as one of the largest and most important European collections of sixteenth-century watercolours depicting natural history on account of both their quantity and their

[1] For an extensive discussion see Egmond F., "Clusius, Cluyt, Saint Omer. The origins of the sixteenth-century botanical and zoological watercolours in Libri Picturati A. 16–30", *Nuncius* 20 (2005) 11–67. All botanical watercolours of the *Libri Picturati* A.18–30 will be published in Koning J. de – Zemanek A. – Zemanek B. (eds.), *Botanical illustrations of the sixteenth century. An enquiry into the Libri Picturati A.18–30* (forthcoming, 2007–2008).

artistic and scientific quality. The botanical volumes are at least on a par with the famous (but much smaller), more or less contemporary botanical watercolours of Conrad Gessner and the so-called *Camerarius Fiorilegium*.[2] The second collection that is relevant to this essay consists of the richly illustrated manuscript albums with the titles *Visboock* (i.e. *Fish Book*, 1577–1581, comprising 125 folios) and *Walvisboock* (i.e. *Whale Book*, 1584–1586, 421 folios) that were written and illustrated by Adriaen Coenen (1514–1587), the self-taught son of a local fisherman from Scheveningen near The Hague, and wholesaler in fish.[3]

The differences seem obvious. The core of the first collection dates from the 1560s, while Coenen produced his albums between the late 1570s and mid 1580s; the former belongs to the Southern Netherlands, the latter to the North – an important distinction in these decades in which the Northern Dutch Provinces had initiated their revolt against Habsburg rule and were in the process of splitting off from the Southern ones. Originally the *Libri Picturati* consisted of a vast collection of loose sheets with watercolours. These sheets were only bound together and (re)organized in the seventeenth century, by a later owner. They mainly focus on botanical subjects and the sheets contain very little annotation. Only the first two volumes (A. 16 and A. 17) specifically concentrate on animals. The majority of the *Libri Picturati* watercolours are of outstanding artistic quality. Several artists must have been involved, and we do not yet know for certain whether the same artists who made the botanical watercolours also produced the zoological ones. Coenen's albums focus almost exclusively on animals living in and near the sea and in some rivers. Coenen painted his watercolours directly in bound albums, accompanying them with texts which present descriptions of the depicted fish and marine creatures, anecdotes, quotations and anything else he considered relevant. Text and pictures are generally closely linked. The voluminous *Visboock* forms a cross between an early encyclopaedia of marine life and a 'paper collec-

---

[2] Both are in the Trew collection in the Library of Erlangen. On this collection see especially Schnalke T. (ed.), *Natur im Bild. Anatomie und Botanik in der Sammlung des Nürnberger Arztes Christoph Jacob Trew* [Exhibition catalogue] (Erlangen: 1995).

[3] On Coenen and his manuscripts see Egmond F. – Mason P., *The Whale Book. Whales and other marine animals as described by Adriaen Coenen in 1585* (London: 2003); and Egmond F., *Het Visboek. De wereld volgens Adriaen Coenen 1514–1587* (Zutphen – The Hague: 2005). The complete *Visboock* is accessible via the website of the Koninklijke Bibliotheek in The Hague (www.koninklijkebibliotheek.nl). All references in this article to folio numbers of the *Visboock* match the *new* numbering made by the Koninklijke Bibliotheek and thus the numbering on the website).

tion', while the *Walvisboock* looks much more like a sixteenth-century 'coffee table book': it contains much less text than the *Walvisboock* and the format is oblong. Coenen, however charming his pictures and accurate his rendering of many fish, was decidedly an amateur artist. The gap between a vast collection belonging to an aristocratic collector of curiosities and the personal albums of a self-taught Dutchman of humble background is considerable. So why do these pictures resemble each other so closely?

A tattooed tuna

Libri Picturati A. 16 contains watercolours of many fish and a smaller number of four-footed beasts. Folio 5 shows an intriguing figure of a fish painted in grey-blue colours with a large pointed nose or beak, facing right [Fig. 1]. It has a dark, round eye, which looks more like a hole than an eye, although the outline of a deep lying eye can be discerned with some difficulty. It has two dorsal fins and an almost symmetrical and smooth tail. The picture does not look like the work of a great artist; it seems stiff and lacks background and shading. In this sense it differs from many other watercolours in the botanical as well as the zoological volumes of the *Libri Picturati*. This fish has one characteristic which makes it completely unique: on its skin dark figures may be discerned which look somewhat like tattoos. The 'tattoos' depict several types of ships: some look like galleys with rowers and occasionally a mast and rolled up sail; others show larger ships with masts and fully deployed sails, but without rowers. The latter can be found along the back of the fish and on its tail and underbelly. Detail is such that even some miniature figures of the rowers in the galleys can be discerned. Curiously, above and to the left of the eye of the fish a 'castle' or turreted gate is depicted in lighter colour on the dark skin of the fish. Just to the left of it, there is another 'black hole', reminiscent of the blow holes of certain whales, although this fish does not resemble a whale at all. There is no annotation, we do not know who the artist is, and the sheet of paper bears a watermark that has not yet been dated or identified.[4]

---

[4] See Ramón-Laca L., "Charles de l'Ecluse and Libri Picturati A. 16–30", *Archives of Natural History* 28 (2001) 195–243, especially its appendix listing watermarks and annotation for all sheets of the *Libri Picturati* volumes A. 16–30.

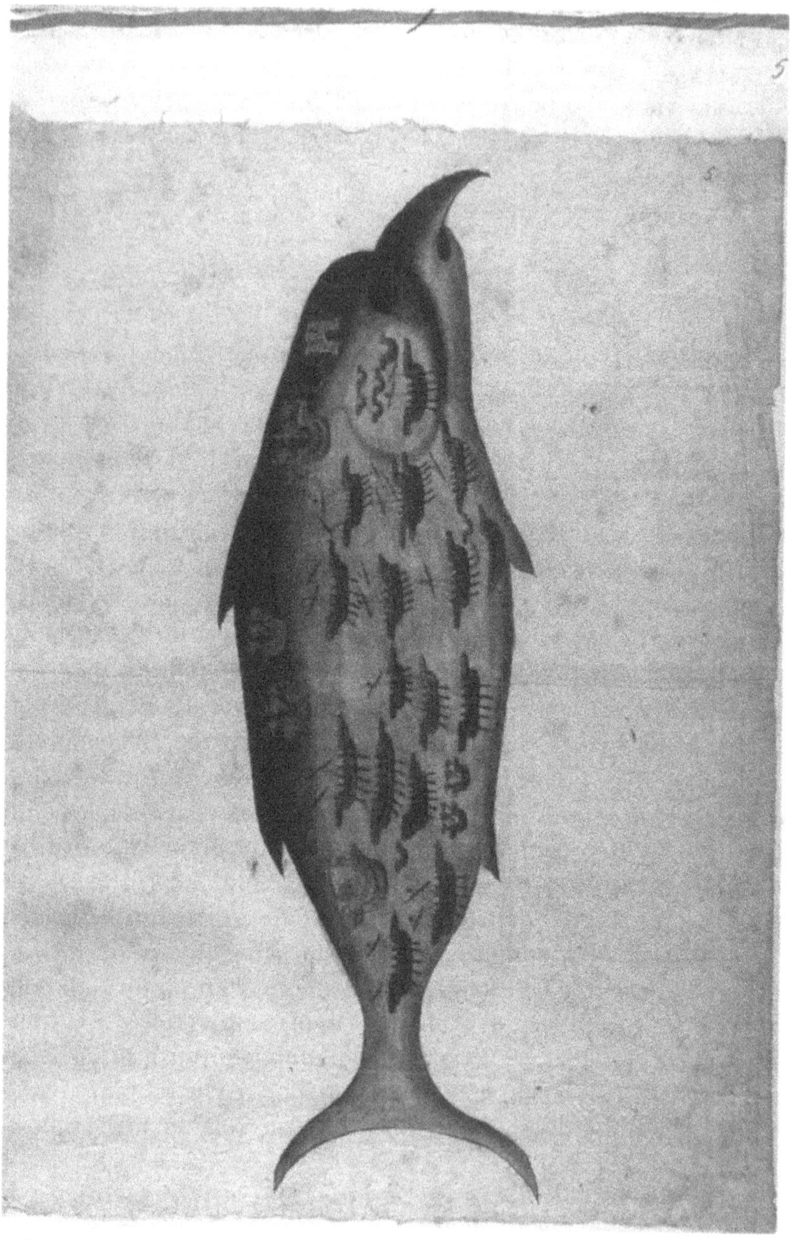

Fig. 1 [Col. pl. VIII]. Tattooed tuna. From *Libri Picturati* A. 16, fol. 5r. Courtesy of Bibliotheka Jagiellonska, Kraków.

The same fish – and it must be the same precisely because of its curious 'tattoos' – is depicted in Coenens *Visboock* (fol. 49v) [Fig. 2] and again, in an almost identical way in Coenens *Walvisboock* (I fol.15r) [Fig. 3]. As the illustration from the *Visboock* shows, Coenens fish faces the other way, to the left, and Coenen has added a purple background, and an attractive decoration at the bottom of the page, showing a sandy beach covered in shells. Coenen's fish is more even in colour (a pale blue), has slightly larger fins, and its beak is slightly more open, while its tail is more frayed, but in other respects the two pictures are extremely similar. Coenen's style of painting is more amateurish but less stiff than in the case of the *Libri Picturati* fish; the way in which he has painted the ships and galleys on the skin of the fish is less detailed and their distribution less well organized than on its counterpart. Yet, Coenen's manner of distributing the two types of ship is roughly the same, and he even depicts the 'blow hole' above and in this case to the right of the fish's eye. Interestingly, the 'castle' seems to be the only 'tattoo' which Coenen depicts in more detail than the *Libri Picturati* artist.

Given the dating, it is tempting to assume that Coenen had somehow had access to the *Libri Picturati* picture and made a copy. Yet, the reversal of the picture poses problems, because such reversals occur generally when a *print* is made after an original. An artisan would cut a copy of the original picture in a wood block (or engrave it in a copper plate), and when printed the picture would automatically be reversed. Here, however, we are dealing with two hand-painted originals. A partial answer can be found in the text that Coenen included in his *Visboock* and at more length in his *Walvisboock*. He explains that he copied the picture and accompanying text from a pamphlet printed in Antwerp by Hans Liefrinck figure cutter [= woodcutter or engraver] at the Lombaerdveste in *t Gulden Turcxhooft*. Coenen had received this pamphlet via a friend and patron who belonged to the lower nobility in Holland: Cornelis Suys, president of the Court of Holland. Suys had received the pamphlet from Brabant, probably from Antwerp where it was printed.[5] Since we know from Coenen's own writings that contacts between Coenen and Suys ceased (for political and religious reasons) around the year 1566 or 1567, it is more than likely that Coenen received this pamphlet not long after it had been printed in 1565. Coenen included the pamphlet itself in

[5] See *Visboock* fol. 49v–50r, and *Walvisboock* I fol. 15r. Cornelis Suys (1514–1580), knight and lord of Rijswijk, was president of the Court of Holland from 1559 to 1572.

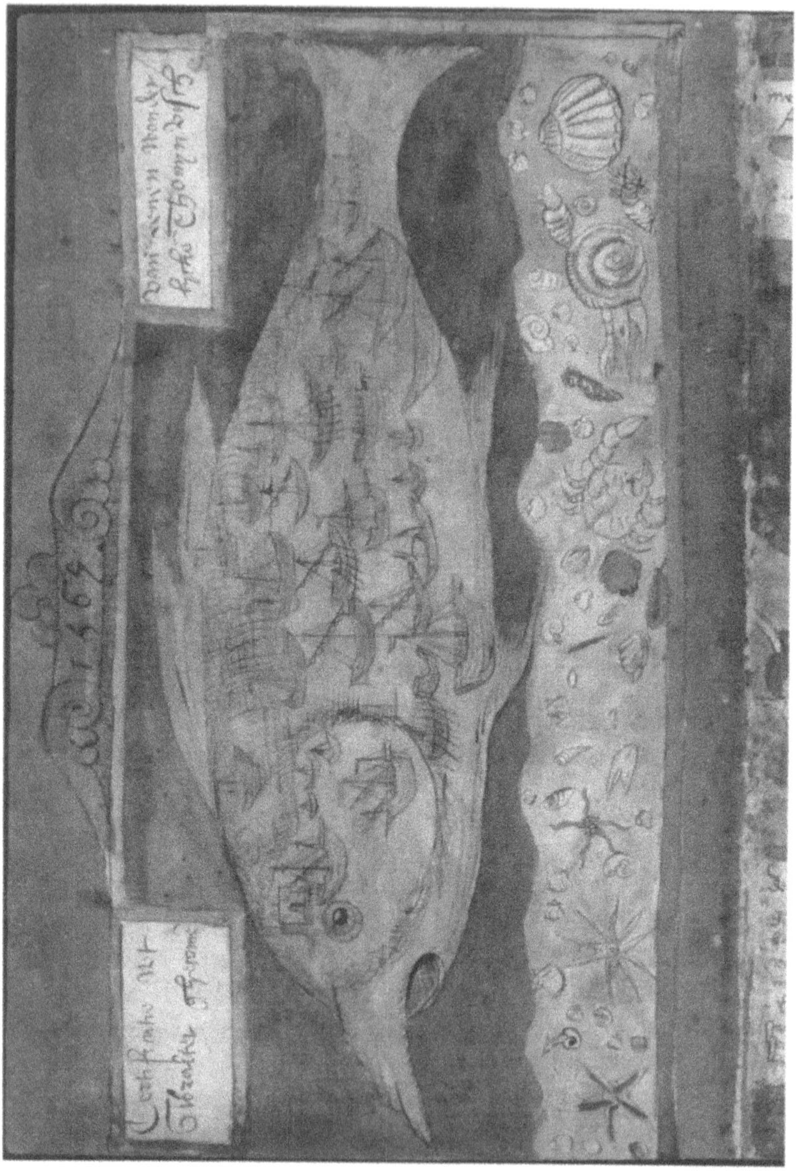

Fig. 2 [Col. pl. IX]. Tattooed tuna. From Adriaen Coenen, *Visboock*, fol. 49v. Koninklijke Bibliotheek, HS. 78 E 54. Courtesy of the Koninklijke Bibliotheek, The Hague.

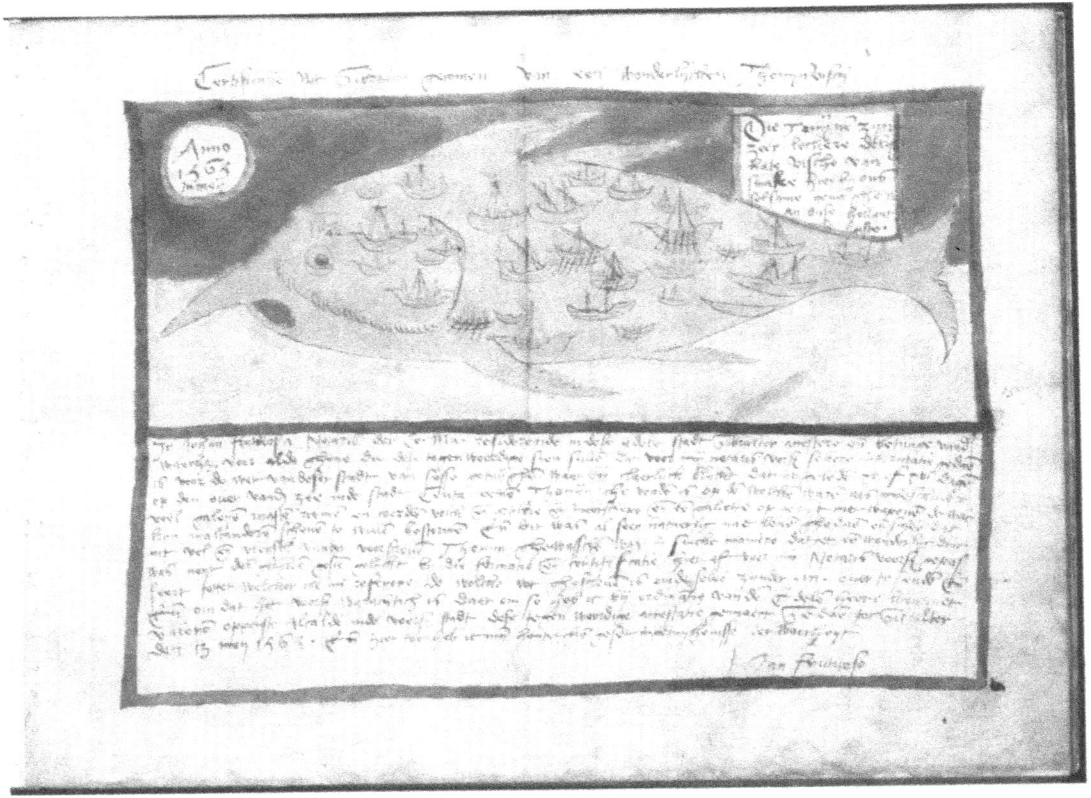

Fig. 3 [Col. pl. X]. Tattooed tuna. From Adriaen Coenen, *Walvisboock* I, fol. 15r. Courtesy of the Koninklijke Maatschappij voor Dierkunde, Antwerp.

an album (now lost) which he was making at that time and which he gave to the leader of the Dutch Revolt, prince William of Orange, in 1574. But he copied both text and illustration for later use, and thus included his own versions of them in his *Visboock* and *Walvisboock*.

There is at present only one copy extant of a German edition of the Dutch pamphlet to which Coenen refers as his source. At the time both versions must have circulated. This single extant copy of the (German) pamphlet was printed in Augsburg and can be found in the sixteenth-century Wickiana collection in Zurich which belonged to the Zurich parson Johann Jacob Wick (1522–1588).[6] It has the woodcut by Matthäus Franck from 1565. It is probably that the Dutch and the German versions of this pamphlet bore the same illustration, and the picture by Franck illustrated here thus was in all likelihood the source of Coenen's picture [Fig. 4]. As could be expected, both the fish in the printed pamphlet and Coenen's fish face left. The two fish look identical in every respect, including the slightly more elaborate 'castle' and the somewhat frayed tail – two very minor points in which they differ from the *Libri Picturati* fish.

Quoting from the printed text in the pamphlet, Coenen relates that it depicts a curious tuna fish which had been caught in the Mediterranean:

> Certificate from Gibraltar of a wondrous tunny, May 1565:
> I, Johan Frutuoso, royal notary, a resident of this noble city of Gibraltar, attest and witness to all who shall see this that it is the truth. Certain information has been given to me, the notary of the court of law of this city, by six witnesses clearly indicating that 15 or 16 days ago a tunny was found on the coast of the sea in the city of Ceuta on which are painted, as it were, many galleys, masts, oars, rowers, artillery and more vessels and an armed galliot which appeared to want to storm one another. This was all done in a very natural and realistic manner as if a drawing had been made in the skin and flesh of this tunny in a wondrous way, never seen before. Done in Gibraltar, 13 May 1565, and below I have set my signature as a witness to its truth, signed Johan Frutuoso.

This rare tuna fish must have been one of the famous natural curiosities of the sixteenth century or it would hardly have deserved a special pamphlet.

---

[6] Pictures of it can be found in Barthelmess K. – Münzing J., *Monstrum horrendum. Wale und Waldarstellungen in der Druckgraphik des 16. Jahrhunderts und ihr motivkundlicher Einfluss* (3 vols., Hamburg: 1991) 64–65; and Strauss W.L., *The German single-leaf woodcut 1500–1600. A pictorial catalogue* (3 vols., New York: 1975) 197. The Wickiana collection now belongs to the Zentralbibliothek in Zurich.

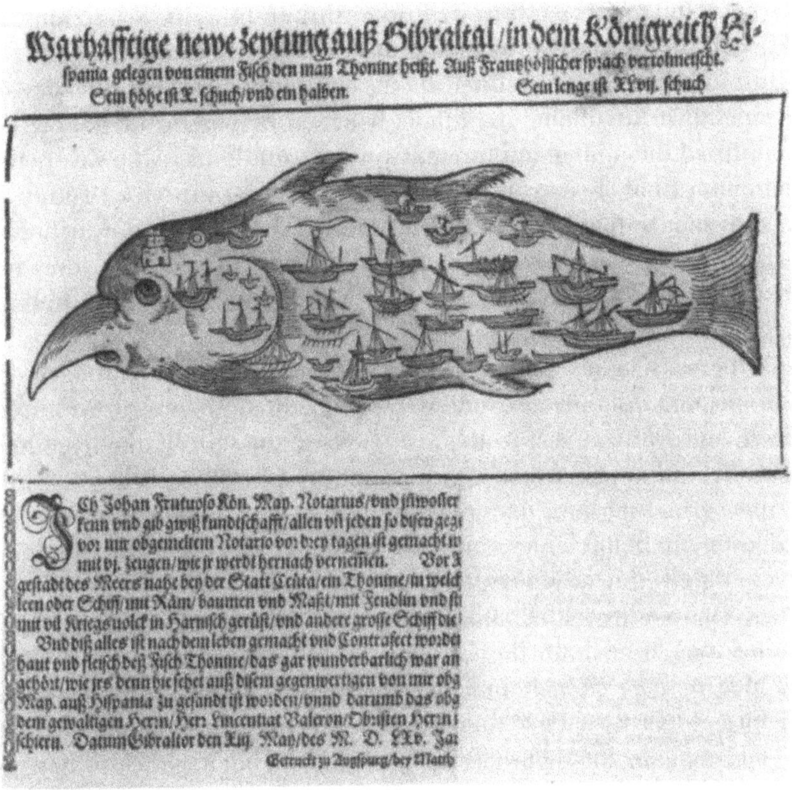

Fig. 4. Tattooed tuna. From a pamphlet dated 1565, with a woodcut by Matthäus Franck, printed in Augsburg. Original now in the Wickiana collection, Zentralbibliothek Zurich.

Via this circuitous route we now know what these curious watercolours in both collections represent. We can also speculate about how a picture of the tuna fish ended up in the *Libri Picturati*. The timing is interesting. Coenen's source, the Antwerp pamphlet, must have been printed soon after the tuna fish was discovered in May 1565. Both the Dutch version of the pamphlet printed in Antwerp and the German version printed in Augsburg probably circulated in various countries. Given the Southern Netherlandish origins of the *Libri Picturati*, it is possible that the *Libri Picturati* watercolour, like Coenen's, was made after the Antwerp version of the pamphlet. It is also conceivable that the very same copy of the pamphlet formed the basis for both, because the self-taught Adriaen Coenen and the aristocratic collector Charles de Saint Omer knew each other personally during the crucial years around 1565. Coenen even occasionally supplied Saint Omer with interesting naturalia for his collection. On 18 June 1565 Coenen spent the sum of six guilders to buy a huge live sea-tortoise from a shrimp fisher in Scheveningen.[7] He kept the tortoise at his home for a while, trying to keep it alive in a tub of water, but it refused to eat and eventually died after having been shown for money by others. Coenen thereupon disembowelled and dried the animal and presented it to Saint Omer, who was spending almost half a year in The Hague for a big court case.[8] Can it be a coincidence that the president of the Court of Holland, which dealt with this case, was Coenen's patron and friend Cornelis Suys, the same man who had obtained the pamphlet with the tuna fish from Brabant and given it to Coenen?

Whereas Coenen's role and the provenance of *his* picture of the curious tuna fish are clear and well documented, we are left with some loose ends with respect to the *Libri Picturati* tuna fish. If the hypothesis sketched above is correct, this tuna too must have been based on the Antwerp or Augsburg pamphlet and must have been painted specifically for the Saint Omer collection around 1565–66 in the Southern Netherlands. But why does the *Libri Picturati* tuna face right? And why does it have a frayed tail and slightly less elaborate castle 'tattoo' – differing thus from both the Coenen and the pamphlet tuna? Could it be that the *Libri Picturati* tuna was the original model for the pamphlet picture, while Coenen's tuna is modelled upon the pamphlet? That would explain the right-left reversal, and it might indicate that the printer or designer of the pamphlet changed some details (such as the frayed tail). On the other hand, it would presuppose a complex and tight-fitting chronology, for if the *Libri Picturati* watercolour was the source for the pamphlet illustration, we could expect it to have been made in or around May 1565 near the Mediterranean and very soon thereafter sent to Northern Europe to be both copied for the pamphlet illustrations and included in the *Libri Picturati*. Perhaps a future identification of the watermark of the *Libri Picturati* sheet will be able to give us an answer. And the situation might actually be considerably more complicated: perhaps the *Libri Picturati* tuna fish itself was painted after yet another – lost – French or Spanish pamphlet. In that case we would

[7] *Visboock* fol. 200v; identical text in Walvisboock II fol. 27r.

[8] For this court case see Nationaal Archief The Hague, Court of Holland, *Civiele Sententies* 537, Book 68 register K, fol. 223r: Charles de Saint Omer versus the heirs of Vincent Cornelisz., 26 January 1563 (1 page); and the more important verdict in National Archive The Hague, Court of Holland, *Civiele Sententies* 538, Book 69 register L, fol. 232r: Anna van Praet, mother of Charles de Saint Omer (and widow of Joos de Saint Omer) versus the heirs of Vincent Cornelisz., 22 December 1564 (32 pp.).

end up with an even more dizzying series of left-right, copy-original and print-watercolour reversals.

A giant moonfish

A second example of related images concerns the beautiful watercolour of a huge moonfish on a double page of *Libri Picturati.* A. 16 (fol. 15r). It bears no annotation but has a watermark dating the paper to 'not before 1554' [Fig. 5].[9] The moonfish is extremely rare in northern waters. It is an easy assumption, therefore, that this fish must have been observed in the Mediterranean. However, something completely different probably happened, and once more Coenen's illustrated manuscripts provide interesting information. In both his *Visboock* and his *Walvisboock* Coenen depicts and describes moonfish [Figs. 6 and 7].[10] He recalls that a local fisherman from Scheveningen caught a small moonfish in 1560, dried it aboard, and brought it back as a curiosity. A few years later, in December 1563, an enormous dead moonfish measuring 8 by 6 feet drifted ashore near the village of Scheveningen:

> Another time, on 13 December 1563, a dead fish of the same type drifted on to the beach of Scheveningen, about half a mile from the village. It was 8 feet in length and 6 feet in height. Once the entrails had been removed I had it loaded onto a wagon – which took the joint forces of eight strong fishermen. I planned to dry the skin and head, but it was a very wet winter and they rotted away. I had the fish portrayed and gave its portrait to the lord of Oosterwijk and the president [of the Court of Holland] master Cornelis Suys, who regarded it as a marvel and wondered very much whether something miraculous would happen – as indeed it did.[11]

There is no incontrovertible proof that the 1563 moonfish depicted by Coenen is the same exemplar as the one depicted in the *Libri Picturati* A. 16, since the fish in these pictures lack any individualizing characteristics. Yet, the extreme rarity with which such fish occur in northern seas combined with the dating of both pictures, makes it likely that they represent the same fish. As Coenen himself tells us, he ordered a painted portrait to be made of the huge moonfish and presented this

---

[9] About the watermark, see Ramón-Laca L., "Charles de l'Ecluse", appendix.
[10] *Visboock* fol. 150r and *Walvisboock* I fol. 14r.
[11] *Walvisboock* I fol. 14r; cf. *Visboock* fol. 151v.

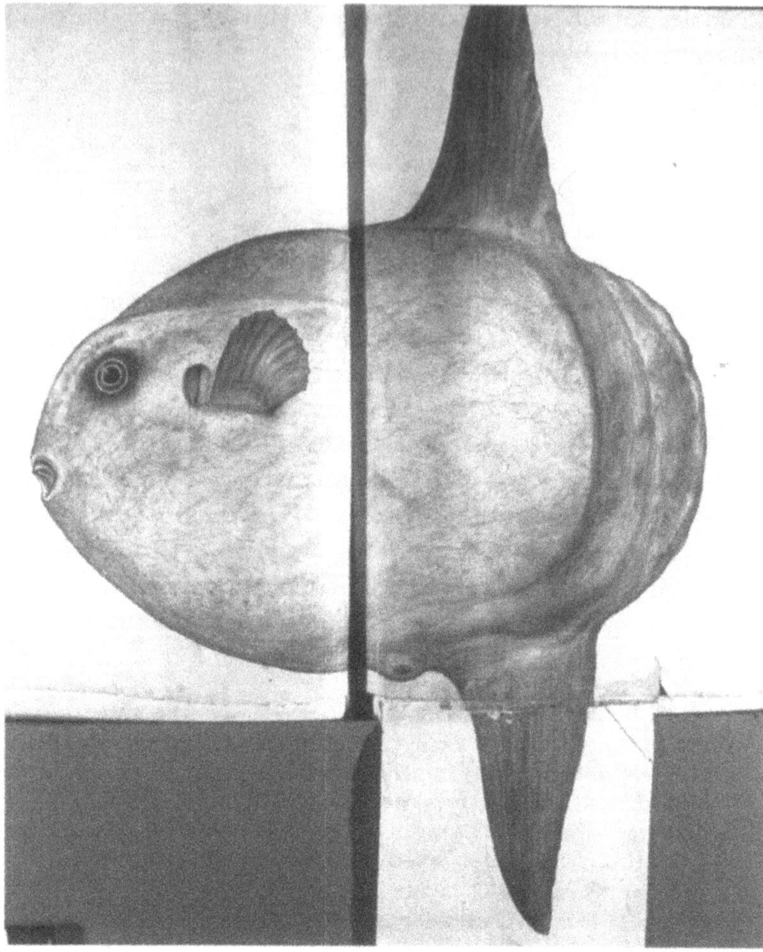

Fig. 5. Moonfish. From *Libri Picturati* A. 16, fol. 15r. Courtesy of Biblioteka Jagiellonska, Kraków.

to Cornelis Suys. Perhaps this portrait or a copy of it reached Saint Omer, who then had it copied by his own artists. The fish stranded in December 1563: it could have been depicted at the earliest in December 1563–January 1564, while Saint Omer's court case in The Hague and the above-mentioned exchanges of naturalia between Coenen, Saint Omer and possibly Suys took place during 1564–65. Coenen's information also demonstrates how the humble collector and practical expert Adriaen Coenen acted – sometimes unintentionally, sometimes self-consciously – as a key in the chain of transmission between aris-

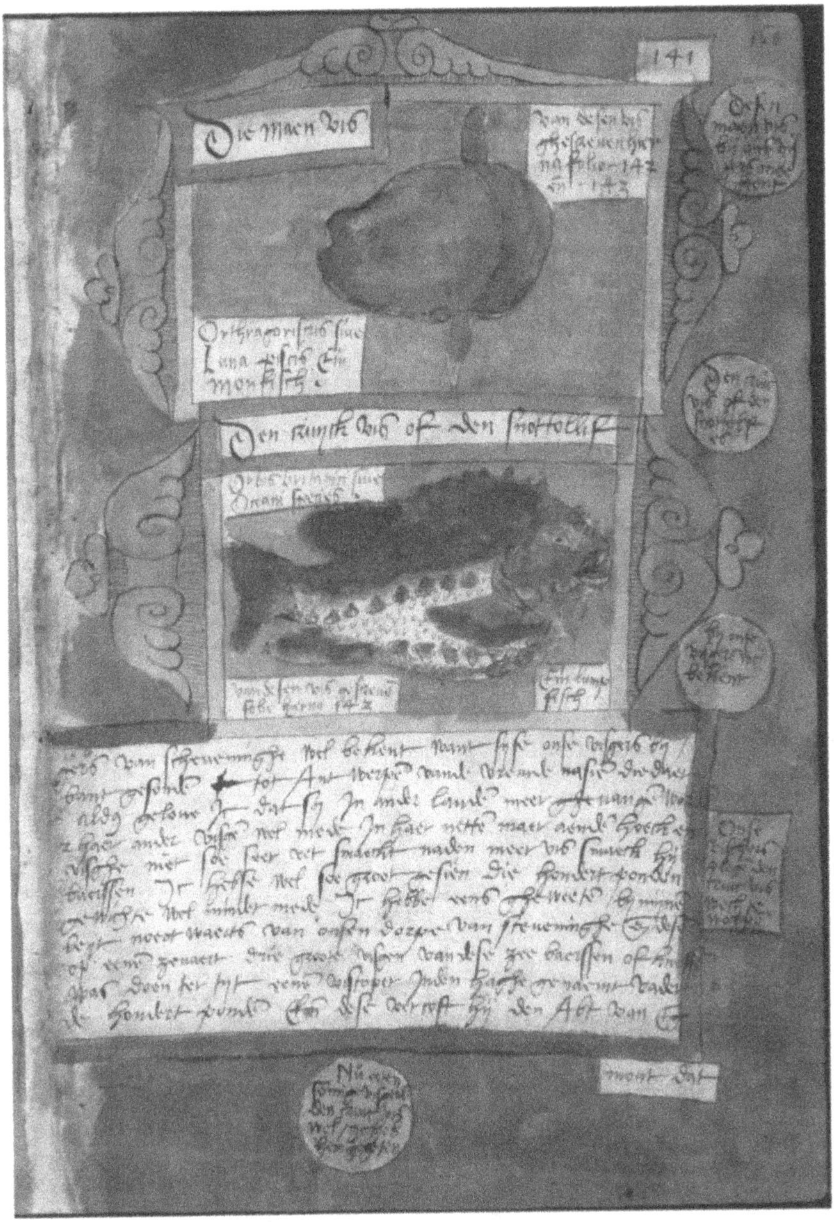

Fig. 6 [Col. pl. XI]. Moonfish. From Adriaen Coenen, *Visboock* fol. 150r. Koninklijke Bibliotheek, HS. 78 E 54. Courtesy of the Koninklijke Bibliotheek, The Hague.

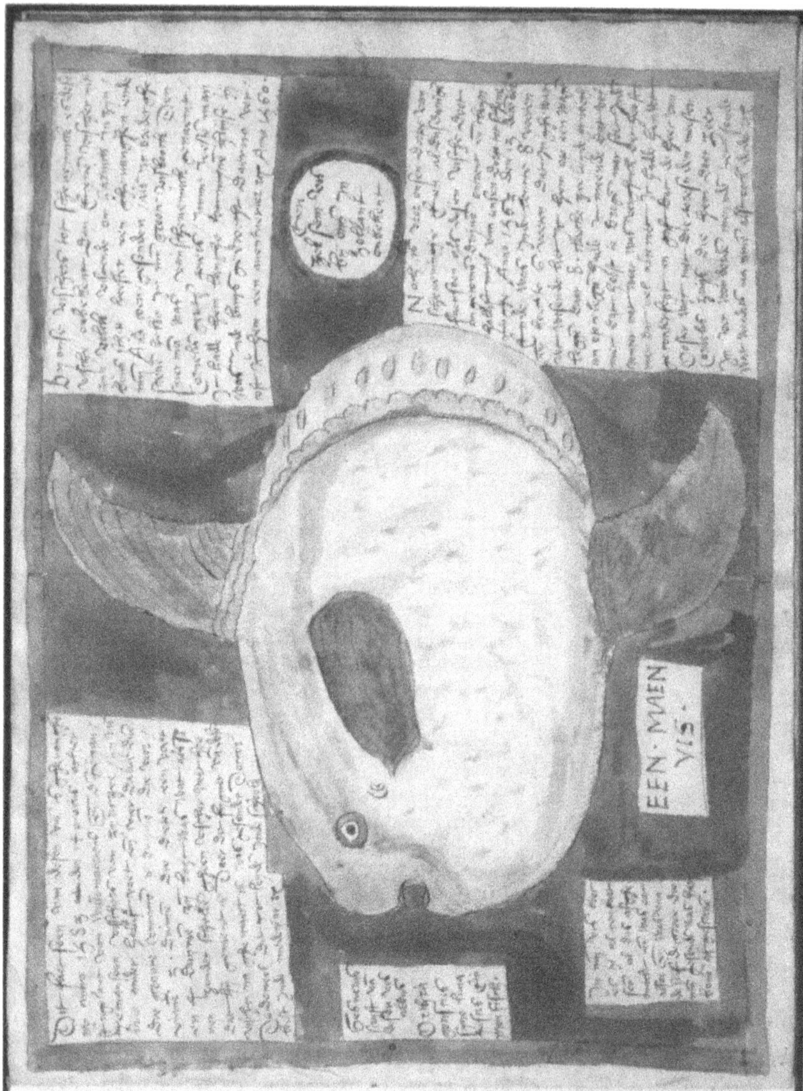

Fig. 7 [Col. pl. XII]. Moonfish. From Adriaen Coenen, *Walvisboeck* I, fol. 14r. Courtesy of Koninklijke Maatschappij voor Dierkunde, Antwerp.

tocratic and princely collectors: Cornelis Suys, Charles de Saint Omer and William Prince of Orange.

While the small picture of the massive moonfish of 1563 in Coenen's *Visboock* looks rather lumpy and brown, lacking detail, the moonfish of the *Libri Picturati* is somewhat more professionally depicted, especially if we look at the details of the fins, the eye and the more scalloped rear. Intriguingly, the moonfish included in Coenen's (later) *Walvisboock* is much more colourful (see fig. 7 above). It has a more pronounced dark fin to the right of the eye, more shading and especially a much more detailed and scalloped rear. The accompanying text makes clear that Coenen is depicting and describing yet another moonfish here – a much smaller one than the one of 1563. This new moonfish was discovered in 1583 off the coast of England by fishermen from the Dutch province of Zealand, several years after Coenen had finished his *Visboock* but just before he started working on his *Walvisboock*:

> This is an illustration of a fish caught on 14 October 1583 behind England by Willemaertens and Dingeman Temanszoon, fishermen from Zierikzee. He was 1.5 feet 4.5 thumbs long, the large fins 9 thumbs, the frontal fins 3 thumbs. The depth 1 foot 4 thumbs. His belly was like silver without scales. I have drawn this portrait from an illustration which was sent to Doctor Dodoens who teaches medicine at the university of Leiden.[12]

As Coenen explains, he exchanged information on unusual fish with the renowned botanist Rembertus Dodonaeus (Dodoens), who had been the physician of the emperors Maximilian II and Rudolph II at Vienna from 1574 to 1578.[13] Dodonaeus spent the last years of his life (1583–85) as a professor of medicine at the recently founded university of Leiden. Coenen too was by then living in Leiden, and both old men must have taken a real pleasure in exchanging information and discussing their lifelong interest in nature. Clearly, Coenen never saw the actual moonfish of 1583 himself, since it was discovered near England and brought to land in Zealand, but he used the picture sent to Dodonaeus as a model for his own drawing.[14]

---

[12] *Walvisboock* I fol. 14r.

[13] For Dodonaeus' life and work, see Meerbeeck P.J. van, *Recherches historiques et critiques sur la vie et les ouvrages de Rembert Dodoens* (Malines: 1841; reprint 1980).

[14] For illustrations and a brief discussion of some drawings owned by Dodonaeus and probably made by Coenen, see the entries by Lodewijk Wagenaar in the exhibition catalogue: Bergvelt E. – Kistemaker R. (eds.), *De wereld binnen handbereik. Nederlandse kunst- en rariteitenverzamelingen, 1585–1735* (2 vols., Zwolle: 1992) 133–134. The exhibition presented drawings/watercolours of an octopus, a moonfish and bottlenose

Coenen must have been an extremely lucky man to encounter three exemplars of such a rare fish in the course of some twenty years. Coenen was also one of the first Europeans to depict them.[15] We know of one, earlier illustration of the moonfish in a printed work by a well-known French expert, which Coenen certainly knew: Guillaume Rondelet's *Libri de Piscibus Marinis*, 1554.[16] Yet, the fish in this printed work is 'anonymous', while each of the Coenen moonfish is a unique and identifiable exemplar thanks to his accompanying text. It is the intimate and indelible connection between text and illustration that turns Coenen's three moonfish into real *portraits* of individual fish – even though they lacked special identifying marks such as the tattooed tuna. The latter is unmistakable on account of its appearance alone.

Depicting animals

Having started with the connections between these pictures of the tattooed tuna and the moonfish, we should widen our inquiry to take a brief look at the immediate context of which they form part – and in particular at the character and making of the two collections.

Perhaps surprisingly, much more is known about the albums of the modest Dutch painter of marine life than about the zoological watercolours of the *Libri Picturati*. This is mainly because Coenen himself provides so much information about his life and manner of working in his texts that we can trace many of his (pictorial and textual) sources. The *zoological* volumes of the *Libri Picturati*, on the contrary, lack almost any annotation other than brief references to the name of the animal. In the volume concerning fish and beasts (A. 16) there are names in French for some of the fish (if any at all), and a few brief remarks concerning the beasts, one of which contains the date 1565 and a

---

whale on loose sheets. We suspect that the latter two and possibly all three were made by Coenen. Cf. Engel H., "Over enkele tekeningen, die eens Dodonaeus toebehoord hebben, waarbij de eerste koningsvisch (lampris luna) van de Europeesche kusten", *Zoölogische Mededelingen* 25 (1945) 9–22.

[15] But not the very first, as has erroneously been claimed by Bergvelt – Kistemaker (eds.), *De wereld binnen handbereik*, II 133.

[16] A second example may occur in François Bossuet, *De Natura Aquatilium Carmen*, 1558, but I have not been able to consult this work and the reference may not be reliable.

reference to André Thevet's *Les Singularitez de la France antarctique, autrement nommée Amerique*, Paris, 1557. The 'bird volume' (A. 17) similarly contains only one direct reference to a date: to Pierre Belon's *L'Histoire de la Nature des Oyseaux* which was first published in French in 1555.[17] The *Libri Picturati* thus offer very few clues that can assist us in the attempt at attribution and identification. This may be one of the reasons why research concerning the zoological volumes of the *Libri Picturati* is far less advanced than that concerning the botanical ones. Both old and recent publications about the *Libri Picturati* almost exclusively concern the botanical volumes, and almost none of the fish and four-footed beasts in Vol. A. 16 or of the large number of beautifully depicted birds in Vol. A. 17 has ever been published.[18] The study of animals and their pictorial representations in history has, in fact, generally attracted far less interest on the part of cultural historians, historians of science and art historians than that of plants. The latter have always been thought to be of greater use as *materia medica*, while the difficulties and costs involved in the preservation of (parts of) animals for purposes of study were much greater than for plants. Modern disciplinary boundaries have played their part as well: the divisions between the humanities, arts and sciences and within the sciences between botany and zoology – not to name their many subdivisions – have hardly helped to maintain the unity of Renaissance 'natural history'.

On the basis of research published more fully elsewhere we can nonetheless say a bit more about these animal volumes of the *Libri Picturati*, making clear why and where they originated, even if not yet by whom they were painted.[19] Again the evidence provided by Coenen plays an important role. Since most of the zoological watercolours of the *Libri Picturati* collection are painted on paper with the same watermark (dateable to not earlier than 1554) as the majority of the botanical ones, it is clear that they originated as part of the same collection. That implies that this collection was conceived from the start as a collection of pictures of *naturalia*. Coenen tells us that he personally presented *some* (dried) seabirds or *procellaria* to Charles de Saint Omer in 1564 or

[17] *Libri Picturati* A.17, fol. 64v. Comparison with other European 'bird' volumes could be interesting. See for instance Kinzelbach R.K. – Hölzinger J. (eds.), *Marcus zum Lamm (1544–1606). Die Vogelbücher aus dem Thesaurus Picturarum* (Stuttgart: 2000).

[18] For an exception see Wille H., "Der Strauss von Joris Hoefnagel", *Scientiarum historia* 21 (1995) 35–41.

[19] See Egmond, "Clusius, Cluyt, Saint Omer".

1565, when the latter spent some months in The Hague in connection with his court case. In this context he mentions that Saint Omer 'was having a book made of all birds'.

> I had these malefijters [procellaria] and sent them to the seigneur of Renouteren, lord of Moerkercke, who was having a book made of all birds, this lord lived in Flanders.[20]

Saint Omer's book comprised 'all' birds and was a typical sixteenth-century encyclopaedic manuscript of the kind that was also being illustrated in painting programmes such as that of the Villa Medici in Rome, thus showing some unexpected similarities with Coenen's more modest encyclopaedic and illustrated albums.[21]

Jacques de Groote's discovery (in 2003–04) of an extensive inventory detailing the possessions of Charles de Saint Omer upon his death in 1569 shows that he also had a menagerie and that he was an extremely important and rich collector.[22] He owned extensive gardens and parks, besides the estate and castle of Moerkercke near Bruges, farms, mills, various feudal landholdings and rights, a town house in Bruges, et cetera. Around the castle of Moerkercke there were gardens, an orchard, and a menagerie with horses, a mule, sheep, goats, swans, pheasants, 'african chickens', an eagle, a stork, a bear, a parrot and some other animals. Indoors a rich wardrobe, jewellery, many tapestries (including verdures), silverware, and paintings testify to his wealth. Possessions specifically pointing to Saint Omer's interests as a collector were his cabinet, containing *singularités*, his collection of weapons and books and the 'cabinet' that is listed explicitly as belonging to his wife Anne d'Ongnies. This cabinet included several (unnamed) paintings, and especially:

---

[20] *Visboock* fol. 113v.

[21] On the painting programme of the Villa Medici see Morel Ph., "Jacopo Zucchi al servizio di Ferdinando de' Medici", in Hochmann M. (ed.), *Villa Medici. Il sogno di un cardinale. Collezioni e artisti di Ferdinando de' Medici* (Rome: 1999) 115–122, especially 117.

[22] De Groote has published major parts of this list (in French) on his website (www.tzwin.be). The original is in Rijksarchief Bruges, Belgium, Family Archive 367. This inventory was presented on 25 July 1569 by Saint Omer's widow Anne d'Ongnies to the court of the Vrije of Bruges (the district surrounding Bruges).

ung couffre ou il y a dedens quatre livres dherbes paintes
ung commenchement dung livre de oyseaulx
ung commenchement dung livre des poissons et aultres animaulx

as well as a large chest in which seeds were kept. This evidence clearly demonstrates that the fish/beasts volume and the bird volume as well as the botanical volumes of the *Libri Picturati* collection originated with Saint Omer and that the two volumes of zoological watercolours were separate from the botanical ones right from the start. Saint Omer's books remained unfinished on account of his untimely death and his plans for publication came to nothing.[23]

Coenen's albums present us with a mixture of an 'encyclopaedic' attempt to describe, chart and (up to a point) classify marine nature, while they form at the same time a 'poor man's collection on paper'. Charles de Saint Omer was a very rich collector. A number of his watercolours may have served as 'documentation' of plants grown in Saint Omer's own garden and of birds and some 'curious' animals in his part of the world. As the evidence of the tuna fish and moonfish shows, however, by no means all watercolours depicted animals or plants that actually belonged to Saint Omer. The watercolours therefore did not simply constitute a catalogue of his own *naturalia*, but formed an integral part of Saint Omer's collection of *naturalia* and as a painted collection on paper complemented his live and dead animals and plants, on the one hand, and his printed works about natural history, on the other. The *Libri Picturati* watercolours may also have served as a kind of home encyclopaedia: a collection in which pictures, copies of pamphlets about curious animals and exotic plants sent to Saint Omer or his friends could be included. Given the fact that Saint Omer's collection comprised art, coins, weapons, and books as well as naturalia in many forms, it was a typical – albeit early – representative of the well known *Kunst- und Wunderkammer*.

Neither Saint Omer's collection of watercolours nor Coenen's collection on paper was created in a simple or single way. As Coenen tells us, several painters formed part of Saint Omer's household. On stylistic grounds the hands of at least five different artists have been distinguished in the *Libri Picturati* collection, but we cannot be certain that all of them were actually living at Saint Omer's court. Almost certainly one of the

---

[23] For a more extensive discussion of the history of this collection see Egmond, "Clusius, Cluyt, Saint Omer".

Libri Picturati painters was Jacques vanden Corenhuyse (born 1529/30), who had been admitted to the St. Lucas guild of Bruges in 1554 and is also mentioned in the register of deaths of Bruges after 1584. The text of Lobelius' famous *Kruydtboeck* (1581) even explicitly informs us that Corenhuyse painted a particular plant under commission to Saint Omer.[24] He must have been very close indeed to the Saint Omer family and may even have been related by kinship.[25] The inventory made up after Saint Omer's death states that Corenhuyse painted the blazons for the funeral. A less strong candidate is the Flemish painter and draughtsman Peter van der Borcht who was discovered in Malines by the botanist Dodonaeus; he worked for Plantin from 1564 and was paid by Plantin to produce more than 3,180 botanical watercolours. Woodcuts after some of these drawings were made by three of Plantin's regular woodcutters whom we know by name: Arnold Nicolaï, and later Gerard van Kampen and Cornelis Muller.[26] These and others must have been instructed to depict many plants and animals in Saint Omer's collection. However, he clearly also included watercolours in his collection which came from elsewhere, but were curious or beautiful enough to fit in, and it is certain that after his early death more watercolours were being added to the core collection.

Coenen, on the other hand, made all of the pictures in his albums himself and he drew many of them after life – or, to be more exact, after dead but real fish and marine animals. But he also frequently made copies after (black and white) illustrations in printed works or pamphlets, usually adding details of his own and always embellishing the picture by adding colour and details. Thereby he turned it into something other than a simply copy. Why Coenen made copies after

[24] These findings are mainly the result of the research by Helena Wille. See especially Wille H., "Der Strauss von Joris Hoefnagel", *Scientiarum historia* 21 (1995) 35–41; eadem, "The discovery of the scientific heritage of Karel van Sint Omaars (1533–1569). Libri picturati A 16–30 in the Jagiellon library in Kraków", *Scientiarum historia* 22 (1996) (67–80) 75–78; and eadem, "The albums of Karel van Sint Omaars (1533–1569) (Libri Picturati A 16–31, in the Jagiellon library in Kraków)", *Archives of Natural History* 24 (1997) 423–437.

[25] Some sixty or seventy years before Charles de Saint Omer's death, a daughter of a certain Maillard van Corenhuyse (the local lord of that town) married the son of a certain Jeanne de Saint Omer, who must be related to Charles. See Gaillard J., *Bruges et le Franc ou leur magistrature et leur noblesse avec des données généalogiques sur chaque famille* (6 vols., Bruges: 1857–64), here II, 161.

[26] For quotations of the textual evidence see Ramón-Laca L., "Charles de l'Ecluse" 205–206.

printed illustrations by other authors even when he was perfectly familiar with the real fish and actually had no need to copy, is not always clear. It does tell us, however, that his pictorial sources were manifold.[27] As briefly indicated above, he occasionally hired a local painter to paint a portrait of a very rare but fast-rotting 'fish' in his possession, such as a moonfish and a seven-foot squid or *poelomp*. It is unusual to find information about such local painters, let alone about their being hired by a person who was himself neither a member of the elite nor a naturalist.[28] Coenen describes how he had the giant squid brought from Scheveningen to his house in The Hague in 1546:[29]

> And the same evening I made a painter by the name of Cornelis Claeszn Sceveninc come to my house in order to portray it. Because it had an exceedingly strange appearance, he desired to have it at his own house in order to portray it all the better and therefore I had it put into a large tub or milkpail and in this manner carried to the painter, who would portray it on canvas as well as he could, but it was stranger than he was able to represent it.

Original paintings of this kind were clearly too big to be included in Coenen's albums; he gave at least one of them away to an important patron. Yet, it is very likely that Coenen's own pictures of these animals were influenced by the portraits he commissioned.

If even a member of the middle class like Adriaen Coenen could commission painters to portray animals, this practice cannot have been an exceedingly uncommon one at the time. Examples usually come from aristocratic circles, however, rather than from middle-class ones. In fact, portraits of individual animals (such as stags, wild boars, bears, dogs or horses) can often still be found as part of the decoration of aristocratic country houses and palaces in Europe, such as Schloss Ambras near Innsbruck, or Schloss Gottorf in Schleswig-Holstein. The hunting lodge Hoflössnitz near Dresden, which was built between 1648 and 1650 on the orders of Johann Georg I Elector of Saxony, is decorated with many 'zoological' paintings. On the ceiling of the central hall 80 Brazilian birds can be seen, while the bedroom and living room of the

---

[27] For a more detailed discussion of Coenen's (visual) sources and manner of presentation see Egmond, *Het Visboek*.

[28] On the sixteenth-century naturalist from Bologna Ulisse Aldrovandi and the painters who worked for him see Olmi G., *L'Inventario del mondo. Catalogazione della natura e luoghi del sapere nella prima età moderna* (Bologna: 1992) especially 61–90.

[29] *Visboock* fol. 46r.

elector of Saxony were decorated with sea-monsters and the portraits of especially impressive hunting trophies (wild boar, colossal beaver, stag, bear) which often detail the relevant date and location as well as further information concerning the animal in text that forms part of the painting.[30] Such zoological paintings are rarely of very high artistic quality, but as displays they can tell us much about the type of interest their owners took in these animals and they certainly should be studied together with other representations of animals in collections.

Contexts

Finally, we should raise the issues of the pertinent visual contexts in which zoological pictures such as the moonfish and tattooed tuna could or should be studied, and the possible relevance of such research. On the one hand, we may look for filiations: can we find more – earlier and later – pictures of these fishes, in order to trace influences in terms of pictorial traditions? On the other hand, we might look at differences and parallels between these pictures and others that belong to the same genre. The latter issue, however, immediately raises the question of genres and their characteristics: to what genre do these pictures actually belong?

A by no means exhaustive search for possible filiations among watercolours and printed illustrations of the sixteenth and seventeenth centuries has as yet yielded few results.[31] The search for moonfishes had to be focused in a *generic* way: although Coenen's three moonfishes can be individually identified thanks to the accompanying text, they do not show any peculiar or unmistakeable visual characteristics to distinguish them from other exemplars of their species. We have not found any pictures of or references to moonfishes at all before Coenen's own lifetime. The illustration of a moonfish in the printed work by Guillaume Rondelet (1554) appears to be the first European representation of a

---

[30] See Beschorner H., *Die Hoflössnitz bei Dresden* (Dresden: 1931). For further information about Ambras and Gottorf, and an analysis of the iconographical programme of the decoration of Hoflössnitz, see Mason P., "Eighty Brazilian birds for Johann Georg", *Folk. Journal of the Danish Ethnographic Society* 43 (2001) 103–121.

[31] An excellent tool for such searches, in so far as published works are concerned, is Barbero Richart M., *Iconografía animal, La representación animal en libros Europeas de Historia Natural de los siglos XVI y XVII* (2 vols., Cuenca: 1999). His text is less reliable.

moonfish.[32] Two more pictures of moonfish can be found in printed works of a later date: Ulisse Aldrovandi, *Serpentum et Draconum Historiae*, 1640 (which is based on his sixteenth-century collection) and Jan Jonston, *Historiae Naturalis de Piscibus* (1650). The latter based his moonfish in fact on Rondelet's picture. It may be a coincidence, but like so many other printed illustrations of animals and plants from the seventeenth century, these two moonfish look more awkward than their sixteenth-century predecessors. A search through the publications concerning collections of original sixteenth-century watercolours (by Albrecht Dürer, Georg Hoefnagel, Anselmus de Boodt and others) has not yet rendered any pictures of moonfish apart from the ones by Coenen and the *Libri Picturati* artist(s).[33] The search for filiations of the tattooed tuna had to be organized differently, precisely because of its non-generic, individual characteristics. Since the fish itself appeared in 1565, there is not much point in looking for illustrations before that year. The only picture we have found of it besides the Coenen and *Libri Picturati* watercolours and the Augsburg/Antwerp pamphlet discussed above, again belongs to the domain of printed illustration: Ulisse Aldrovandi included an identical tuna fish with tattoos of ships on its skin (facing right) in his *De Piscibus Libri V*, which was posthumously published in 1640. Its caption actually identifies it as the tuna of 1565.

The issue of pictorial sources and filiations in the early modern period is clearly a complicated and fascinating one, especially if we want to compare original watercolours or drawings rather than printed illustrations. However provisional, the present findings provide food for thought. By its sudden appearance in printed specialist works of the mid sixteenth-century, the moonfish seems to have jumped out of nothing straight into the domain of scientific illustration, to be 'individually' portrayed in colour only for the first time by Coenen and the *Libri Picturati* artists during the 1560s–1580s. It thus appears to have followed

[32] See Guillaume Rondelet, *Libri de piscibus marinis* (Lyon: 1554). An abbreviated French edition appeared in Lyon in 1558 under the title *L'histoire entière des poissons*.

[33] On De Boodt see Maselis M.C. – Balis A. – Marijnissen R.H., *De Albums van Anselmus de Boodt (1550–1632). Geschilderde natuurobservatie aan het Hof van Rudolf II te Praag* (Tielt: 1989). On Hoefnagel see Vignau-Wilberg Th., *Archetypa Studiaque Patris Georgii Hoefnagel. Natur, Dichtung und Wissenschaft in der Kunst um 1600/Nature Poetry and Science in Art around 1600* (Munich: 1994). For Dürer see Koreny F., *Albrecht Dürer and the Animal and Plant Studies of the Renaissance* (New York–Boston: 1988). Often, by no means all of the works in collections of watercolours have been published, so this is no more than a provisional result.

an inverted version of the 'normal' process, in which representations of animals found their way from original drawings or watercolours into print. In fact, it is more than likely that the printed illustration in Rondelet was made after an original drawing which is no longer extant or may even still await rediscovery. The fact that our search has as yet only led to *printed* illustrations of the tuna and moonfish probably reflects no more than the simple reality that access to non-printed visual material is much more difficult. The watercolour of the tattooed tuna in Coenen's work also demonstrates that it is no longer self-evident (if it ever was) that watercolours are always 'original' models for mechanically reproduced representations of animals. Drawings or watercolours can be copies too, albeit unique ones.

The question of relevant contexts and the uses of studying and comparing representations of animals are closely linked. On a purely practical level the examples of the moonfish and tuna show that a comparison at the level of the individual sheets can help enormously with the dating and origins of the collection of *Libri Picturati*. Vice versa, the *Libri Picturati* watercolours may throw light on other sixteenth-century naturalia and their pictures as well. Given the size of the collection in Kraków and the fact that most of the zoological watercolours have never been published, a comparison with the botanical and zoological representation of the sixteenth century has been virtually impossible up to now. It promises to be extremely interesting. Besides the moonfish and tuna fish, some other striking but as yet unexplained similarities can be observed: for instance between some pictures of four-footed beasts in the *Libri Picturati* and the Anselmus de Boodt collection.[34] And the comparison should obviously be extended to the works of Jacques de Gheyn II and Hendrick Goltzius.[35] Such comparisons may, however, not only serve to resolve practical problems concerning specific collections, their origins, background and attribution; they can also help to throw light on more general issues such as the history of collecting, relations between and practices of artists, scholars and collectors or patrons, and the history of visual representation.

---

[34] This especially concerns the lama with a collar, which looks like an inverted image of the one depicted by De Boodt, while a giraffe and coati(?) likewise wear collars. There is also a hare reminiscent of Dürer. De Boodt came from an elite Bruges family and it is likely that he would have known (of) Saint Omer during his young years. See Maselis – Balis – Marijnissen, *De Albums van Anselmus de Boodt*.

[35] For examples of their work see *Dawn of the Golden Age: Northern Netherlandish Art 1580–1620* [Exhibition Catalogue Rijksmuseum Amsterdam] (Zwolle: 1993).

Finally, the particular type of picture that has been discussed in this essay – watercolours and other animal portraits which offer potential links with 'paper collections', printed illustrations and *Kunst und Wunderkammern* – are also eminently suited to investigate matters of categorization and genre. It seems to be generally accepted that scientific illustration of the natural world is (or has become) 'scientific' precisely because it attempts to depict the generic and not the individual, while 'art' is usually associated with individualizing representation and 'portrait'.[36] Where plants and animals are concerned, scientific illustrations generally show the characteristics typical of that species (often presenting an ideal type) and deliberately attempt *not* to depict individual variation. In the process the link with 'reality' – in the sense of depicting a situation or state existing at a given moment in time – is often lost: plants can be depicted with flowers and fruits at the same time; the colour patterns of furs or feathers can be stylized and simplified. Many representations of animals and plants – such as the famous hare or 'large piece of turf' by Dürer – are classified as 'art' rather than scientific illustration on account of both their quality and the fact that they actually depict highly individual characteristics. Some of the sixteenth-century watercolours discussed in this essay are definitely individual portraits, while others from the same collections belong to the generalizing type. The artists (or the persons who instructed them) seem to alternate between generic and individualizing ways of depicting animals, and this should make us wonder about the categories of art and scientific illustration themselves. This is all the more relevant since the genre of scientific illustration only came into

[36] There is a considerable literature on this issue. To mention only a few stimulating publications: Ackerman J.S., "Early Renaissance 'naturalism' and scientific illustration", in Ellenius A. (ed.), *The natural sciences and the arts. Aspects of interaction from the Renaissance to the 20th century* [...] (Uppsala: 1985) 1–17; idem, "The involvement of artists in Renaissance science", in David Hoeniger F. – Shirley J.W. (eds.), *Science and the Arts in the Renaissance* (Cranbury – London: 1985) 94–129; Hall B.S., "The didactic and the elegant; some thoughts on scientific and technological illustrations in the Middle Ages and the Renaissance", in Baigrie B.S. (ed.), *Picturing knowledge. Historical and philosophical problems concerning the use of art in science* (Toronto: 1996) 3–39, and several of the other essays in the same volume; Ashworth W.B. Jr., "Natural history and the emblematic worldview", in Lindberg D.C. – Westman R.S. (eds.), *Reappraisals of the Scientific Revolution* (Cambridge: 1990) 303–332; and Ashworth W.B. Jr., "Emblematic natural history of the Renaissance", in Jardine N. – Secord J.A. – Spary E.C. (eds.), *Cultures of natural history* (Cambridge: 1996) 17–37. Two interesting books on natural history illustration (by Sachiko Kusukawa and Brian Ogilvie) had been announced but not yet published when this essay was written.

being during the sixteenth century while the notion itself of natural science as a scientific discipline was being constructed. The genre did not emerge all at once with all of its (later) characteristics present, and this is precisely why comparison of representations such as the Coenen watercolours, the *Libri Picturati* watercolours, animal portraits of the Hoflössnitz type, the Gessner watercolours, the Weiditz watercolours in the Platter herbarium[37] as well as the more famous ones by Dürer, De Boodt, Hoefnagel and others can turn out to be extremely revealing with regard to the emergence of a new genre.

[37] See the brilliant Rytz W., *Das Herbarium Felix Platters: ein Beitrag zur Geschichte der Botanik des 16. Jahrhunderts* (Basel: 1933). The Felix Platter herbarium is an exceptional case in which watercolours and (hand coloured) woodcuts are included in a late sixteenth-century set of herbarium volumes with dried plants. These watercolours are the work of Hans Weiditz and the original models of the printed illustrations in the famous *Herbarum vivae eicones* (Strasbourg: 1530) by Otto Brunfels. As Rytz has pointed out, many of these watercolours are highly individual portraits of plants.

Selective Bibliography

ACKERMAN J.S., "Early Renaissance 'naturalism' and scientific illustration", in Ellenius A. (ed.), *The natural sciences and the arts. Aspects of interaction from the Renaissance to the 20th century.* [...] (Uppsala: 1985) 1–17, republished with an addendum in Ackerman J.S., *Distance Points: Essays in Theory and Renaissance Art and Architecture* (Cambridge Mass.: 1991).

—— "The involvement of artists in Renaissance science", in David Hoeninger F. – Shirley J.W. (eds.), *Science and the Arts in the Renaissance* (Cranbury – London: 1985) 94–129, republished with an addendum in Ackerman J.S., *Distance Points: Essays in Theory and Renaissance Art and Architecture* (Cambridge Mass.: 1991).

ALDROVANDI ULISSE, *Serpentum et draconum historiae* (Bologna: 1640).

ASHWORTH W.B. Jr., "Natural history and the emblematic worldview", in Lindberg D.C. – Westman R.S. (eds.), *Reappraisals of the Scientific Revolution* (Cambridge: 1990) 303–332.

——, "Emblematic natural history of the Renaissance", in Jardine N. – Secord J.A. – Spary E.C. (eds.), *Cultures of natural history* (Cambridge: 1996) 17–37.

BARBERO RICHART M., *Iconografia animal, La representación animal en libros Europeas de Historia Natural de los siglos XVI y XVII* (2 vols., Cuenca: 1999).

BARTHELMESS K. – MÜNZING J., *Monstrum horrendum. Wale und Waldarstellungen in der Druckgraphik des 16. Jahrhunderts und ihr motivkundlicher Einfluss* (3 vols., Hamburg: 1991).

BELON PIERRE, *L'Histoire de la Nature des Oijseaux* (Paris: 1555; 1st Latin edition 1557).

BERGVELT E. – KISTEMAKER R. (eds.), *De wereld binnen handbereik. Nederlandse kunst- en rariteitenverzamelingen, 1585–1735* (2 vols., Zwolle: 1992).

BESCHORNER H., *Die Hoflössnitz bei Dresden* (Dresden: 1931).

Dawn of the Golden Age: Northern Netherlandish Art 1580–1620 [Exhibition Catalogue Rijksmuseum Amsterdam] (Zwolle: 1993).

EGMOND F., "Clusius, Cluyt, Saint Omer. The origins of the sixteenth-century botanical and zoological watercolours in *Libri Picturati* A. 16–30", *Nuncius* 20 (2005) 11–67.

——, *Het Visboek. De wereld volgens Adriaen Coenen 1514–1587* (Zutphen–The Hague: 2005).

—— – MASON P., *The Whale Book. Whales and other marine animals as described by Adriaen Coenen in 1585* (London: 2003).

ENGEL H., "Over enkele tekeningen, die eens Dodonaeus toebehoord hebben, waarbij de eerste koningsvisch (lampris luna) van de Europeesche kusten", *Zoölogische Mededelingen* 25 (1945) 9–22.

GAILLARD J., *Bruges et le Franc ou leur magistrature et leur noblesse avec des données généalogiques sur chaque famille* (6 vols., Bruges: 1857–64).

HALL B.S., "The didactic and the elegant; some thoughts on scientific and technological illustrations in the Middle Ages and the Renaissance", in BAIGRIE B.S. (ed.), *Picturing knowledge. Historical and philosophical problems concerning the use of art in science* (Toronto: 1996) 3–39.

JONSTON JAN, *Historiae naturalis de piscibus* (Frankfurt a.M.: 1650).

KINZELBACH R.K. – HÖLZINGER J. (eds.), *Marcus zum Lamm (1544–1606). Die Vogelbücher aus dem Thesaurus Picturarum* (Stuttgart: 2000).

KONING J. de – ZEMANEK A. – ZEMANEK B. (eds.), *Botanical illustrations of the sixteenth century. An enquiry into the Libri Picturati A.18–30* (forthcoming, 2007–2008).

KORENY F., *Albrecht Dürer and the Animal and Plant Studies of the Renaissance* (New York – Boston: 1988).

MASELIS M.C. – BALIS A. – MARIJNISSEN R.H., *De Albums van Anselmus de Boodt (1550–1632). Geschilderde natuurobservatie aan het Hof van Rudolf II te Praag* (Tielt: 1989).

MASON P., "Eighty Brazilian birds for Johann Georg", *Folk. Journal of the Danish Ethnographic Society* 43 (2001) 103–121.

Meerbeeck P.J. van, *Recherches historiques et critiques sur la vie et les ouvrages de Rembert Dodoens* (Malines: 1841; reprint 1980).
Morel Ph., "Jacopo Zucchi al servizio di Ferdinando de' Medici", in Hochmann M. (ed.), *Villa Medici. Il sogno di un cardinale. Collezioni e artisti di Ferdinando de' Medici* (Rome: 1999) 115–122.
Olmi G., *L'Inventario del mondo. Catalogazione della natura e luoghi del sapere nella prima età moderna* (Bologna: 1992).
Ramón-Laca L., "Charles de l'Ecluse and Libri Picturati A. 16–30", *Archives of Natural History* 28 (2001) 195–243.
Rondelet Guillaume, *Libri de piscibus marinis* (Lyon: 1554; abbreviated French edition *L'histoire entière des poissons* [Lyon: 1558]).
Rytz W., *Das Herbarium Felix Platters: ein Beitrag zur Geschichte der Botanik des 16. Jahrhunderts* (Basel: 1933).
Schnalke T. (ed.), *Natur im Bild. Anatomie und Botanik in der Sammlung des Nürnberger Arztes Christoph Jacob Trew* [Exhibition catalogue] (Erlangen: 1995).
Strauss W.L., *The German single-leaf woodcut 1500–1600. A pictorial catalogue* (3 vols., New York: 1975).
Tevet Auguste, *Les Singularitez de la France antarctique, autrement nommée Amerique* (Paris: 1557).
Vignau-Wilberg Th., *Archetypa Studiaque Patris Georgii Hoefnagel. Natur, Dichtung und Wissenschaft in der Kunst um 1600/Nature Poetry and Science in Art around 1600* (Munich: 1994).
Wille H., "Der Strauss von Joris Hoefnagel", *Scientiarum historia* 21 (1995) 35–41.
——, "The discovery of the scientific heritage of Karel van Sint Omaars (1533–1569). *Libri picturati* A 16–30 in the Jagiellon library in Kraków", *Scientiarum historia* 22 (1996) 67–80.
——, "The albums of Karel van Sint Omaars (1533–1569) (*Libri Picturati* A 16–31, in the Jagiellon library in Kraków)", *Archives of Natural History* 24 (1997) 423–437.

FROM BRAZIL TO EUROPE: THE ZOOLOGICAL DRAWINGS OF ALBERT ECKHOUT AND GEORG MARCGRAF

Rebecca Parker Brienen

Discovered in 1500 by the Portuguese, Brazil was almost immediately christened the 'land of parrots', with early travellers claiming that blue skies would darken as if night had fallen when huge, raucous flocks of the birds flew overhead. But the Europeans who spent time in this Portuguese (and, for a short time, Dutch) colony were not satisfied with simply looking at the wondrous, colourful, and odd animals around them; they often brought specimens back with them, including parrots, armadillos, coati, lizards, and even small monkeys. The overseas journey, however, spelled certain death for most of the animals: even if they weren't eaten by hungry sailors, several months of inadequate care, inappropriate food, and a radical change in climate meant that only the most robust and adaptable animals could survive.[1] Nonetheless, dried animal remains were highly sought after by many collectors of *naturalia* throughout Europe, and it was well-known that harbour cities such as Amsterdam were excellent places to obtain such materials.

In his overview of cabinets of curiosities assembled by Dutch burghers between 1585 and 1735, Roelof van Gelder describes the obvious enthusiasm for *naturalia*, the majority of it non-native, demonstrated by the contents of such collections, with most of the specimens coming from areas of VOC and WIC trade and colonization, including the East Indies, South Africa, Brazil, Surinam, and New Netherland.[2] Of

---

[1] Jean de Léry describes his arduous trip back to France from Brazil, during which monkeys and parrots, intended to be exotic pets, were instead consumed as food. See his 1578 *History of a Voyage to the Land of Brazil*, trans. J. Watson (Berkeley: 2000) 208–216. For other fascinating accounts of the difficulties encountered when importing exotic species into Europe, see Robbins L., *Elephant Slaves and Pampered Parrots: Exotic Animals in Eighteenth-Century Paris* (Baltimore: 2002) 9–37.

[2] Gelder R. van, "De wereld binnen handbereik: Nederlandse kunst- en rariteitverzameling, 1585–1735", in Bergvelt E. and Kistemaker R. (eds.), *De wereld binnen handbereik. Nederlandse kunst- en rariteitenverzamelingen, 1585–1735* (Zwolle – Amsterdam: 1992). See also, Smit P. (ed.), *Hendrik Engel's Alphabetical List of Dutch Zoological Cabinets and Menageries* (Amsterdam: 1986).

course, the Dutch were not the only collectors interested in flora and fauna from the Americas; as noted by Wilma George in her analysis of printed catalogues of European zoological cabinets from the second half of the seventeenth century, the armadillo—or rather its dried carapace—was the most common animal in such collections, and after Europe, the majority of the specimens came from South America.[3] This does not mean, however, that everyone was satisfied with dried specimens from the New World. Living South American animals were highly desirable because of their unusual appearance, rarity, or beauty and, for obvious reasons related to wealth and position, they frequently ended up in the menageries of European nobility.

Collections of *naturalia* often also included albums of drawings or single images of flora and fauna that served to complement preserved remains, but only the highest ranking members of the European elite were able to commission artists to make images after the living, non-European animals in their own private zoos.[4] Holy Roman Emperor Rudolf II's 'book of animals', or *Museum*, from the late sixteenth century, is a well-known example of this type of visual production. That this was not an isolated example is made clear by the images of South American animals belonging to the Grand Duke of Tuscany, including parrots and an agouti, which were created by Italian artist Jacopo Ligozzi in the 1580s, or the oils studies on canvas after the exotic animals in Louis XIV's menagerie at Versailles by Flemish artist Pieter Boel. Although such images were put to a variety of uses, their contemporary value seems clear: while stylistically divergent, these drawings and paintings preserved the exterior appearance, including the colour and form, of the exotic animal. As such, they functioned as a permanent visual record of both the animal and, with respect to the corpus of images, the menagerie as a whole. They furthermore highlighted the collector's ability to assemble and possess wondrous and exotic creatures from afar. But how are we to assess the value of images like these if the menagerie, the noble collector, and his artists, were not in Europe, but in the New World as part of a European colonial presence?

---

[3] George W., "Alive or Dead: Zoological Collections in the Seventeenth Century", in Impey O. – MacGregor A. (eds.), *The Origins of Museums: the Cabinet of Curiosities in Sixteenth- and Seventeenth-Century Europe* (Oxford: 1985) 179–187.

[4] See, for example, discussion of William Courten's collection practices and his patronage of scientific illustration in England in the late seventeenth century in Gibson-Wood C., "Classification and value in a seventeenth-century museum: William Courten's collection", *Journal of the History of Collections* 9,1 (1997) 61–77.

This brings us to the case of the German count Johan Maurits van Nassau-Siegen, who was governor-general of Dutch Brazil from 1637–1644. As soon as he was appointed to this position by the directors of the WIC, Johan Maurits determined that the investigation, collection, and representation of Brazil would be a priority during his governorship. Before he left the Dutch Republic, he hired the painters Frans Post and Albert Eckhout and then brought them with him to Brazil to work as his court artists.[5] By 1638, they had been joined by the scientists Willem Piso and Georg Marcgraf (who also had artistic training), both of whom worked as 'nature describers' for the count; their collaborative work would be published in 1647 as *Historia naturalis Brasiliae*, the first natural history of Brazil, a work that remained authoritative until the nineteenth century.[6] Although scholarship on the works of art created at Johan Maurits's court has tended to focus on the oil paintings on canvas by Post and Eckhout, the artists and naturalists at Johan Maurits's Brazilian court also produced over 700 natural history drawings and oil studies on paper. These unique works, which were used for a variety of decorative and scientific purposes, were highly valued by contemporaries in Europe and became part of an elite gift exchange after Johan Maurits returned to the Dutch Republic in 1644.[7]

For the rest of this essay, I will focus on the zoological drawings created in Brazil by Dutch painter Albert Eckhout and the German naturalist-illustrator Georg Marcgraf, whose images make up the majority of this visual corpus. As will beome clear, Eckhout's oil studies and Marcgraf's watercolours drawings belong to separate visual traditions

[5] Literature on the work of these artists includes: Thomsen T., *Albert Eckhout, ein niederländischer Maler und sein Gönner Johan Maurits der Brasilianer* (Copenhagen: 1938); Larsen E., *Frans Post, interprète du Brésil* (Amsterdam: 1962); Boogaart E. van den (ed.), *Johan Maurits van Nassau-Siegen, 1604–1679. A Humanist Prince in Brazil and Europe* (The Hague: 1979); Whitehead P.J.P. – Boeseman M., *A portrait of Dutch 17th century Brazil: Animals, plants and people by the artists of Johan Maurits of Nassau* (Amsterdam: 1989); Brienen R., *Visions of Savage Paradise: Albert Eckhout, Court Painter in Colonial Dutch Brazil* (Amsterdam: 2006). The exhibition catalogue, Buvelot Q. (ed.), *Albert Eckhout: A Dutch Artist in Brazil* (The Hague: 2004) contains excellent reproductions and a useful overview of the literature by Buvelot.

[6] Regarding the importance of Marcgraf's zoological descriptions in the *Historia*, see Ashworth W., "Remarkable Humans and Singular Beasts", in Kenseth J. (ed.), *Age of the Marvelous* (Hanover: 1993) 113–144.

[7] For those who cannot travel to Poland, study of the drawings has been made more accessible by their publication in Ferrao C. – Soares J.P.M. (eds.), *Brasil-Holandês/Dutch-Brasil*, 5 vols. (Rio de Janeiro: 1995), which includes all of the Brazilian images, although the quality of the reproduction is quite variable.

and demonstrate diverging priorities in their encounter with and reproduction of Brazilian nature. Marcgraf, who may have been trained as a miniature painter, followed the descriptive style of the nature study, while Eckhout employed the dynamic and naturalistic vocabulary of a seventeenth-century Netherlandish still-life painter.[8] By addressing the conditions of production and the colonial function of their drawings, I hope to shed light on the continuties as well as the differences between Johan Maurits and his contemporaries in Europe with respect to their patronage of natural history illustration. This essay will close with a discussion regarding the use of the images by Marcgraf and Eckhout in Europe after 1644, when the Count returned from Brazil to the Dutch Republic.

The Brazilian Drawings: from Mauritsstad to Kraków

Created by Marcgraf and Eckhout in Brazil between 1637 and 1644, the images addressed in this essay formed part of a larger gift presented by Johan Maurits to his patron, Friedrich Wilhelm, the Elector of Brandenburg, in 1652, eight years after the Count's return from Brazil. The Brazilian drawings became part of the *Libri picturati* collection of natural history illustrations in the Elector's library, which later became the Royal Library in Berlin. During WWII, the *Libri picturati* collection and other important books and manuscripts were removed from Berlin, with the Brazilian volumes sent to Silesia and later Grüssau (Krzeszów) for safekeeping. They were transported to Kraków after the war and eventually found their way into the Jagiellon University Library. Outside of Poland, the Brazilian volumes of the *Libri picturati* and other important works from the Royal Library, including music manuscripts by Bach and Mozart, were thought to have been lost until the late 1970s, when they were incorporated into the Jagiellon library's general collection and made accessible to readers both from Poland and abroad.[9]

The first document to address these images is the anonymous 18 September 1652 contract for the transfer of Johan Maurits's gift to

---

[8] It is not known who trained either one of these artists.

[9] For a more complete discussion, see Whitehead – Boeseman, *A portrait of Dutch 17th century Brazil* 33–35 and Whitehead P.J.P., "The Treasures at Grüssau", *New Scientist* 94 (1982) 226–231. The Brazilian drawings have never travelled outside of Poland and are in desperate need of conservation.

Friedrich Wilhelm, which includes a complete inventory. Items fourteen and fifteen are described as follows:

> 14. A large book in royal folio and another somewhat smaller (in folio), in which everything that can be seen and found in Brazil (people, quadrupeds, birds, worms, fish, trees, herbs, flowers) with miniatures is artfully represented after life [*nach dem Leben*] added to which are (descriptions, names, and characteristics) names, qualities, and characteristics.
>
> 15. Yet more than (several) hundred other (Indian paintings of animals and all sorts of other subjects with oil paints) paintings with oil paint on paper, unbound.[10]

Between 1660 and 1664, the German scientist Christian Mentzel (1622–1701) gathered most of the loose images, described in item 15 above as including several hundred 'Indian' [i.e., West Indian] paintings of animals and other subjects in oils, into four large folio volumes, to which he gave the collective title *Theatrum Rerum Naturalium Brasiliae* (Theater of the Natural Things of Brazil). Mentzel was a highly learned botanist as well as the personal physician to the Elector of Brandenburg.[11]

Arranging the images into separate volumes by subject, Mentzel pasted the drawings onto new sheets of paper, sometimes cutting off parts of the original representations in his effort to fit them into the new format. Volume I (*Icones Aquatilium*) includes images of fish; volume II (*Icones Volatilium*) is limited to birds; volume III (*Icones Animalium*) includes images of people and quadrupeds; and volume IV (*Icones Vegetabilium*) is devoted to images of plants. The vast majority of the studies and chalk drawings in these four volumes can be attributed to Albert Eckhout, although it is clear that other hands are also present.[12] Mentzel's source

---

[10] See the inventory of the items („Verzeichnis derer im vorgedachten Akkord Uns von Sr. Lbd. Ueberlassenen Stuecken"), No. 14. Ein gross Buch in Royal-Folio und eins etwas kleiner (in Folio), worin alles was in Brasilien (von Menschen, vierfuessigen Thieren, Gevoegel, Gewuermen, Fische, Baeume, Kraeuter, Blumen) zu sehen und zu finden ist, mit Miniaturen kuenstlich nach dem Leben abgebildet ist, mit beigefuegten (Beschreibung, Nahmen und Eigenschaften) Nahmen, Qualitaeten und Eigenschaften. No. 15. Noch ueber (etzliche) hundert andere (Indianische Schildereyen von Thieren und allerhand Sachen mit Oelfarben) Schildereyen mit Oelfarben auf Papier, so nicht zusammen gebunden'. This inventory is reproduced by Larsen, *Frans Post, interprète du Brésil* 252–253, doc. No. 50, which is based on Driesen L., *Leben des Fürsten Johann Moritz von Nassau-Siegen* (Berlin: 1849), Beilage A 356. Cf. Driesen's discussion, 107–109. There is another version of this inventory in the KHA in The Hague.

[11] For Mentzel, see Artelt W., *Christian Mentzel: Leibarzt des Grossen Kurfürsten* (Leipzig: 1940).

[12] For a history of the attribution of these images to Eckhout, see Brienen, *Visions of Savage Paradise*.

for the Brazilian names, which he wrote at the top of the page above many of the images, as well as his basic system of classification, was Marcgraf and Piso's *Historia*, both the 1648 and the 1658 editions, in which the natural world is separated into plants, quadrupeds, birds, insects, and fishes and sea animals, a fairly standard system of classification by the seventeenth century.

In the 1652 inventory reproduced above, item 14 lists two volumes of 'miniatures artfully represented after life'. These books are immediately recognizable as the *Libri Principis*, also called the *Handbooks*. Unlike the oil studies and drawings from Brazil, the watercolour and body colour images in the *Handbooks* had already been gathered into separate volumes before Johan Maurits presented them to the Elector. These images are also unsigned, but Marcgraf's authorship of them has become well-recognized in the literature.[13] In 1652 these books included 'everything that can be seen and found in Brazil'. Today, however, they comprise a more limited selection of subjects, chiefly Brazilian birds, fishes, and other animals, and a handful of botanical images, which suggests that the images of people mentioned in the inventory were removed by Mentzel. It is nonetheless certain that no drawings from the *Handbooks* were incorporated into the *Theatrum*. As Mentzel stated with approval in his preface to *Theatrum* volume I, 'Very correctly, His Serene Highness the Elector forbade that such water colours were placed with our larger oil-painting collection, determining that some here and there would constitute their own "Theatre" separately'.[14] Watercolour images could create their own 'Theater of Brazilian Nature', but it was considered unacceptable to mix them with the drawings in oil on paper for reasons that will be addressed in greater detail at the end of this essay.

[13] See Brienen R.P., "Georg Marcgraf (1610–ca. 1644) A German Cartographer, Astronomer, and Naturalist-Illustrator in Colonial Dutch Brazil", *Itinerario* 20,1 (2001) 85–112.

[14] 'Tandem accedebat et hoc, quod in Celsissimi Principis Iconibus aliae, et praecipue duorum Librorum, coloribus quidem aquariis, sed egregie per punctulorum minutissima distinctae essent, quas coniungi cum illis maioribus nostris, olearibus coloribus depictis, summo accuratoque indicio vetabat, Serenissimus ELECTOR: Illas ergo, quae colorum essent oleariorum, THEATRO suo non minus separatim atque illas aquariorum colorum mandaret', Mentzel C., preface to *Theatrum rerum Naturalium Brasiliae*, vol I., as reproduced both in translation and in the original Latin (in facsimile) in *Brasil-Holandês/Dutch-Brazil*, vol. IV 22–23.

Johan Maurits's Brazilian Menagerie and Cabinet of Curiosities

In order to fully understand the history of these drawings, it is necessary to examine the conditions of their creation in Dutch Brazil. Although Mentzel asserts in his introduction to the *Theatrum* that the Brazilian oil studies were 'pictures of the things that Nature hath produced, nourished and nurtured, formed in their own original habitat and painted in colours exactly *after life* [ad vivum] in order to reproduce nature itself as perfectly as possible', it is nonetheless clear that the representations by Marcgraf and Eckhout do not record encounters with wild animals in their natural environment.[15] In fact, Johan Maurits's artists, like their European contemporaries, did not need to participate in dangerous expeditions into the interior to study Brazilian nature; they could instead examine plants and animals (certainly in greater numbers and varieties than could be found in Europe) at the governor's Brazilian residences Boa Vista and Vrijburg Palace, in his menagerie, cabinet of curiosities, and botanic gardens.

During the early modern period, menageries of exotic animals became an essential part of princely self-fashioning throughout western Europe. In the fifteenth century the powerful unofficial ruler of Florence, Lorenzo de' Medici, numbered parrots, apes, peacocks, and even a giraffe in his famous collection of animals.[16] Collections like this one were not simply for pleasure, although that certainly played a part, they were also a physical manifestation of the reach of one's empire and a reminder of political affiliations. Generally speaking, only the very highest ranking members of the European elite, from the Medici and Emperor Rudolf II in the sixteenth century to Archdukes Albert and Isabella and Louis XIV in the seventeenth century, possessed menageries with large numbers of exotic animals.[17]

---

[15] '[...] THEATRUM BRASILIENSIUM scilicet RERUM, quas NATURA produxit, aluit atque enutrivit, ICONES ipsis in natalium suorum incunabulis ac patria sua BRASILIA nempe coloribus ad vivum exactissime depictas', Mentzel, preface to *Theatrum rerum naturalium Brasiliae*, part 1, *Icones Aquatilium*, as reproduced in translation and the original Latin (in facsimile) in *Brasil-Holandês/Dutch-Brazil* vol. IV 18–19. In this translation, *ad vivum*, or after life, was translated into English as 'bright'. I have restored this important term to the sentence, which will be discussed in more detail later in this essay.

[16] Comito T., *The Idea of the Garden in the Renaissance* (New Brunswick: 1978) 8.

[17] As discussed by Smith and Findlen in a recent essay on commerce and images of nature, the Indian rhinoceros that arrived in Lisbon in 1515, most famously pictured by Albrecht Dürer, started its long journey to western Europe as a diplomatic

By building a zoo of South American and African animals during his governorship in Brazil, Johan Maurits demonstrated his desire for affiliation with other European noble collectors of exotic and native animals, whose ranks included members of his extended family, such as Landgrave Wilhelm IV and the Dutch stadhouders Prince Maurits and Prince Frederik Hendrik.[18] As mentioned by Caspar Barleaus, the Count's historian and principle apologist for Dutch Brazil, animals arrived in Recife via ships from Africa, the East Indies, and other areas for the Count, and the presence of these exotic imports, mixed in with indigenous Brazilian specimens, is documented in both the *Theatrum* and the *Handbooks*.[19] Eyewitness accounts by the Portuguese friar Manuel Calado and the German WIC servant (later governor of the Cape colony) Zacharias Wagener suggest that many of the Brazilian animals in the Count's menagerie were gifts, given to curry favor with Johan Maurits.[20] According to Calado, a frequent visitor at the Vrijburg court, the Count 'brought thither every kind of bird and animal that he could find; and since the local *moradores* [colonists] knew his taste and inclination, each one brought him whatever rare bird or beast he could find in the back-lands'.[21]

Not all of the animals that made their way into the Count's possession were still alive. Many of these creatures were preserved (presumably dried and stuffed) and thereafter incorporated into Johan

gift from Sultan Muzafar II to Alfonso d'Albuquerque, the governor of Portuguese India. Alfonso shipped this exceptional creature to his king, Don Manuel, who in turn recycled the gift by sending it on to the Medici pope Leo X in Rome. Unfortunately, the rhinoceros did not make it to Rome alive, because its ship sank off the coast of Italy. See Smith P. – Findlen P., "Commerce and the Representation of Nature in Art and Science", in *Merchants and Marvels: Commerce, Science, and Art in the Early Modern Period* (London: 2002).

[18] Although Maurits and Frederik Hendrik owned exotic animals, including Maurits's famous talking parrot, the later menagerie of Stadholder Willem V may be the best documented of these Dutch princely collections. See Sliggers B.C. – Wertheim A.A. (eds.), *Een vorstelijke dierentuin: De menagerie van Willem V* (Zutphen: 1994).

[19] Caspar Barlaeus, *Rerum per octennium in Brasilia et alibi nuper gestarum sub preafectura illustrissimi comitis I. Mauritii Nassoviae* [...] *historia* (Amsterdam: 1647). This work has been translated into Dutch as *Nederlandsch Brazilie onder het bewind van Johan Maurits Grave van Nassau 1637–1644*, ed. and trans. S.P. L'Honore (The Hague: 1923).

[20] Under his illustration of a Taÿim bugh (taibia), Wagener states, 'in 1640, a Portuguese called Antonio Robero brought one alive for his excellency', as reproduced in translation in *The "Thierbuch" and "Autobiography" of Zacharias Wagener*, trans. Á.A. Bragança Júnior, *Dutch Brazil* (Rio de Janeiro: 1997) 120. This is a facsimile of Wagener's *Thierbuch* and all subsequent references will be to this translation.

[21] See *O valeroso lucideno e triunfo da liberdade* (1648), as quoted in translation by Boxer C.R., *The Dutch in Brazil, 1624–1654* (Oxford: 1957) 116.

Maurits's cabinet of curiosities. Artists, apothecaries, professors, and princes throughout western Europe demonstrated their keen interest in natural history and the accumulation of exotic and unusual objects by creating cabinets of curiosities or *Wunderkammeren*, during the early modern period.[22] In addition to his private zoo, Rudolf II had impressive *Kunst- und Wunderkammer*, which 'embraced all the human arts, all the branches of human knowledge, and all the different realms of nature'.[23] Although his collection was exceptional, the Emperor's natural history holdings paled in comparison to the thousands of specimens amassed by Ulisse Aldrovandi (1522–1605), an Italian professor of medicine and a key figure in early modern science.[24] As described by Barlaeus, the Count's Brazilian 'museum' included both *artificialia*, such as non-European ethnographic objects like clothing and weapons, in addition to *naturalia* in the form of shells and dried plants and animals.

Unlike the vast collections of objects, specimens, and finely-crafted items from around the world amassed by many of his elite contemporaries, Johan Maurits did not attempt eycyclopedic coverage. Rather his menagerie and cabinet reflected, and I would argue, reinforced, his position as a governor-general for the Dutch West India Company by highlighting animals and specimens from Brazil and western Africa, the primary areas of WIC trade. Given the physical location of Dutch Brazil, far removed from 'civilization', Johan Maurits clearly felt the need to adopt European forms of princely authority in order to make visible his claims to power. These manifestations of princely authority included his new city Mauritsstad (Recife), his various architectural projects, including Vrijburg palace and the Boa Vista pleasure house, in addition to his gardens, menagerie, and cabinet of curiosities. The intended audience for these projects and collections was not limited to servants of the WIC, colonists, and native allies, but also included foreign interests, such as the Portuguese in nearby Olinda and the west

---

22 The literature on collecting is vast. A useful introduction to the materials can be found in Kenseth J. (ed.), *Age of the Marvelous* (Hanover: 1993). Other important works include Impey O. – MacGregor A. (eds.), *The Origins of Museums: the Cabinet of Curiosities in Sixteenth- and Seventeenth-Century Europe* (Oxford: 1985); Bergvelt – Kistemaker (eds.), *De Wereld Binnen Handbereik* and Findlen P., *Possessing nature: museums, collecting, and scientific culture in early modern Italy* (Berkeley – Los Angeles: 1994).

23 Kenseth J., "A World in One Closet Shut", in idem (ed.), *The Age of the Marvelous* 85.

24 See Swan C., "From Blowfish to Flower Still Life Paintings: Classification and Its Images, circa 1600", in Kenseth (ed.), *Merchants and Marvels* 109–137.

African ambassadors who made Maurtisstad the goal of their diplomatic missions during this period.

Johan Maurits's Patronage of Natural History Illustration

But what of Johan Maurits's motivations with respect to his Brazilian patronage of natural history drawings? As argued by Lee Hendrix, Rudolf II's *Museum*, or *Thierbuch*, included images of the 'the natural world as it was embodied in and defined by his [Rudolf II's] own collections', which makes it an important precedent for the drawings by Eckhout and Marcagraf in the *Theatrum* and the *Handbooks*.[25] The images in the *Museum*, which include large-scale oil paintings on parchment, documented the living animals present in the Emperor's zoo as well as the curiosities and preserved specimens of *naturalia* from his *Wunderkammer*. In the same way, the animals represented in the *Theatrum* and the *Handbooks* mirrored the contents of Johan Maurits's personal collections. The *Museum* functioned as a record of Rudolf II's impressive ability to gather together the new and the unusual from all over the world, as emblems of influence and empire. In a similar fashion, the *Theatrum* and the *Handbooks* reproduced the contents of the Count's Brazilian menagerie and cabinet of curiosities; one may even argue that these images stood in for all of the animals and plants that fell within Johan Maurits's larger sphere of influence in South America and Africa.

Among the Brazilian drawings by Eckhout and Marcgraf we find images of young animals (which were easier to catch and train), chained monkeys, and parrots on perches, demonstrating captivity as well as domestication [Fig. 1]. Other animals, including a bird of paradise (here pictured with feet, but lying stiffly on its side), were clearly dead before being drawn [Fig. 2]. This suggests that Marcgraf, the drawing's author, had access to dried specimens in the Count's cabinet of curiosities; birds of paradise, imported from New Guinea, had figured prominently in such collections in Europe since the sixteenth century.[26] Often their feet were removed, which added credence to claims that

---

[25] Hendrix L., "Natural History Illustration at the Court of Rudolf II", in Fucíková E. et al. (eds.), *Rudolf II and Prague: the imperial court and the residential city as the cultural and spiritual heart of central Europe* (Prague: 1997) 162.

[26] Koreny F., *Albrecht Dürer and the animal and plant studies of the Renaissance* (Boston: 1985) 100–112.

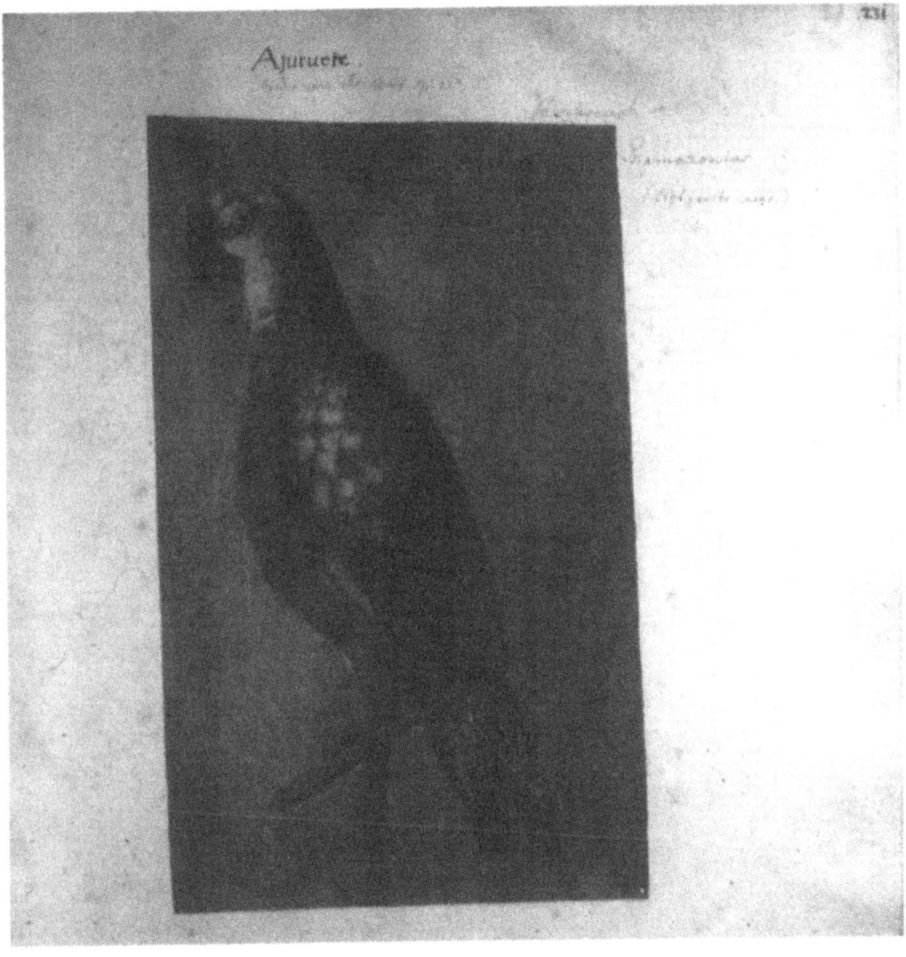

Fig. 1 [Col. pl. XIII]. Albert Eckhout, Amazonian parrot (*Amazona aestiva*). Oil on paper. *Theatrum* II fol. 231r, *Libri picturati* A33 (Jagiellon University Library, Krákow).

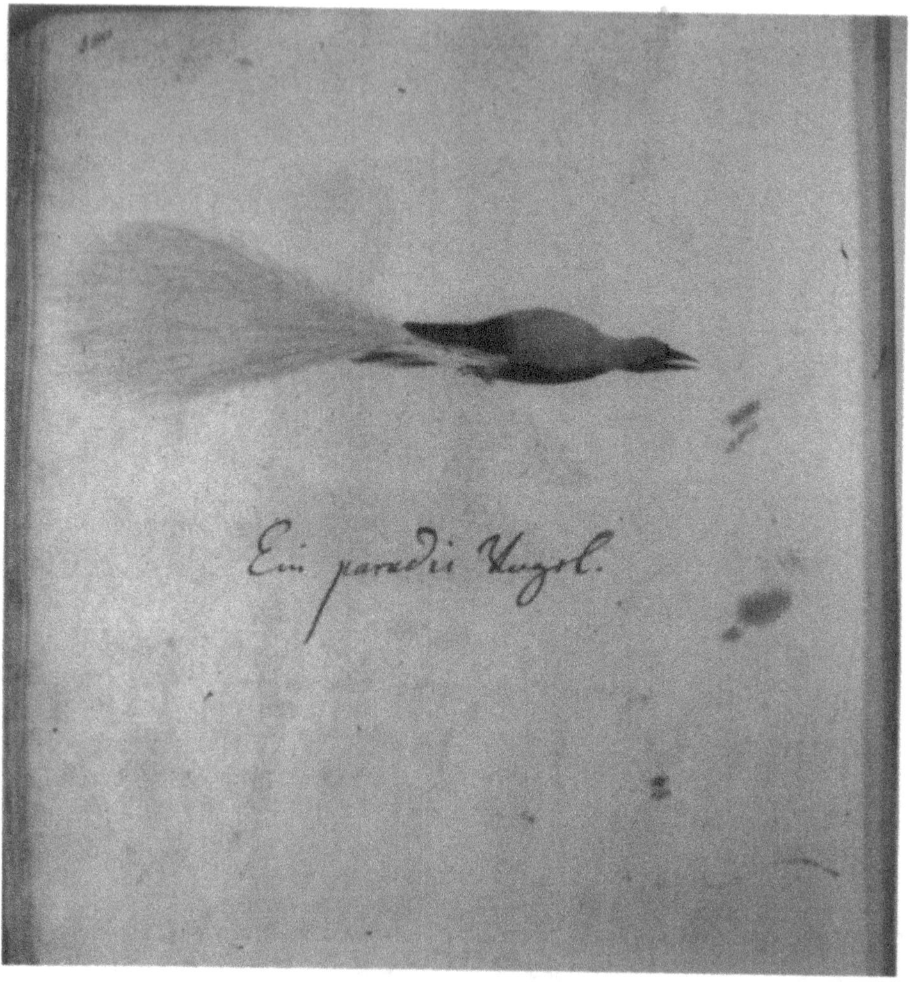

Fig. 2 [Col. pl. XIV (detail)]. Georg Marcgraf, Bird of paradise (*Paradisaea minor*), ca. 1640. Watercolour and body colour on paper, 31 × 20 cm. *Handbook* I fol. 180r, *Libri picturati* A36 (Jagiellon University Library, Krákow).

they lived their entire life in the air.[27] Of the native Brazilian species, it can be inferred that all of the representations of aquatic animals in these volumes were created after dried or recently caught specimens, with some coming from the ponds on the grounds of Vrijburg palace and others, such as this flying fish, obtained by fishermen and sailors off the coast of the colony [Fig. 3].

Beyond documenting the Count's Brazilian collection, it seems clear that Eckhout's role as Johan Maurits's court artist was also to create a comprehensive group of realistic, large-scale, and authoritative images of exotic flora and fauna, which could then be used as reference materials and models for decorative projects—from tapestries, to ceiling paintings, to paintings on canvas—both in Brazil and in Europe. Eckhout's works on paper, which generally lack annotation and remained unbound until the 1660s, were not originally intended for educational, interactive use. Unlike these loose oil studies on paper, Marcgraf's drawings were already bound into two volumes in Brazil with a specific, hands on, function in mind. While Marcgraf may have intended these drawings to serve as models for the illustrations to his treatise on zoology in the *Historia naturalis Brasiliae*, in Brazil the *Handbooks* appear to have functioned as Johan Maurits's personal natural history guides and notebooks. They include his handwritten descriptions and observations about various animals, providing evidence for the Count's personal engagement with the study of Brazilian nature and his deliberate use of animals as princely and diplomatic gifts.

By writing his own descriptions underneath many of the images, Johan Maurits became a 'princely practitioner' of New World natural history as well as the ruler of all he surveyed.[28] Under a simple black and white watercolour drawing of a jaguar cub (*Panthera onca*), the Count writes: 'a young tiger, reaching such a size, that, when full grown, its back is level with the height of a table; very fierce'.[29] Johan Maurits's annotations give a keen sense of the interest in and interaction with some of the more unusual Brazilian animals in the colony, which were clearly objects of fascination for the Europeans. In describing

---

[27] Ashworth W.B. jr., "Remarkable Humans and Singular Beasts", in Kenseth (ed.), *Age of of the Marvelous* 125–126.

[28] Moran B.T., "German Prince-Practitioners: Aspects in the Development of Courtly Science, Technology, and Procedures in the Renaissance", *Technology and Culture* 22 (1981) 253–274.

[29] *Liber Principis*, vol. I, fol. 58, as reproduced in translation in *Brasil-Holandês/Dutch-Brasil* vol. II 21.

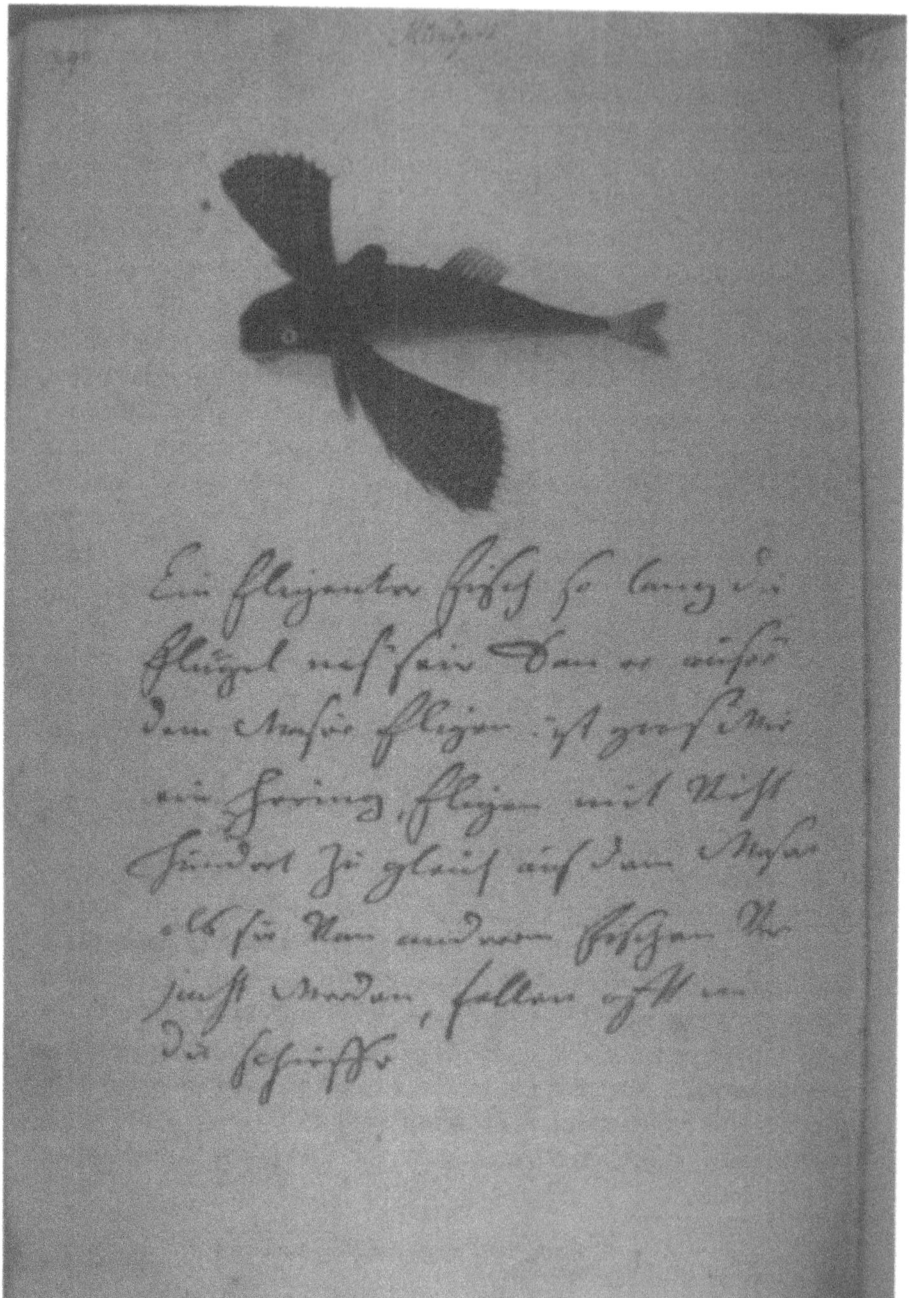

Fig. 3. Georg Marcgraf, Flying Fish (*Dactylopterus volitans*), ca. 1640. Watercolour and bodycolour on paper, 21 × 30 cm. *Handbook* I fol. 390r, *Libri picturati* A36 (Jagiellon University Library, Krákow).

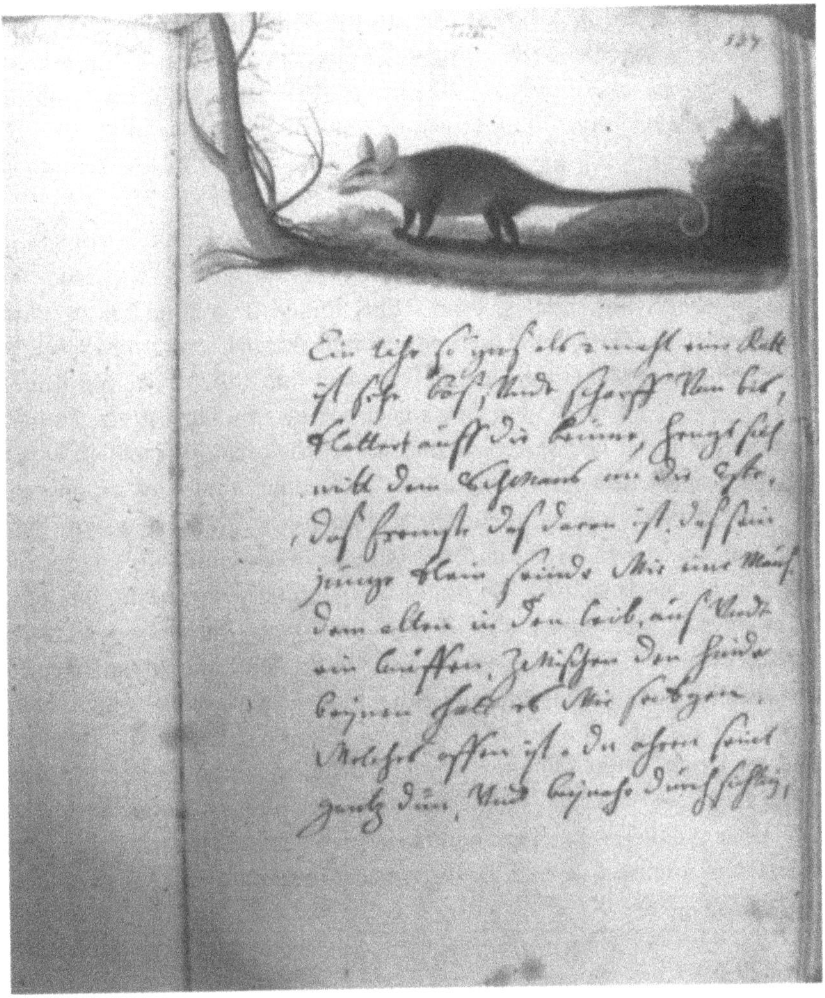

Fig. 4 [COL. PL. XV (detail)]. Georg Marcgraf, Taibi (*Didelphis marsupialis*), ca. 1640. Watercolour and body colour on paper, 31 × 20 cm. *Handbook* I fol. 30r, *Libri picturati* A36 (Jagiellon University Library, Krákow).

a taibi (*Didelphis marsupialis*), a sort of South American opossum, he writes: 'An animal, twice as large as a rat, very fierce and with a sharp bite. It climbs on trees, hanging by the tail from the branches [...] its young, as small as a mouse, get in and out of the adult's belly...the ears are thin and almost transparent' [Fig. 4].[30] Unlike the taibi, sloths

[30] *Liber Principis*, vol. I fol. 137, as reproduced in translation in *Brasil-Holandês/ Dutch-Brasil* vol. II 41 (with some adjustments to the English).

had been featured in a variety of European accounts of the Americas beginning in the sixteenth century. Under this drawing of three-toed sloths (*Bradypus tridactylus*), one climbing a tree and the other lying on the ground, he notes 'The Germans call this animal a sloth since it walks no more than fifty steps a day. It feeds on old leaves. Its hair is thick, and it reaches about two feet long' [Fig. 5].[31]

In addition to the 'tiger' mentioned above, the Count's collection included other animals traditionally associated with royalty, such as game animals and horses. Under this image of a graceful, reddish brown deer drinking from a stream, Johan Maurits describes it as: 'A game animal smaller than a roe. The stag has antlers like the European ones [...] I have sent live specimens to His Majesty (S. Hoog., probably refering to Frederik Hendrik) but they died of cold' [Fig. 6]. The princely context of this volume is further reinforced by images of two Brazilian horses, including a dapple-gray horse of a type that also appears in Jan Brueghel the Elder's paradise landscapes from the early seventeenth century. As noted by Arianne Faber Kolb, this type of horse was 'the highest class of horse and [was] therefore associated with royalty'.[32] This is also made clear by the fact that Johan Maurits specifically sent this animal and other fine riding horses to 'S. Hoch. Prins. Henderig' (Stadhouder Frederik Hendrik, Prince of Orange) and 'S. Hoch. Prins Wilhem' (Frederik Hendrik's son, Prince William II of Orange). The fact that *Handbook* II originally included a treatise on horse medicine, perhaps handwritten by Johan Maurits, similarly betrays its practical as well as its princely function.

Style and Pictorial Mode in Marcgraf's Brazilian Natural History Drawings

In his preface to the *Theatrum,* Mentzel calls Marcgraf's watercolours in *Handbook* I and II 'meticuously painted', and he notes that in each of the two volumes 'the Prince described in his own hand, in our own German language, the true grandeur and dimensions of the creatures portrayed, so that these books must be admired and appreciated as

[31] *Liber Principis*, vol 1, fol. 112, as reproduced in translation in *Brasil-Holandês/ Dutch-Brasil* vol. II 34. There are three different types of three-toed sloths; this appears to be the pale-throated variety.

[32] Kolb A.F., *The Entry of the Animals into Noah's Ark* (Los Angeles: 2005) 63.

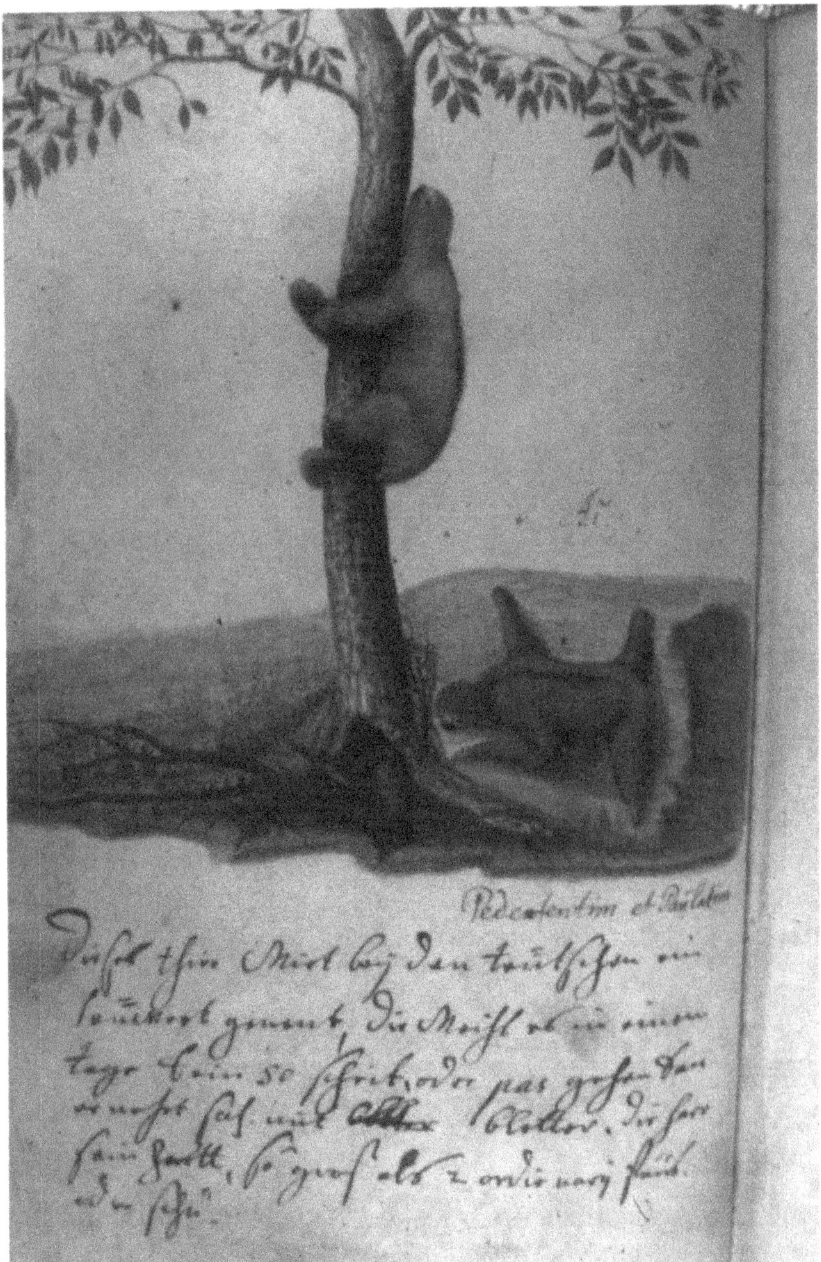

Fig. 5 [Col. pl. XV (detail)]. Georg Marcgraf, Three-toed Sloths (*Bradypus tridactylus*), ca. 1640. Watercolour and body colour on paper, 31 × 20 cm. *Handbook* I fol. 30r, *Libri picturati* A36 (Jagiellon University Library, Krákow).

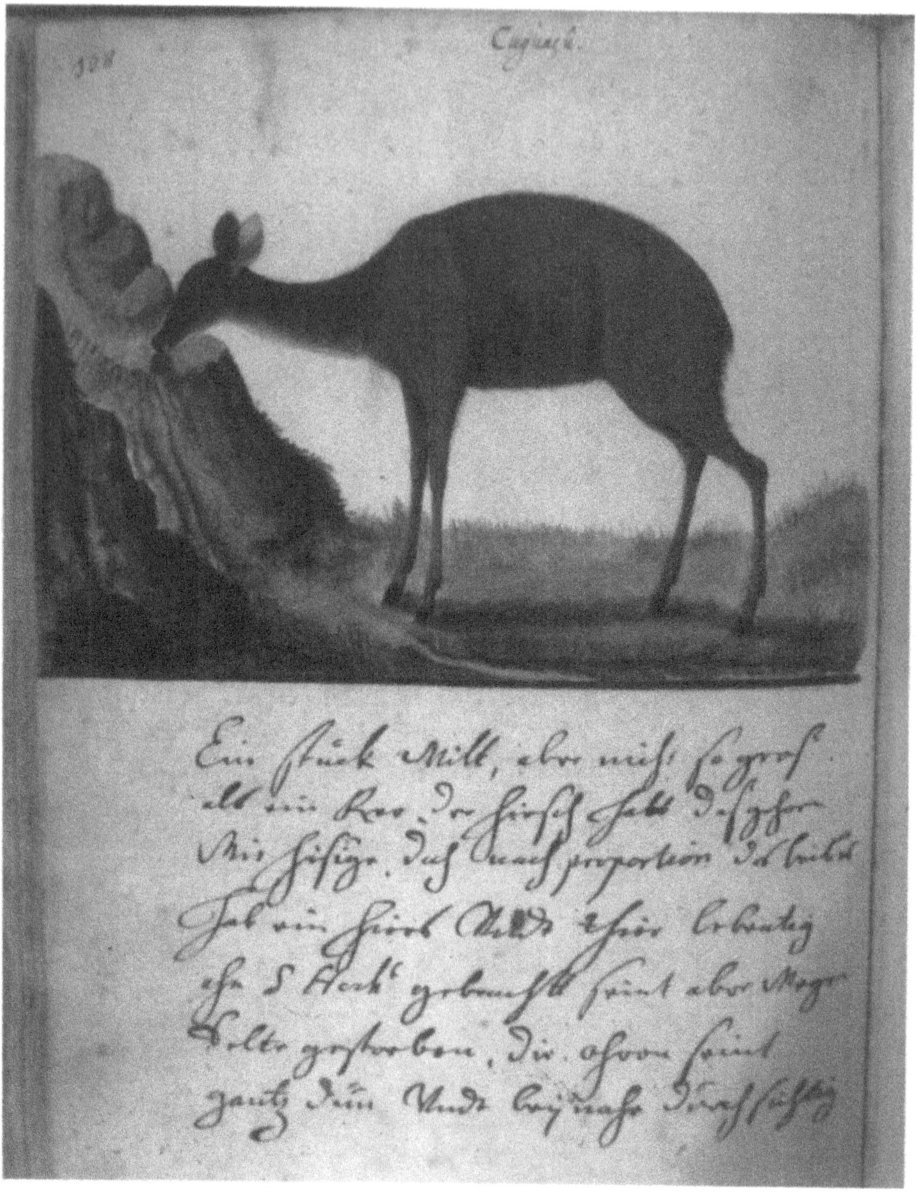

Fig. 6 [Col. pl. XVI (detail)]. Georg Marcgraf, Brazilian Deer (çuguaçû?), ca. 1640. Watercolour and body colour on paper, 31 × 20 cm, *Handbook* I fol. 108r, *Libri picturati* A36 (Jagiellon University Library, Krákow).

authentic and authorized records of the birds, and land and sea animals'.[33] It is clear that the scientific labels written by Marcgraf above each animal, the Count's witness to visual accuracy in his written descriptions, and Marcgraf's 'meticulous' style were understood by contemporaries, including Mentzel as well as *Historia* editor Johannes de Laet, to be evidence that these images were sources of visual knowledge.[34] It is also probable that the very small size of these works, which not only reinforce the viewer's sense of viewing something precious and unique, but also connect the images to an established pictorial tradition for model book drawings and images of nature, made such an interpretation possible.

Marcgraf's drawings demonstrate their relationship to this visual tradition not only by their small size, but also because of their lack of animation, consistent use of the profile pose, clear and careful manner of drawing, watercolour and gouache media, and use of a fixed position (often in profile) for the animal.[35] As argued by Otto Pächt in his seminal article on Italian nature studies, adoption of this artificial mode of representation, in which the subjects 'lack the full breath of life', does not, however, rule out the possibility that such images were produced during an encounter with the living animal.[36] Rather, he argues that this mode is actually necessary for a nature study. In Robert Scheller's work on model book drawings, he further suggests that artists may also have chosen to produce a stylized and less naturalistic image in order to ensure the long-term status of their drawing as an

---

[33] 'Ipse talium curiosissimus Princeps duobus istis quos dixi libris, qui figuras depictas coloribus ut plurimum punctim applicatis continent, veram molem ac dimensionem rei deliniatae manu sua nostra Teutonum lingua adscripserat, ut hos libros, tanquam authenticos, pro Canone diiudicandorum animalium tam aëreorum quam terrestrium et marinorum, inspicere ac venerari ubique possit', Mentzel, preface to *Theatrum rerum Naturalium Brasiliae*, vol I., in *Brasil-Holandês/Dutch-Brazil*, vol. IV 22–23.

[34] For discussion of the use of images to aid study and taxonomy, see Freedberg D., "Cassiano, Ferrari, and other Drawings of Citrus Fruits", in *Citrus Fruit*, vol. I, *The Paper Museum of Cassiano dal Pozzo. A Catalogue Raisonné. Drawings and Prints in the Royal Library at Windsor Castle, the British Museum, the Institut de France and other Collections* (London: 1997) 45–84 and Freedberg D., *Fossil Woods and other Geological Specimens*, vol. III, *The Paper Museum of Cassiano dal Pozzo. A Catalogue Raisonné. Drawings and Prints in the Royal Library at Windsor Castle, the British Museum, the Institut de France and other Collections* (London: 2000).

[35] See Scheller R.W., *Exemplum: Model-Book Drawings and the Practice of Artistic Transmission in the Middle Ages (ca. 900–ca. 1470)* (Amsterdam: 1995) 62.

[36] Pächt O., "Early Italian Nature Studies and the Early Calendar Landscape", *Journal of the Warburg and Courtauld Institutes* 13 (1950) 13–47.

exemplum.[37] Given the eventual use of Marcgraf's drawings as models for the illustrations in the *Historia*, the adoption of such a visual mode would have been highly desirable.

Marcgraf's watercolours belong to the same representational family as the well-known 'animal portrait' of a hoopoe, attributed to the fifteenth-century Flemish minaturist Simon Marmion [Fig. 8]. This finely rendered, completely static watercolour drawing was in the collection of Rudolf II as part of an album that included nature studies by a variety of artists.[38] The preference for watercolour and body colour for images of nature has been well documented: figures as different as the German naturalist Conrad Gesner and the Netherlandish artist Joris Hoefnagel favored the medium—the artist for its precision and delicacy and the scientist for its transportabilty, ease of use, and ability to dry quickly. Other early modern northern European masters of the nature study include Albrecht Dürer, Hans Hoffman, Hans Bol, and Jacob de Gheyn II. Dutch specialists in this genre from the seventeenth century also include Pieter Holsteyn II, whose largely unstudied work includes watercolour drawings of anteaters and other South American animals.

Marcgraf's highly contained watercolours remain within the bounds of the two-dimensional space of the paper, emphasizing exterior description over three-dimensional, realistic representation. Seventeenth-century botantical images often display images of fruit that is cut open to reveal the seeds within; zoological imagery did not yet favor this type of 'anatomical' representation, although dissection was standard practice among natural historians. The inventory cited above describes Marcgraf's style as 'artful', perhaps refering to the highlights of gold and silver paint around the eyes and on the scales of his fish and the inclusion of small landscapes for some of the birds and quadrupeds [Figs. 3–7]. Marcgraf's attention to detail in the *Handbooks* demonstrates that his images are not simple or schematic field studies, made to quickly capture the basic form and colours of an animal, but rather represent later, more finished versions of earlier images.

Marcgraf's study of a giant anteater (*Myrmecophaga tridactyla*) is a good example of one of his more fully worked up drawings in the *Handbooks* [Fig. 7]. Marcgraf began by making a simple pencil sketch

---

[37] Scheller, *Exemplum* 62.
[38] Koreny, *Albrecht Dürer* 28–29.

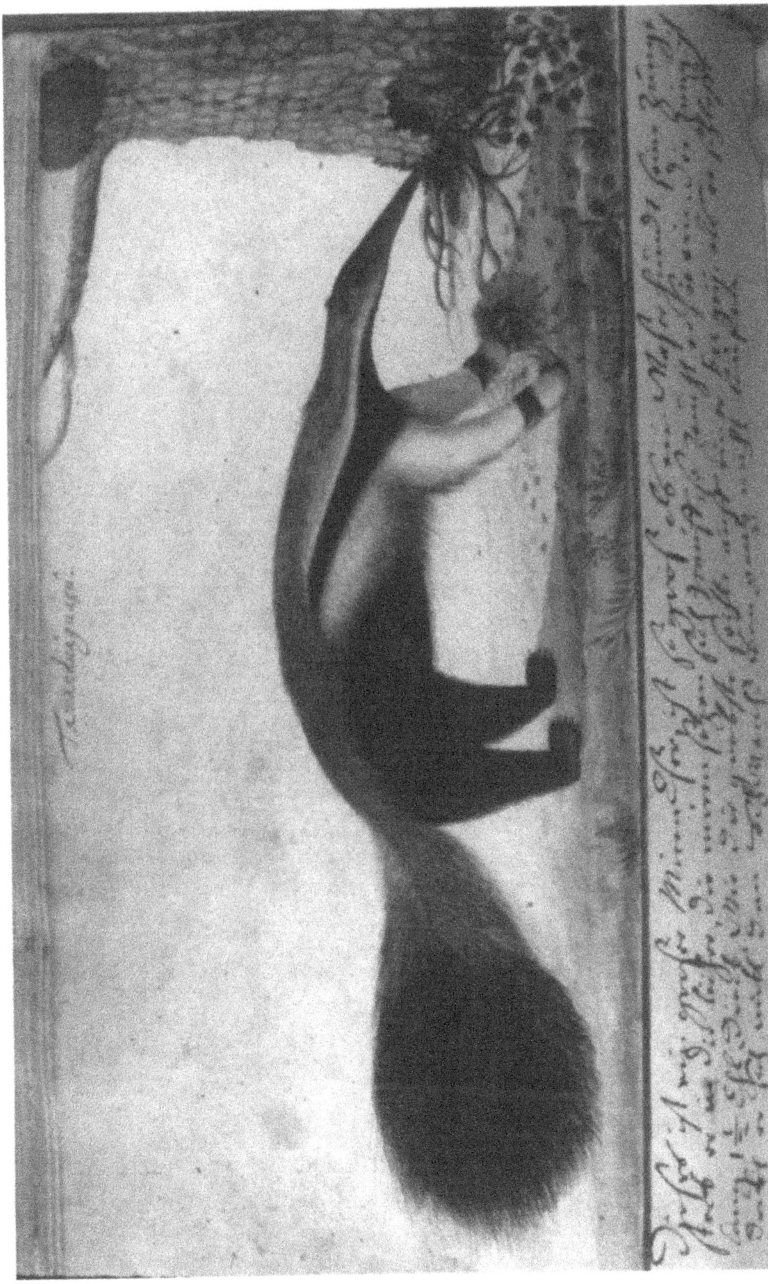

Fig. 7 [Col. pl. XVII]. Georg Marcgaf, Giant anteater (*Myrmecophaga tridactyla*). Watercolour and body colour on paper, 31 × 20 cm. *Handbook* I fol. 84r, *Libri picturati* A36 (Jagiellon University Library, Krákow).

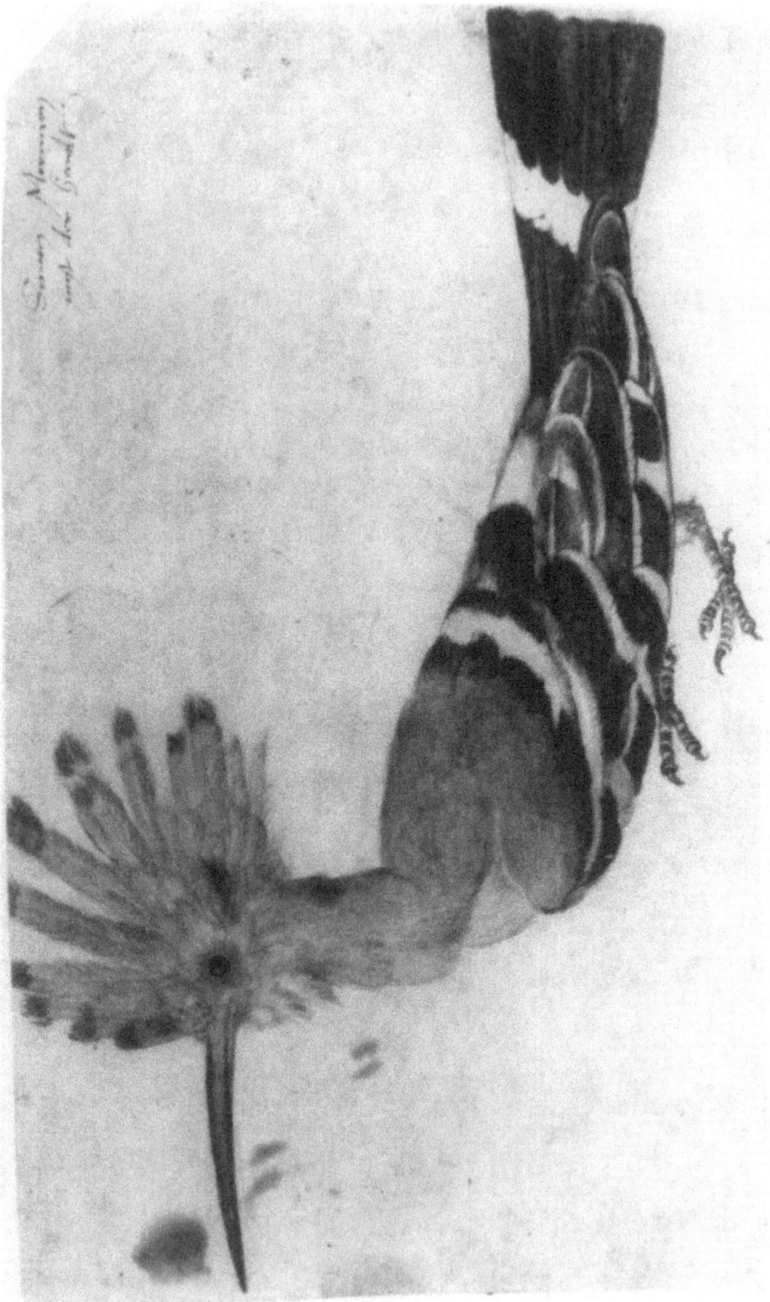

Fig. 8. Simon Marmion, Hoopoe. Watercolour and body colour on paper, 17 × 28.2 cm. Vienna, Österreichische Nationalbibliothek, Handschriften- und Inkunabelsammlung, Cod. min. 42, fol. 55r.

on paper, thereafter applying a gray watercolour wash, which allowed him to establish the forms and patterns for the subsequent application of details in body colour or gouache. Using very fine, highly linear brushstrokes, Marcgraf carefully recorded the anteater's colouring and the pattern of its fur, neatly representing the individual hairs of its tail.[39] As demonstrated by this example, Marcgraf's drawings are contained, relatively two-dimensional, and static; as an artist he rejected the sensuality possible with oils in favor of the clarity of line and form that characterize watercolour and body colour. This image conveys essential information (ie, that which enabled one to recognize an animal and distinguish it from similar species, such as its general form, colouring, type of feet, number of toes, length of tail, etc.), although Johan Maurits's annotations provide key information on sizes. In this image, the anteater stands in strict profile, which is characteristic of nature studies as well as illustrations of animals found in early modern natural history texts. Marcgraf's inclusion of additional information regarding food and habitat in his representations of quadrupeds would become standard in the eighteenth century with naturalist-illustrators like Mark Catesby. In Marcgraf's drawing, the anteater is posed before a tree, from which it has removed a round section of bark to expose an ant nest--its main food. Although Marcgraf's rendering is quite careful, the back feet are stiff and unconvincing, and demonstrate a poor understanding of perspective. Marcgraf's *tabia*, another delicately rendered work and perhaps his most animated animal, is pictured with teeth bared before its burrow [Fig. 4]. The tabia's toothy grimace accords well with Johan Maurits's desciption of the animal as 'fierce'. In other images, brown capuchin monkeys (*Cebus apella*) and the three-toed sloths climb trees, demonstrating both habitat and characteristic behavior [Fig. 5]. Marcgraf's main concern in these images appears to have been the transmission of visual information about the animals, which could then be studied by his patron Johan Maurits in Brazil and later reproduced in printed form in the *Historia*.

By his use of the phrase 'nach dem Leben', the anonymous author of the 1652 inventory mentioned at the beginning of this essay gave recognition to the fact that Marcgraf's images fit the criteria for scientific

---

[39] For a similar approach to the rendering of hair, see the badger (ca. 1570) included in the Lambert Lombard Album (Rijksprentenkabinet, Amsterdam) and the watercolour and body colour drawing of a beech marten, attributed to Hans Verhagen den Stomme of Antwerp. Both images are reproduced in Koreny, *Albrecht Dürer* 130.

(especially botanical) illustrations as they began to develop in the sixteenth century. Images so intended had to downplay the presence of the artist and his inventive powers and instead demonstrate a direct, one-to-one, relationship with nature. As discussed in recent literature on Northern European art, during the sixteenth century it became increasingly common for images to be produced and labelled as sources of accurate visual information rather than examples of artistic invention.[40] In particular such representations included, but were not limited to, portraits, images of natural or preternatural phenonmenon, natural history representations, and ethnographic illustrations in travel accounts.[41] In his detailed analysis of the use of the term *contrafactum* on printed images in the sixteenth century, Peter Parshall has argued that such informational images (which ranged widely in quality, style, and degree of verisimilitude) were not distinguished by their use of 'conventionally' realistic representation, but rather by the vocabulary used to describe them. The designation of an image as a true 'counterfeit' vouched for its correspondence to the original. As he argues, 'the type or class of portrayal designated by *contrafactum* came to be determined primarily by the intention to convey some particle of information deemed transmissable through a picture'.[42] Especially in cases where the original was unavailable for purposes of verification, such a designation was deemed necessary to assure the viewing public of the representation's accuracy. In this respect it is not insignificant that the nature study of the hoopoe addressed above includes the handwritten text, 'Simon Mormion myt der handt' (Simon Marmion [made this] with his hand), which makes the same kind of claim to eyewitness accuracy.[43]

Focusing on the sixteenth and early seventeenth centuries, Claudia Swan has explored the implications of the term *ad vivum* and its vernacular cognates, such as the German *nach dem Leben* and the Dutch *naer het leven*, coming to many of the same conclusions as Parshall with respect to the designation *counterfactum*.[44] As she notes, 'Like images iden-

[40] Parshall P., "Imago Contrafacta: Images and Facts in the Northern Renaissance", *Art History* 16 (1993) 555 and Landau D. and Parshall P., *The Renaissance Print 1470–1550* (New Haven: 1994).

[41] Parshall, "Imago Contrafacta" 556.

[42] Parshall, "Imago Contrafacta" 555.

[43] Koreny, *Albrecht Dürer* 28–29. See also Swan C. *Art, Science and Witchcraft in Early Modern Holland: Jacques de Gheyn II (1565–1629)* (Cambridge: 2005) 37.

[44] See Swan C., "*Ad Vivum*, naer het leven, from the life: Defining a Mode of Representation", *Word and Image* 11 (1995) 353–372 and Swan, *Art, Science and Witchcraft*.

tified as *gheconterfeyt*, those said to have been done *naer het leven* promise a verifiable conformity with their subject, an iconic correspondence'.[45] When identified as having been made 'after life', the image took on the status of a document. As Swan argues, 'The specification of how the drawing (or portion of it) was made – *naer t'leven*—invests it with a functional value neither its style nor its subject could'.[46] Although Marcgraf does not write this type of description directly on his images in the *Handbooks*, he does specifically state in the preface to his section of the *Historia* that he produced images *ad vivum* of Brazilian nature.

Like Parshall, Swan focuses on images that were designated 'true counterfeits' or accurately produced 'after life' by the artists who created them or by those who commissioned or published them. Swan nonetheless also brings attention to the important artist and theorist Karel van Mander and his use of the phrase 'gheconterfeyt nae t' leven' to describe portraits and natural history images in *Het Schilderboek* (1604). She argues that such an assessment required Van Mander to make a 'leap of faith', because he was neither familiar with the original nor had he witnessed the image's creation. Regarding the visual characteristics that allowed an observer like Van Mander to make such an assessment, Swan suggests that an image's 'lifelike effect' appears to have been understood by Van Mander as evidence of the image's production *naer het leven*.[47] Certainly this 'life-like' effect was created in part by Van Mander's favourable recognition of pictorial naturalism and the realistic use of colour.[48] Apart from all too brief discussions like this, there has been little investigation regarding how particular combinations of visual and physical characteristics (apart from linguistic evidence) may have made a viewer believe that the image was a trustworthy bearer of information.

[45] Swan, "'*Ad vivum*'" 356.

[46] Ibidem 358.

[47] For discussion of Van Mander and *naer het leven*, see Swan, "'*Ad vivum*'" and chapter one in Swan, *Art, Science and Witchcraft*, especially 36–39.

[48] Regarding Van Mander's use of *naer het leven*, Swan notes 'he often combines a description of the wrok done *nae t'leven* with its characterization as *natuerlijck* ('naturalistic'). For example, he commends Pieter Aertsen for having painted "kitchens with all sorts of goods and foods after the life, so accurately capturing all of the colors that it appeared to be natural."' See Swan, *Art, Science and Witchcraft* 38.

Artistic Authority and Oil Paint: Eckhout's Zoological Images in the Theatrum

Although Mentzel calls Marcgraf's gouache and watercolour drawings 'authentic and authorized records', he also forcefully asserts that 'no less authority must be given to the paintings [Eckhout's works on paper], fruits of the same land, made with the same care and accuracy by those who gazed upon them with their own eyes'.[49] In fact, he carefully draws attention to the status of these images, calling them 'pictures of the things that Nature hath produced, nourished and nurtured, formed in their own original habitat and painted in exact colours after life [*ad vivum*]'.[50] For the anonymous writer of the 1652 inventory, the label *nach dem Leben* verified the accurate content of Marcgraf's drawings, and established a functional distinction between the oil studies and the watercolour and gouache drawings. Given this assessment it is somewhat surprising that only eleven years later Mentzel, while recognizing the informational content of the watercolours, would nonetheless turn this hierarchy on its head by valuing Eckhout's oils on paper more highly as complete visual statements.[51]

Eckhout's oil studies are painted on paper, a practice common in the Southern Netherlands, but rather uncommon among Dutch artists in the seventeenth century and highly unusual for images of nature.[52] Unlike watercolour, oil paint was employed exclusively by trained artists, not scientists. In her work on natural history illustration at Rudolf II's court, Lee Hendrix suggests that the use of oils for images of nature 'should also be seen as part of the larger dominance of oil painting as an artistic goal in Northern and Central Europe during the late sixteenth and early seventeenth centuries'.[53] That images of nature in oils on paper had a higher status than body colour on paper is sug-

---

[49] 'Quanquam caeterae oleaginae, ut ita dicam, picturae non minoris autoritatis censendae, quandoquidem eodem jussu curaque, imo eodem simul natali solo prodierunt, oculorum fide commendatissimae' Mentzel, preface to *Theatrum rerum Naturalium Brasiliae*, vol. IV 22–23.

[50] See note 15.

[51] This interest suggests that Mentzel was the one responsible for having Eckhout's images varnished, giving them a glossier, more life-like effect.

[52] The only other contemporary Dutch artist who also made oil studies on paper with any frequency was Dirk Hals, the brother of Frans Hals.

[53] Hendrix, "Natural History Illustration at the Court of Rudolf II" 165. It is possible that by the seventeenth century, some German princely collectors favored the use of oil for nature studies as a way of distinguishing between those images that were

gested by the Great Elector's response to the Brazilian drawings when they came into his possession in 1652, and his directions to Mentzel to separate the images by medium.

Careful examination of these images reveals Eckout's working method. He first made a simple sketch in chalk on a prepared piece of paper, covered with what now appears to be a pinkish-brown ground layer of paint. Over his chalk sketch he painted in oils, for the most part following the technique of a painter on canvas. Like Marcgraf's watercolours, Eckhout's studies record accurate details of the specimens under consideration, but he does not represent habitat and only rarely food. For example, Eckhout's image of two batfish, 'Guacucua', (*Ogocephalus vespertilio*), seen from the bottom and the top, equals his extraordinary study of a lizard (a young iguana?) both in its drama as well as in its verisimilitude [Fig. 9]. Here an impressively life-like green lizard, painted quite thickly with prominent yellow highlights, presents a convincingly three-dimensional form, with dark shadows attesting to its physical presence. One has the sense that the reptile has paused for just a moment, but will soon skitter off the page. Other beautifully painted, realistic works, which demonstrate a similarly 'scientific' still-life aesthetic, include this representation of a green lizard and two lobsters, again represented from the bottom and the top [Fig. 10].

In contrast to Marcgraf's watercolours, which are intimate in size and carefully analytic, Eckhout's oil studies are large, robust, and project a three-dimensional presence. In his works it is not the detail of each brushstroke that attracts one's attention; rather it is the body of the living animal that draws the viewer in. Eckhout's images never hide his brushstrokes; he uses thick layers of paint to create texture and protruding drops of paint for the highlights. Eckhout also employs black outlines, which are usually fine and delicate. Twenty-five of the thirty largest images (all measuring around 50 by 20 cm) in volume III (*Icones Animalium*) can be attributed to Eckhout, a pattern also seen in other volumes of the *Theatrum*. Eckhout's skillful rendering of many of the animals in his oil studies, from crabs and lobsters to parrots and lizards, is similar to what we find in the work of his contemporaries in the northern and southern Netherlands. Eckhout's pair of red and glistening cooked lobsters addressed above would not be out of place in

intended to first have an aesthetic function (in a collector's cabinet) and only second to carry information necessary for scientific classification.

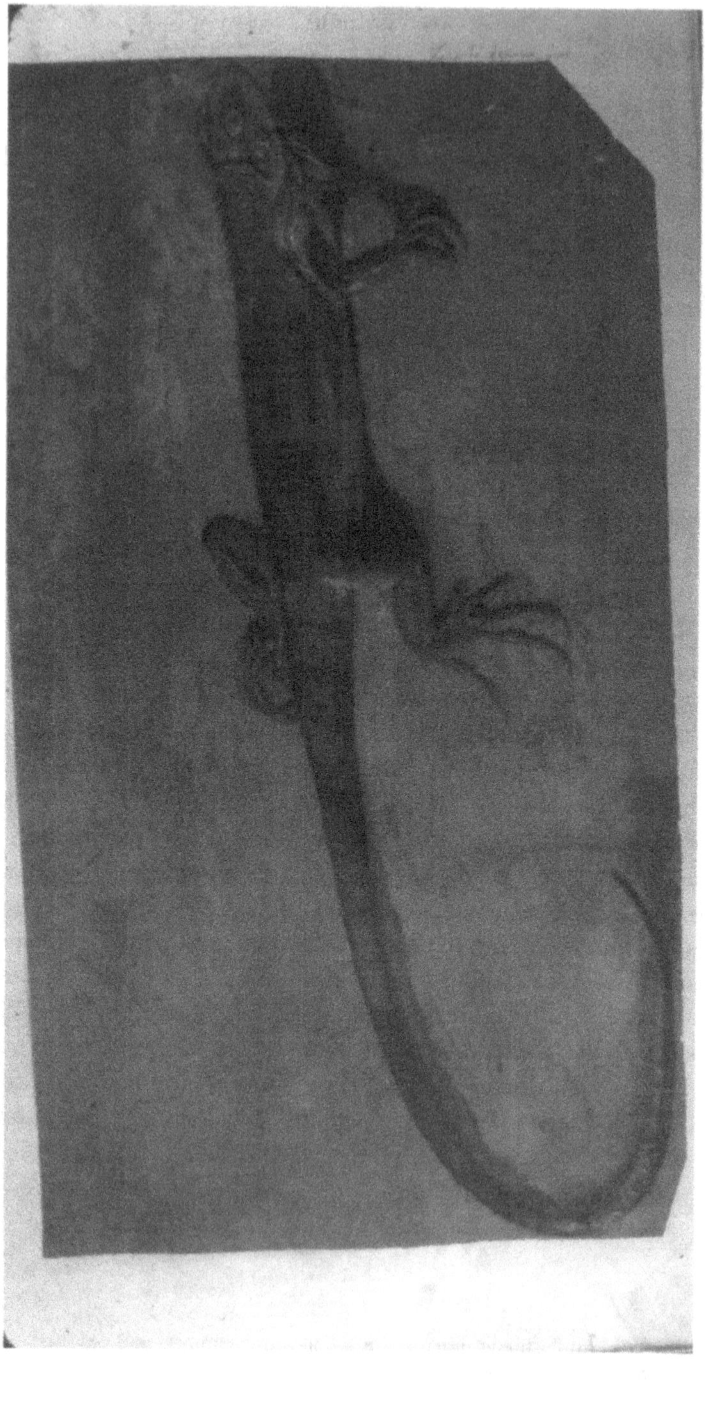

Fig. 9 [Col. pl. XVIII]. Albert Eckhout, Green Lizard (Young Iguana?). Oil on paper, 47 × 27 cm. *Theatrum rerum naturalium*, vol. III fol. 168r, *Libri picturati* A34 (Jagiellon University Library, Krákow).

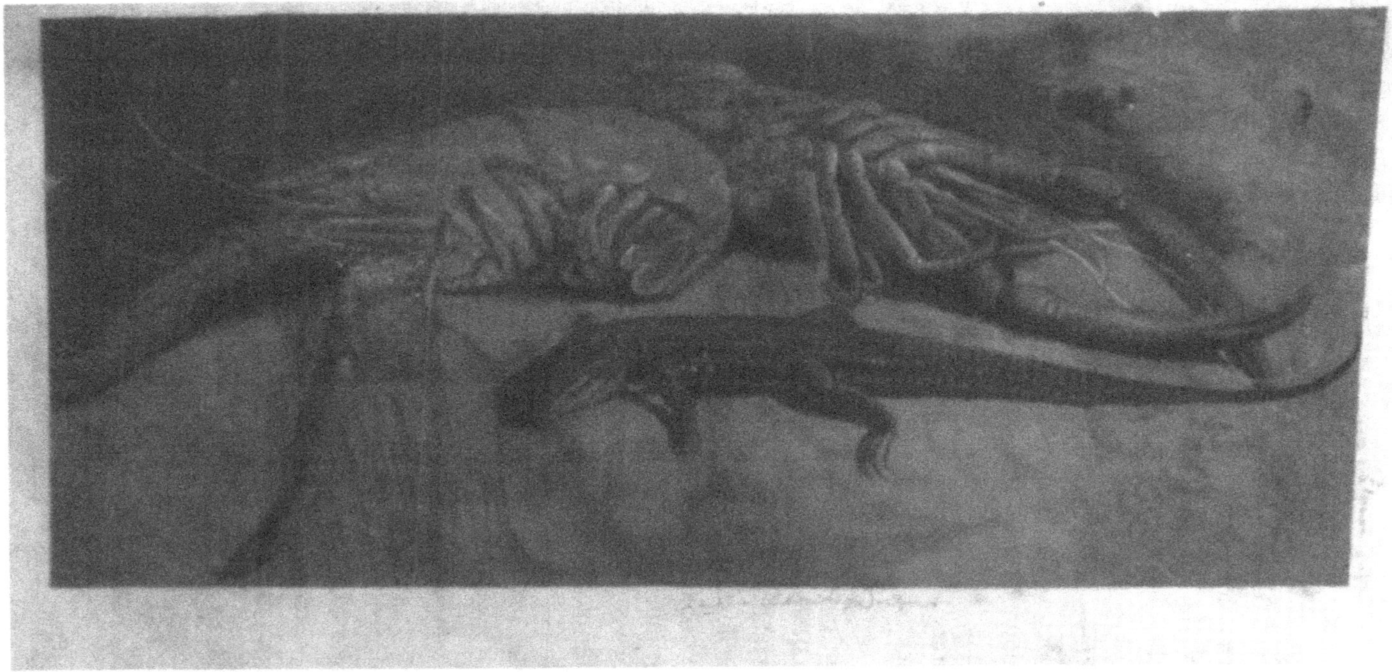

Fig. 10 [Col. pl. XIX]. Albert Eckhout, Cooked lobsters and a lizard. Oil on paper, 38.5 × 17.9 cm. *Theatrum rerum naturalium*, vol. I fol. 323r, *Libri picturati* A32 (Jagiellon University Library, Krákow).

this ca. 1640 banquet piece by Jan Davidsz. De Heem [Fig 11]. Even Eckhout's parrots compare favorably in terms of skill and naturalism with the South American parrot pictured by De Heem in this elaborate 'pronk' still life.

Because of their large size, subject matter, naturalism, and dynamic, painterly technique Eckhout's works are also in keeping with the images produced at the court of Louis XIV by the Flemish painter Pieter Boel. Boel's lively oil studies on canvas of animals from the king's zoo at Versailles were specifically created as designs for the Gobelins tapestry works.[54] There are, nonetheless, some important differences, among them the fact that Eckhout always represents the entire animal, never the head alone. Eckhout also positions his animals in much more formal and static poses than Boel, in this way suggesting his closer affinity with contemporary botanical and zoological prints.

Nonetheless, a comparison of illustrations made by Eckhout and Marcgraf after the same specimen, in this case a yellow and green Amazonian parrot, demonstrates their fundamental representational differences [Figs. 1 and 12]. Although both images include the same basic information on color and form, Eckhout's painterly execution and life-size format produce a much more dynamic visual impact than Marcgraf's small, flat, and linear production. But even though the parrot in Eckhout's image is stiffly posed, it is much more physically present. The parrot even appears to turn towards the viewer, with the light reflecting off of its eye giving the feeling of movement. In contrast, the static quality of Marcgraf's drawing provides a curious contrast with Johan Maurits's lively description, in which he asserts that this parrot was able to answer and ask questions, making some believe it was a mouthpiece for the devil.[55] Similarly, while the olive-yellow and brown body of this ocelot (*Leopardus pardalis*) is painted in a fairly flat manner, Eckhout paints the head of the animal turned out, with the eyes gazing straight at the viewer [Fig. 13]. He achieves the greatest degree of animation in the face, which is dominated by large and liquid golden eyes. No such illusion of life is conveyed in Marcgraf's representation

---

[54] Pieter Boel's work for the king has been the subject of a recent exhibition at the Louvre in Paris. See Foucart-Walter E., *Pieter Boel: Painter of Louis XIV's Animals* (Paris: 2001).

[55] *Handbook* I, reproduced as *Libri Principis*, vol. I, in *Brasil-Holandês/Dutch-Brazil*, vol. II 69.

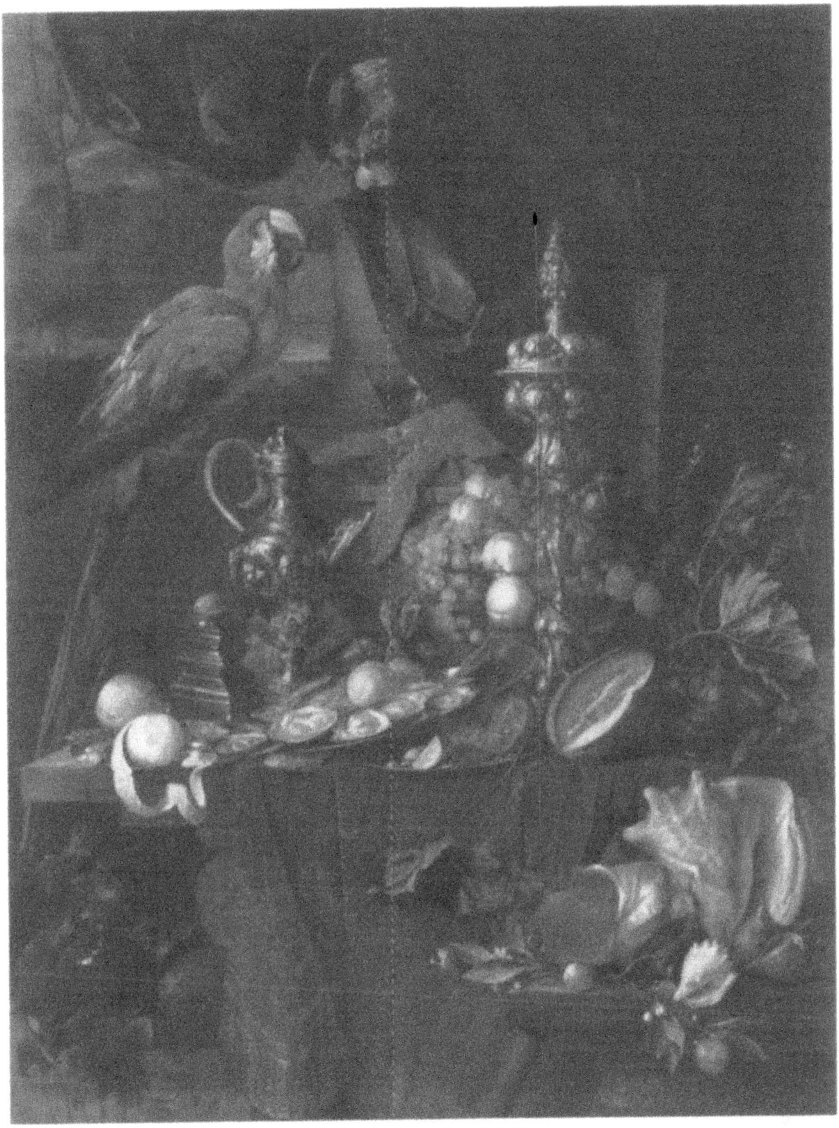

Fig. 11. Jan Davidsz. De Heem, Elaborate tabletop still life with foodstuffs, exotic animals, and precious objects. Oil on canvas, 150.49 × 117.74 cm. Ringling Museum of Art, Sarasota, Florida.

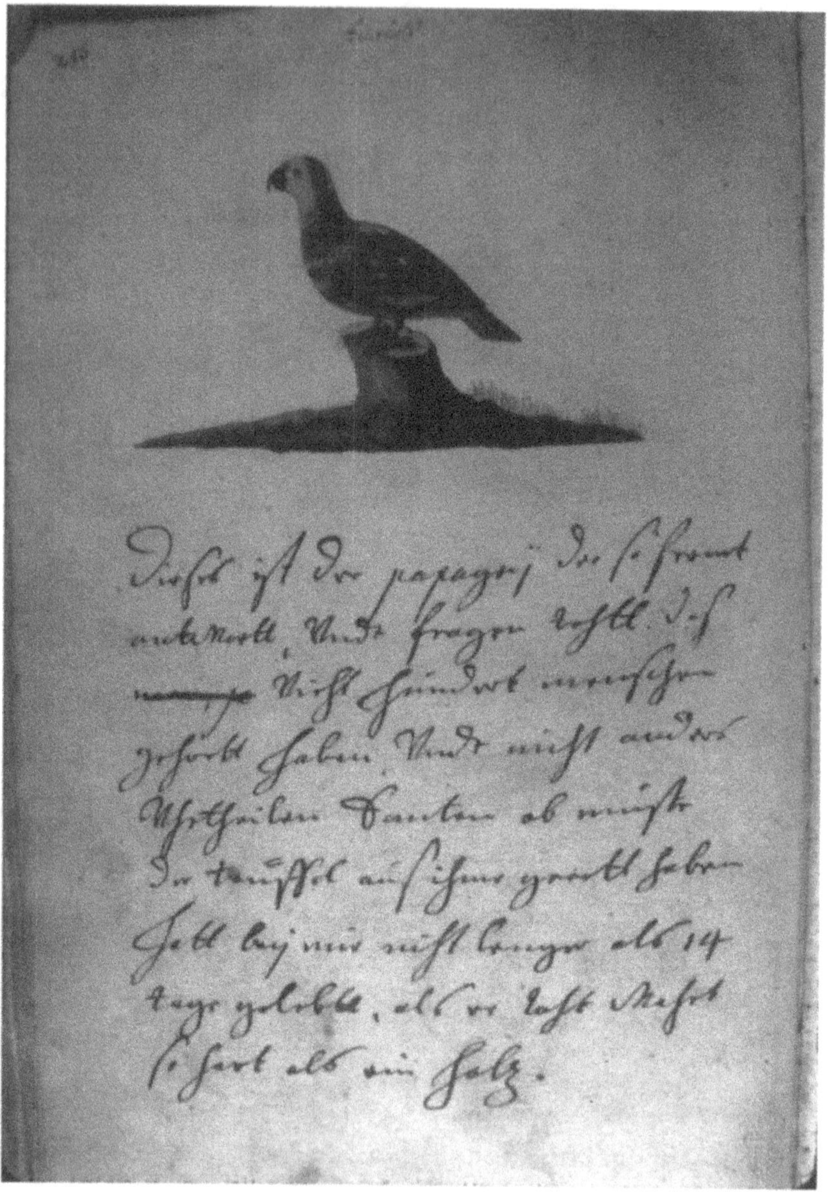

Fig. 12 [Col. pl. XIV (detail)]. Georg Marcgraf, Amazonian parrot (*Amazona aestiva*), ca. 1640. Watercolour and body colour on paper, 21 × 20 cm. *Handbook* I fol. 216r, *Libri picturati* A36 (Jagiellon University Library, Krákow).

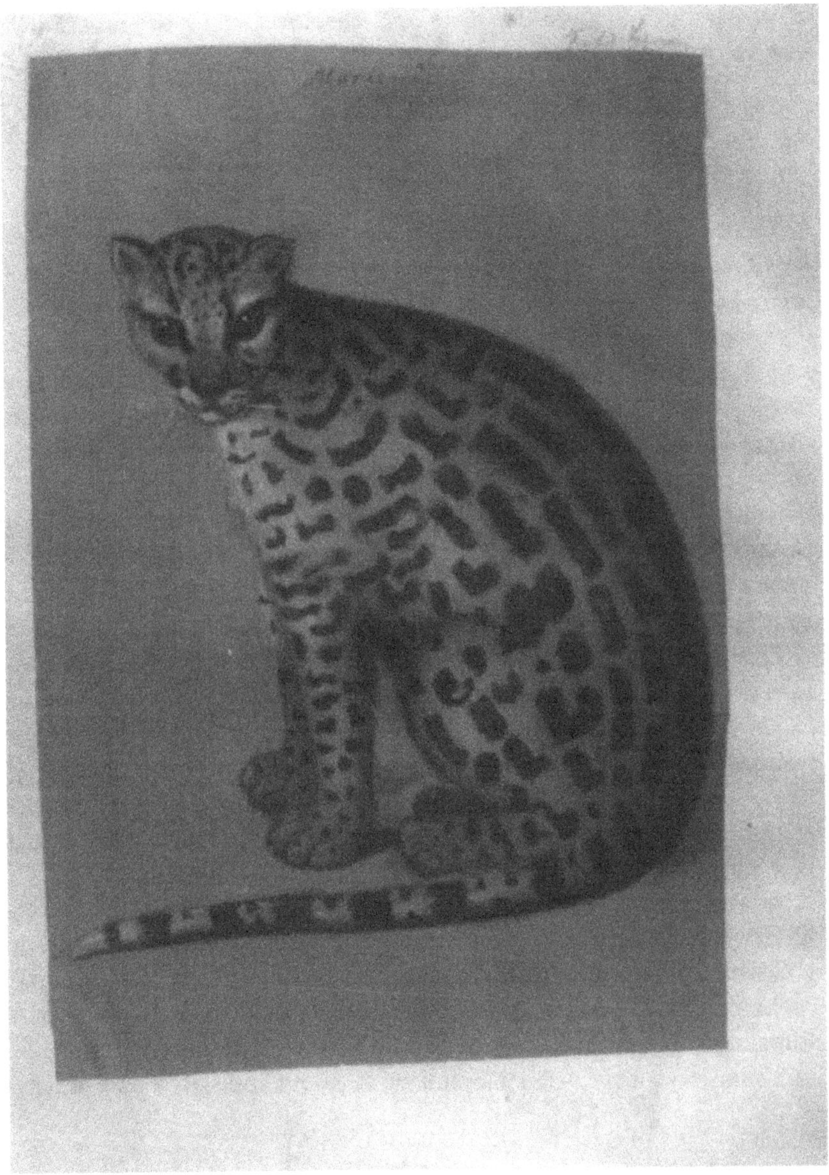

Fig. 13 [Col. pl. XX]. Albert Eckhout, Ocelot. Oil on paper, 28 × 18.1 cm. *Theatrum rerum naturalium*, vol. III fol. 55r, *Libri picturati* A34 (Jagiellon University Library, Krákow).

of young 'tiger'—there the clear and silent recording of the animal's exterior was his primary task.

The Brazilian Drawings in The Dutch Republic, 1644–1652

As mentioned above, even before Johan Maurits returned to the Dutch Republic in 1644, he had already started sending back horses, game animals, in addition to parrots and paintings (presumably by Eckhout, Post, or Marcgraf) as presents to individuals (such as the stadhouder) and institutions (such as the anatomy theatre in Leiden) interested in owning South America exotica.[56] While most of the Count's Brasilian collection of drawings, paintings, and other types of exotic products would eventually be given away, he first made them available to WIC Director and amateur naturalist Johannes de Laet. De Laet had become the editor of the *Historia naturalis Brasiliae* (1647) following Marcgraf's untimely death in Africa around 1644. Given the fact that many of the animals and especially plants reproduced by Eckhout were unknown in Europe and had never been illustrated, it is not surprising that de Laet used a large number of Eckhout's oil studies (especially botanical images) to supplement the images by Marcgraf as models for the woodcut illustrations to this work [Fig. 14].[57] The frontispiece, illustrated here, borrows liberally from the works of both Marcgraf and Eckhout in its depiction of Brazil as a lush, tropical paradise filled with curious animals and friendly natives.

Although Eckhout's paintings and drawings supplemented this scientifc project, we find immediate and clear quotations from his images in a variety of 'high art', princely decorative projects. It is, for example, probable that the paintings of birds documented on the ceiling of a room in the Mauritshuis circa 1645 were based on his Brazilian drawings and may even have been painted by Eckhout himself. These images

---

[56] In his autobiography, Wagener discusses how he was charged with bringing writings (documents), paintings, and parrots to individuals in a number of different Dutch cities. See Whitehead – Boeseman, *A portrait of Dutch 17th century Brazil* 25.

[57] Given the count's interest in natural history and his support for this project, Eckhout may even have been enlisted to create images, especially for Piso's botanical portion of the *Historia*, while still in Brazil. The majority of the botanical images in the *Historia* are based on Eckhout's oils studies in the *Theatrum*. With respect to the zoological images, Marcgraf's drawings are nearly always prefered to Eckhout's as models for the woodcuts in cases where an image by each artist of the same animal was available.

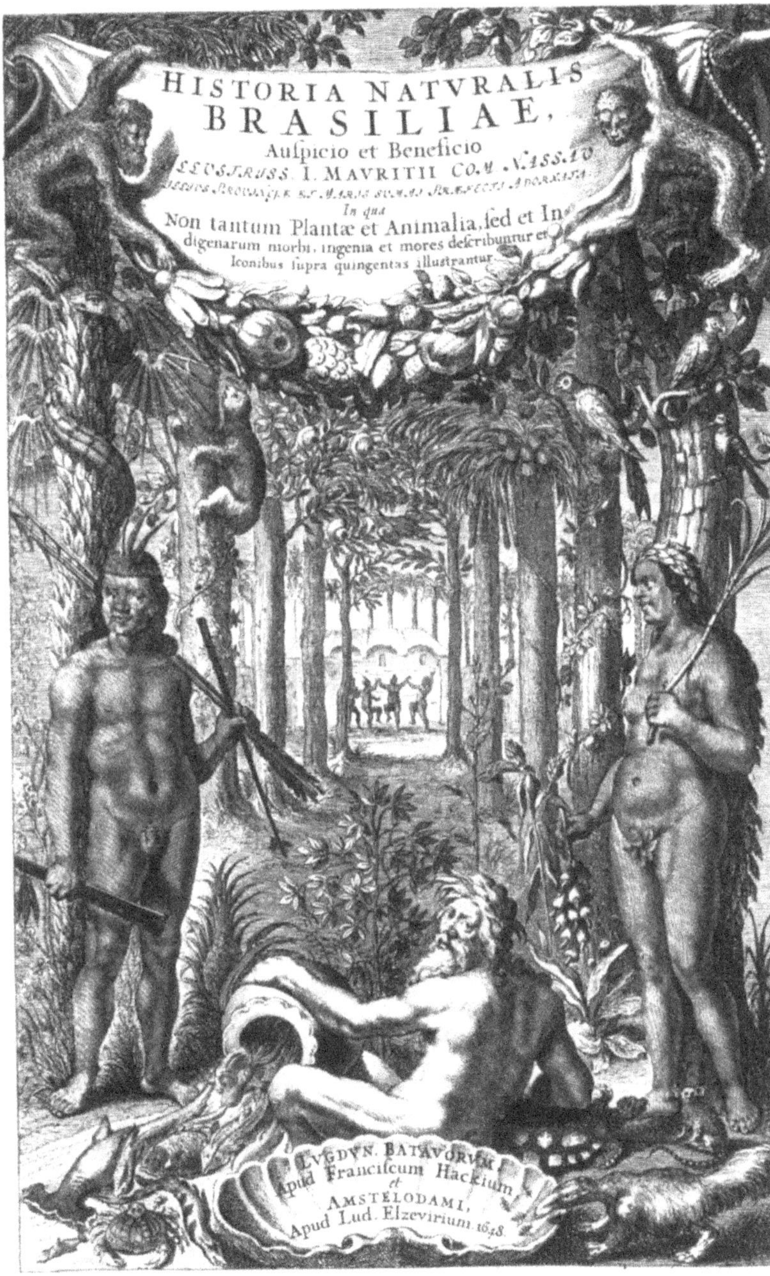

Fig. 14. Georg Marcgraf and Willem Piso, *Historia naturalis Brasiliae*, *Frontispiece*, 1648. Courtesy of the Linda Hall Library, Kansas City.

were undoubtedly similar in format to the ceiling paintings of Brazilian birds painted at Hoflössnitz lodge (Weingergschloss), Radebeul, near Dresden, Germany, during Eckhout's 10 years (1653–1663) there as court artist to Johan Gregor II, the future Elector of Saxony.[58]

That Eckhout's drawings were also a source of inspiration to other painters in the Dutch Republic is demonstrated by Jacob van Campen's work from the 1650s, which borrows Brazilian motifs from Eckhout.[59] Van Campen, was one of the leading artists of Dutch classicism, and the architect (with Pieter Post) of Johan Maurits's Dutch palace, the Mauritshuis. Surely some of the Brazilian drawings mentioned as in the possession of Van Campen in 1646 were by Eckhout; the two artists also appear to have collaborated around 1650 on a series of still lifes at Het Hoogerhuis in Amersfoort, which included Brazilian motifs drawn directly from Eckhout's works on paper and his paintings.[60]

Conclusion: Christopher Mentzel, Eckhout's Oil Studies, and the Theatrum

By 1652, Johan Maurits's collection of Brazilian natural history drawings had already been put to a number of uses. They had served as models for zoological and botanical illustrations for the *Historia* and cartoons had been created after them for a tapestry series produced by Gücht in Delft (now lost). They had also served as inspiration for a series of ceiling paintings of birds at the Mauritshuis and they had been quoted in paintings by Dutch artists such as Jacob van Campen. Given the reality that the Dutch were in the process of losing hold of their Brazilian colony and the fact that the drawings by Eckhout and Marcgraf were no longer quite as novel as they had been in 1644, it is not surprising that Johan Maurits chose this moment to present them to his patron, Wilhelm Frederich, the Great Elector of Brandenburg. It is the Elector's physician Christian Mentzel who is responsible for the current division of the drawings into seven volumes comprising the *Handbooks* (2 vols.),

[58] In Johan Maurits's letter of recommendation for Eckhout, sent to Johann Georg in 1653, he specifically mentions (surely as a selling point) that the painter would bring Brazilian drawings with him.

[59] See Buvelot Q., "Jacob van Campen als schilder en tekenaar" in Huisken J. et al. (eds.), *Jacob van Campen: het klassieke ideaal in de Gouden Eeuw* (Amsterdam: 1995).

[60] Buvelot, "Albert Eckhout a Dutch artist in Brazil", in *Albert Eckhout* (2004) 35 and note 150.

the *Theatrum* (4 vols.) and the *Misc. Cleyeri* (1 vol.). As curator of the Elector's collection of Brazilian natural history illustrations, Mentzel was especially interested in Marcgraf and Eckhout's drawings because of their status as unique representations of exotic and largely unknown species of animals and plants that were made *in situ*.

As addressed above, there was a movement in the early modern period towards defining an appropriately 'scientific' style and manner of working with respect to images (especially botanical) intended to have informational value. The 'nature study mode', here exemplifed by Marcgraf's work, with its small size, linearity, and lifelessness, had not been embraced by everyone as the ideal. here exemplified by Marcgraf's work, were not yet embraced as essential criteria, and there was quite a bit of uncertainty regarding what was appropriate for zoological representation. For the famous botanist Carolus Clusius in the sixteenth century, an 'image "drawn after the plant itself" and "in natural colours" was deemed capable of supplying the necessary information' to classify the specimen—he does not mention style, format, or a prefered medium.[61] Sixteenth-century German botanist Leonard Fuchs emphasized clarity of form and lack of artifice for scientific illustration; of Marcgraf and Eckhout, he clearly would have favoured the work of the former.[62] Although a scientist by training, Mentzel by contrast appears to have responded to and judged images much in the same way that Van Mander did; Eckhout's paintings appeared more "lifelike" than Marcgraf's because of their naturalism and vivid, realistic use of color. Although Mentzel's eye was far from 'innocent', he did not know the artists and he had never been to Brazil. Unfamilar with living examples of its indigenous flora and fauna, Mentzel could not judge the drawings with respect to their visual accuracy beyond what he could infer from the images themselves.

In his dedication to the *Theatrum*, Mentzel writes: 'When our Prince Maurice of Nassau, conqueror of Brazil, did bring with him a collection of paintings of Brazil's natural things, incomplete, however, and in detail lacking, Thou [Friedrich Wilhelm], not content with thy splendid feats, wished, as a complement to thy achievements, that all

[61] Swan C., "Jacques de Gheyn and the Representation of the Natural World" (Dissertation, Columbia: 1997) 131.

[62] See the preface to Leonhard Fuchs, *Historia stirpium* (Basel: 1542); Landau – Parshall, *Renaissance Print* 254.

be compiled and classified in those four volumes'.[63] The desire to sort, classify, apply scientific labels, and arrange the 'raw' nature of the unsorted images into neatly-bound and organized volumes was the essence of this enterprise. According to Mentzel, the Elector planned to make the collection available to 'students of Humanities and Arts in the sanctuary of his library'.[64] The acquisition of the South American natural history images and the creation of a 'theater' of Brazilian nature furthermore allowed the Elector to ally himself with a scientific and colonial enterprise at a time when he was contemplating the establishment of overseas trading companies for economic expansion into Africa and the New World.[65]

Despite Mentzel's collation of Eckhout's oils into a set of study volumes on the natural history of Brazil, it is striking that his images lack anything more than a label giving the name of the plant or animal. Yet it is precisely this lack of written description that is essential for understanding Mentzel's view of scientific illustration. The naturalism possible with oils and the rich, realistic colours and textures of Eckhout's images made Mentzel believe that each oil study could stand alone as a complete description, making verbal annotations superfluous. He describes the oil studies as visual substitutes for the real object, conveying the information of a written and a visual description in one. Like many other early modern scientists, Mentzel believed in the power and importance of images in transferring vital information—information that could not be expressed by words alone.[66] Aldrovandi believed that images could replace words in their descriptive capabilities.[67] In a letter to Charles V (1500–1558), Holy Roman Emperor and King of

---

[63] 'Quidquid enim rerum naturalium Brasiliae primitus depictum, sed indigestum atque incompletum, princeps noster Nassavius Brasiliae domitor redux quondam attulerat, stupendis tuis factis non contentus concinne id ordinari ac perfici quatuor his tomis voluisti bonarum actionum complementum." (Note: this part of the preface is written in all capital letters) Mentzel, preface to the *Theatrum Rerum Naturalium Brasiliae*, vol. I, as reproduced both in translation and in the original Latin (in facsimile) in *Brasil-Holandês/Dutch-Brazil* vol. IV 16.

[64] *Theatrum rerum naturalium Brasiliae*, vol. I, as reproduced both in translation and in the original Latin (in facsimilie) in *Brasil-Holandês/Dutch-Brazil* vol. IV 18.

[65] A Brandenburg-African company was finally established in 1682, although the Elector had been interested in forming such a company as early as 1647. See Schmitt E., "The Brandenburg Overseas Trading Companies in the Seventeenth Century", in Blussé L. – Gaastra F. (eds.), *Companies and Trade* (Leiden: 1981) 159–176.

[66] Freedberg D., "The Failure of Colour," in *Sight and Insight: Essays on Art and Culture in Honour of E.H. Gombrich at 85* (London: 1994).

[67] Swan, "'*Ad Vivum*'" 359.

Spain, the famous anatomist Andreas Vesalius (1514–1564) expressed a similar sentiment when he wrote: 'Pictures greatly aid the understanding of these anatomical matters, and how much more accurately they put things before the eyes than even the clearest language'.[68] In the following quote, Mentzel suggests a similar point of view, which also calls upon the paragone, or rivalry between the arts in their efforts to reproduce nature:

> So behold the magnificence of this 'Theatre', its splendour, its recommendation: that the powers immediately perceive at a single glance the truth and authenticity of the things, without the need for long detailed descriptions, and that this subject leaves nothing to be desired. Indeed, even to the most demanding pen would it be impossible to describe with perfection the particularities of the colours and distinguish them, as well as the number and division of the parts, details that may be seen in the picture at a single glance. It would be in vain to strive to describe the things of natures since the Creator's skill and art are far superior to the phantasy of any writer. In practice, such greater worth hath painting than eloquence, principally when the reproductions are authentic and created by skilled hands, which do not intend to surpass natural beauty.[69]

It is ironic that Marcgraf's images, which adopted the formal language of the model book drawing, were created by a naturalist-illustrator, and were specifically described by both the artist and at least one contemporary as having been produced 'after life', were judged by Mentzel (perhaps under the influence of Wilhelm Frederich) to be less effective or desirable natural history images because of their use of watercolour, which resulted in an image that was less 'life-like' than Eckhout's oils on paper. Mentzel's preference for Eckhout's dynamic images would not, however, prove typical for naturalists of later generations. Colour would be rejected because of its mutability, and a painterly manner like Eckhout's would be abandoned in favor of the linear style of

[68] As quoted in Whitehead P.J.P., "The original drawings for the *Historia naturalis Brasiliae* of Piso and Marcgrave (1648)", *Journal of the Society of Bibliography of Natural History* 7 (1976) 409–422.

[69] 'Isthaec enim THEATRI huius magnificencia, iste splendor, ista commendatio est, ut, quo nihil in historia divinius putatur, Veritatem ac Certitudinem Rerum, ipso statim aspectu, non longis atque ambitiosis descriptionibus assequare. Neque enim vel mimima ista colorum inter se discrimina aut partium membrorumque competentiae ac numeri, ullius vel disertissimi hominis eloquio exprimi sic possunt, quam solo obtutu et statim ac simul hauriuntur: frustraque in enumerandis rerum aspectabilium figuris desudatur, ubi pictoris ars atque industria omnem circumscribentis phantasiam prevenit ac superavit', Mentzel, preface to *Theatrum rerum naturalium Brasiliae*, part I, as reproduced both in translation in Latin (facsimile) in *Brasil-Holandês/Dutch-Brazil*, vol. IV 22–23.

Marcgraf's representations. Indeed, natural historians such as Carolus Linneaus in the eighteenth century would favor the written word over representation. The Brazilian drawings and oil studies by Eckhout and Marcgraf were created during a period when artists and scientists alike were attempting to create an appropriate visual language for the representation of animals and nature more broadly. Produced in the same colonial environment for the same princely patron, and sometimes even making pictures after the same animal, the drawings of Marcgraf and Eckhout nonetheless present distinct visual responses to the challenge of naturalism in zoological representation.

Selective Bibliography

Artelt W., *Christian Mentzel: Leibarzt des Grossen Kurfürsten* (Leipzig: 1940).
Ashworth W., "Remarkable Humans and Singular Beasts", in Kenseth J. (ed.), *Age of the Marvelous* (Hanover: 1993) 113–144.
Bergvelt E. – Kistemaker R. (eds.), *De wereld binnen handbereik. Nederlandse kunst- en rariteitenverzamelingen, 1585–1735* (Zwolle – Amsterdam: 1992).
Boogaart E. van den (ed.), *Johan Maurits van Nassau-Siegen, 1604–1679. A Humanist Prince in Brazil and Europe* (The Hague: 1979).
Boxer C.R., *The Dutch in Brazil, 1624–1654* (Oxford: 1957).
Brienen R.P., *Visions of Savage Paradise: Albert Eckhout, Court Painter in Colonial Dutch Brazil* (Amsterdam: 2006).
——, "Georg Marcgraf (1610–ca. 1644) A German Cartographer, Astronomer, and Naturalist-Illustrator in Colonial Dutch Brazil", *Itinerario* 20,1 (2001) 85–112.
Buvelot Q. (ed.), *Albert Eckhout: A Dutch Artist in Brazil* (The Hague: 2004).
Ferrao C. – Soares J.P.M. (eds.), *Brasil-Holandês/Dutch-Brasil*, 5 vols. (Rio de Janeiro: 1995).
Findlen P., *Possessing nature: museums, collecting, and scientific culture in early modern Italy* (Berkeley – Los Angeles: 1994).
Foucart-Walter E., *Pieter Boel: Painter of Louis XIV's Animals* (Paris: 2001).
Gelder R. van, "De wereld binnen handbereik: Nederlandse kunst- en rariteitverzamelingen, 1585–1735", in Bergvelt – Kistemaker (eds.), *De wereld binnen handbereik*.
George W., "Alive or Dead: Zoological Collections in the Seventeenth Century", in Impey O. – Macgregor A. (eds.), *The Origins of Museums: the Cabinet of Curiosities in Sixteenth- and Seventeenth-Century Europe* (Oxford: 1985) 179–187.
Hendrix L., "Natural History Illustration at the Court of Rudolf II", in Fucikova E. et al. (eds.), *Rudolf II and Prague: the imperial court and the residential city as the cultural and spiritual heart of central Europe* (Prague: 1997).
Impey O. – MacGregor A. (eds.), *The Origins of Museums: the Cabinet of Curiosities in Sixteenth- and Seventeenth-Century Europe* (Oxford: 1985).
Kenseth J. (ed.), *Age of the Marvelous* (Hanover: 1993).
Kolb A.F., *The Entry of the Animals into Noah's Ark* (Los Angeles: 2005).
Koreny F., *Albrecht Dürer and the Animal and Plant Studies of the Renaissance* (Boston: 1985).
Larsen E., *Frans Post, interprète du Brésil* (Amsterdam: 1962).
Moran B.T., "German Prince-Practitioners: Aspects in the Development of Courtly Science, Technology, and Procedures in the Renaissance", *Technology and Culture* 22 (1981) 253–274.
Pächt O., "Early Italian Nature Studies and the Early Calendar Landscape", *Journal of the Warburg and Courtauld Institutes* 13 (1950) 13–47.
Parshall P., "Imago Contrafacta: Images and Facts in the Northern Renaissance", *Art History* 16,4 (1993) 554–579.
Robbins L., *Elephant Slaves and Pampered Parrots: Exotic Animals in Eighteenth-Century Paris* (Baltimore: 2002).
Sliggers B.C. – Wertheim A.A. (eds.), *Een vorstelijke dierentuin: De menagerie van Willem V* (Zutphen: 1994).
Smit P. (ed.), *Hendrik Engel's Alphabetical List of Dutch Zoological Cabinets and Menageries* (Amsterdam: 1986).
Smith P. – Findlen P. (eds.), *Merchants and Marvels: Commerce, Science, and Art in the Early Modern Period* (London: 2002).
Swan C., *Art, Science and Witchcraft in Early Modern Holland: Jacques de Gheyn II (1565–1629)* (Cambridge: 2005).

——, "*Ad Vivum*, naer het leven, from the life: Defining a Mode of Representation", *Word and Image* 11 (1995) 353–372.
——, "From Blowfish to Flower Still Life Paintings: Classification and Its Images, circa 1600", in Kenseth (ed.), *Merchants and Marvels* 109–137.
——, "Jacques de Gheyn and the Representation of the Natural World" (Diss., Columbia: 1997).
THOMSEN T., *Albert Eckhout. Ein niederländischer Maler und sein Gönner Johan Maurits der Brasilianer* (Copenhagen: 1938).
WAGENER ZACHARIAS, *The "Thierbuch and Autobiography" of Zacharias Wagener*, trans. Á.A. Bragança Júnior, *Dutch Brazil* (Rio de Janeiro: 1997).
WHITEHEAD P.J.P. – BOESEMAN M., *A portrait of Dutch 17th century Brazil: Animals, plants and people by the artists of Johan Maurits of Nassau* (Amsterdam: 1989).
WHITEHEAD P.J.P., "The original drawings for the *Historia naturalis Brasiliae* of Piso and Marcgrave (1648)", *Journal of the Society of Bibliography of Natural History* 7 (1976) 409–422.

www.ingramcontent.com/pod-product-compliance
Lightning Source LLC
LaVergne TN
LVHW010624300926
849162LV00001B/50

* 9 7 8 9 0 0 4 1 3 1 8 8 0 *